Chemistry of the Solid-Water Interface

Chemistry of the Solid-Water Interface

Processes at the Mineral-Water and Particle-Water Interface in Natural Systems

WERNER STUMM

Professor, Swiss Federal Institute of Technology, ETH Zurich,
Institute for Water Resources and Water Pollution Control (EAWAG)

with contributions by
LAURA SIGG (Chapter 11)
BARBARA SULZBERGER (Chapter 10)

A Wiley-Interscience Publication
John Wiley & Sons, Inc.
New York / Chichester / Brisbane / Toronto / Singapore

Library of Congress Cataloging in Publication Data:

Stumm, Werner
    Chemistry of the solid-water interface : processes at the mineral-
water and particle-water interface in natural systems / Werner
Stumm : with contributions by Laura Sigg (chapter 11), and Barbara
Sulzberger (chapter 10).
      p. cm.
    ''A Wiley-Interscience publication.''
    Includes bibliographical references and index.
    ISBN 0-471-57672-7 (pbk.)
    1. Water chemistry. 2. Surface chemistry. I. Sigg, Laura.
II. Sulzberger, Barbara. III. Title.
GB855.S79 1992
551.46—dc20   92-9701
                                           CIP

Printed in the United States of America

10 9 8 7 6 5 4 3 2 1

Printed and bound by Malloy Lithographing, Inc..

Contents

Preface

The aim of this book is to provide an introduction to the chemistry of the solid–water interface. Of primary interest are the important interfaces in natural systems, above all in geochemistry, in natural waters, soils, and sediments. The processes occurring at mineral–water, particle–water, and organism–water interfaces play critical roles in regulating the composition and the ecology of oceans and fresh waters, in the development of soils and the supply of plant nutrients, in preserving the integrity of waste repositories, and in technical applications such as in water technology and in corrosion science.

This book is a teaching book; it progresses from the simple to the more complex and applied. It is addressed to students and researchers (chemists, geochemists, oceanographers, limnologists, soil scientists and environmental engineers). Rather than providing descriptive data, this book tries to stress surface chemical principles that can be applied in the geochemistry of natural waters, soils, and sediments, and in water technology.

Interface and colloid science has a very wide scope and depends on many branches of the physical sciences, including thermodynamics, kinetics, electrolyte and electrochemistry, and solid state chemistry. Throughout, this book explores one fundamental mechanism, the interaction of solutes with solid surfaces (adsorption and desorption). This interaction is characterized in terms of the chemical and physical properties of water, the solute, and the sorbent. Two basic processes in the reaction of solutes with natural surfaces are: 1) the formation of coordinative bonds (surface complexation), and 2) hydrophobic adsorption, driven by the incompatibility of the nonpolar compounds with water (and not by the attraction of the compounds to the particulate surface). Both processes need to be understood to explain many processes in natural systems and to derive rate laws for geochemical processes.

The geochemical fate of most reactive substances (trace metals, pollutants) is controlled by the reaction of solutes with solid surfaces. Simple chemical models for the residence time of reactive elements in oceans, lakes, sediment, and soil systems are based on the partitioning of chemical species between the aqueous solution and the particle surface. The rates of processes involved in precipitation (heterogeneous nucleation, crystal growth) and dissolution of mineral phases, of importance in the weathering of rocks, in the formation of soils, and sediment diagenesis, are critically dependent on surface species and their structural identity.

The dynamics of particles, especially the role of particle–particle interactions (coagulation) is critically assessed. The effects of particle surfaces on the catalysis of redox processes and on photochemically induced processes are discussed, and it is shown that the geochemical cycling of electrons is not only mediated by micro-organisms but also by suitable surfaces, and is thus of general importance at particle–water interfaces.

The chemical, physical, and biological processes that are analyzed here at the micro level influence the major geochemical cycles. Understanding how geochemical cycles are coupled by particles and organisms may aid our understanding of global ecosystems, and on how interacting systems may become disturbed by civilization.

Acknowledgements

Chemistry of the Solid-Water Interface covers many subjects where I have been involved in my own research. This research has been stimulated by many colleagues. Regretfully, only casual recognition of some of their papers can be made because this book does not review the literature comprehensively; but I should like to acknowledge the great influence of Paul W. Schindler (University of Berne) on the ideas of surface coordination, and the significant contributions of Garrison Sposito to the surface chemistry of soils. My own research could not have been carried out without the help and the enthusiasm of a number of doctoral students. My own scientific development owes a great deal to Elisabeth Stumm-Zollinger. The prolonged association with James J. Morgan (California Institute of Technology, Pasadena), Charles R. O'Melia (Johns Hopkins University, Baltimore), and Laura Sigg (Swiss Federal Institute of Technology, EAWAG, Zurich) has been of great inspiration.

I greatly acknowledge my colleagues Laura Sigg and Barbara Sulzberger for contributing the two last chapters in this book.

In the preparation of this book I am greatly indebted to Lilo Schwarz who skillfully carried the manuscript through its many revisions to camera-readiness. Heidi Bolliger drew and redrew most of the illustrations. I am also grateful to Sonja Rex and Gerda Thieme.

Many of my colleagues, including M. Blesa, J. J. Morgan, F. M. M. Morel, C. R. O'Melia, Jerry L. Schnoor, L. Sigg, and B. Sulzberger, have made valuable suggestions for corrections and improvements. Rolf Grauer (Paul Scherrer Institute, Switzerland) deserves special credit for giving critical and constructive advice on many chapters.

December 1991   Werner Stumm

Chemistry of the Solid-Water Interface

Chapter 1

Introduction
Scope of Aquatic Surface Chemistry

The various reservoirs of the earth (atmosphere, water, sediments, soils, biota) contain material that is characterized by high area to volume ratios. Even the atmosphere contains solid-water-gas interfaces. There are trillions of square kilometers of surfaces of inorganic, organic and biological material that cover our sediments and soils and that are dispersed in our waters. Very efficient interface chemistry must occur to maintain appropriate atmospheric chemistry and hydrospheric chemistry. Mineral-based assemblages and humus make up our soil systems that provide the supply of nutrients and support our vegetation. The action of water (and CO_2 and organic matter) on minerals is one of the most important processes which produce extremely high surface areas and reactive and catalytic materials in the surface environments. The geological processes creating topography involve erosion by solution and particle transport. Such processes provide nutrient supply to the biosphere. The mass of material eroded off the continents annually is thus of an order of magnitude similar to that of the rate of crust formation and subduction (Fyfe, 1987). Human activity is greatly increasing erosion; and soil erosion has become a most serious world problem. The oceanic microcosmos of particles – biological particles dominate the detrital phases – plays a vital role in ocean chemistry.

The actual natural systems usually consist of numerous mineral assemblages and often a gas phase in addition to the aqueous phase; they nearly always include a portion of the biosphere. Hence natural systems are characterized by a complexity seldom encountered in the laboratory. In order to understand the pertinent variables out of a bewildering number of possible ones it is advantageous to compare the real system with idealized counterparts, and to abstract from the complexity of nature.

Adsorption

Adsorption, the accumulation of matter at the solid-water interface, is the basis of most surface-chemical processes.
1) It influences the *distribution of substances* between the aqueous phase and particulate matter, which, in turn, affects their transport through the various reservoirs of the earth. The affinity of the solutes to the surfaces of the "conveyor belt" of the settling inorganic and biotic particles in the ocean (and in lakes) regulate their (relative) residence time, their residual concentrations and their

ultimate fate (Fig. 1.1). Adsorption has a pronouced effect on the speciation of aquatic constituents.

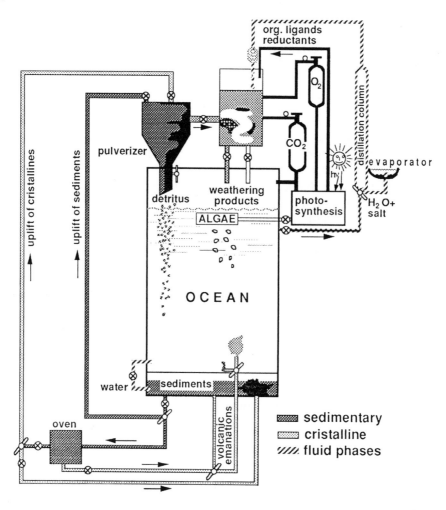

Figure 1.1

Circulation of rocks, water and biota. Steady state model for the earth's surface geochemical system likened to a chemical engineering plant. The interaction of water with rocks in the presence of photo-synthesized organic matter continuously produces reactive material of high surface area in the surface environment. This process provides nutrient supply to the biosphere and, along with biota, forms the array of small particles (soils). Weathering imparts solutes to the water and erosion brings particles into surface waters and oceans. A large flux of settling detrital and biogenic particles continuously runs through the water column. The steady state conveyor belt of settling particles which are efficient sorbents of heavy metals and other trace elements regulates their concentrations in the water column. The sediments are the predominant sink of trace elements.

(Modified from Siever (1968)

2) Adsorption affects the *electrostatic properties* of suspended particles and colloids, which, in turn, influences their tendency to aggregate and attach (coagulation, settling, filtration).

3) Adsorption influences the reactivity of surfaces. It has been shown that the rates of processes such as precipitation (heterogeneous nucleation and surface precipitation), dissolution of minerals (of importance in the weathering of rocks, in the formation of soils and sediments, and in the corrosion of structures and metals), and in the catalysis and photocatalysis of redox processes, are critically dependent on the properties of the surfaces (surface species and their structural identity).

Atoms, molecules and ions exert forces upon each other at the interface. In this book, adsorption reactions are discussed primarily in terms of intermolecular interactions between solute and solid phases. This includes: 1) *Surface complexation reactions* (surface hydrolysis, the formation of coordinative bonds at the surface with metals and with ligands). 2) The *electric interactions* at surfaces, extending over longer distances than chemical forces. 3) *Hydrophobic expulsion* (hydrophobic substances) – this includes non-polar organic solutes – which are usually only sparingly soluble in water, tend to reduce the contact in water and seek relatively non-polar environments and thus may accumulate at solid surfaces and may become absorbed on organic sorbents. 4) *Adsorption of surfactants* (molecules that contain a hydrophobic moiety). (interfacial tension and adsorption are intimately related through the Gibbs adsorption law; its main message – expressed in a simple way – is that substances that tend to reduce surface tension, tend to become adsorbed at interfaces). 5) *The adsorption of polymers* and of polyelectrolytes – above all humic substances and proteins – is a rather general phenomenon in natural waters and soil systems that has far-reaching consequences for the interaction of particles with each other and on the attachment of colloids (and bacteria) to surfaces.

Surface Coordination

One of the more important generalizations emphasized in this book is that the solids can be considered as inorganic and organic polymers, the surfaces of which can be looked at as *extending structures* bearing surface functional groups. These functional groups contain the same donor atoms which are found in functional

groups of solute ligands such as $-OH$, $-SH$, $-SS$, $-C{\overset{O}{\underset{OH}{\diagup}}}$ etc. Such functional

groups provide a diversity of interactions through the formation of *coordinative bonds*. Fig. 1.2 illustrates three possible adsorption mechanisms of metal ions on an oxide-water interface as well as sorption through the formation of a surface precipitate. In a similar way ligands can replace surface OH groups (ligand exchange) to form ligand surface complexes. The concept of active sites has been a highly

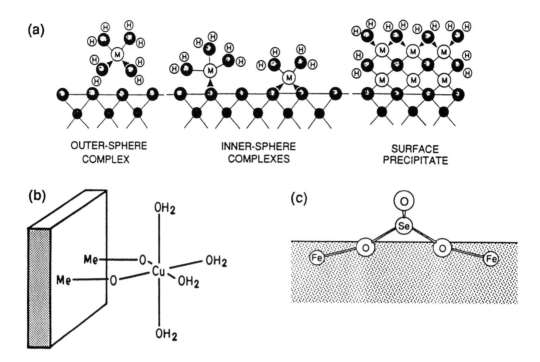

Figure 1.2

Structural arrangements in surface complexes at an oxide surface.

a) Definition of possible sorption complexes at the solid/water interface, which is represented by the horizontal line. The solid substrate is below the line and the solution is above the line. The circles labeled M represent sorbed metal atoms in various types of sorption complexes. The larger shaded spheres in the solid substrate and surrounding the metal in the solution phase are oxygens. The smaller dark spheres in the solid substrate are metal ions, as are the spheres labeled M in the sorption complexes and surface precipitate. (From Brown, 1990)

b) Surface complex of Cu(II) on $\delta-Al_2O_3$ (structure inferred from EPR measurements). (From Motschi, 1987)

c) Proposed structure for SeO_3^{2-} coordinated with Fe atoms of goethite based on Extended X-ray Absorption Fine Structure (EXAFS) spectroscopy. (From Hayes, Roe, Brown, Hodgson, Leckie, and Parks, 1981)

productive one in understanding catalysis by enzymes and coenzymes. Although surface functional groups at solid-water interfaces are often characterized by less specificity than that of enzymes, they form an array of surface complexes, whose reactivities determine the mechanism of many surface controlled processes. We know from research on nucleation and biomineralization that the specific surface sites can extent "molecular recognition"; they determine in nucleation not only what allotropic modification of the solid phase is formed but also the morphology of the new phase. It is only today that we are discovering some of the basic mechanistic steps. Many mechanisms can readily be described in terms of Brønsted acid sites

Table 1.1   Coordination Chemistry of the Solid-Water Interface: Concepts and important Applications in natural and technical Systems

Surface Complex Formation	Applications: Distribution of Solutes between Water and Solid Surface	Applications: Rates depend on Surface Speciation
Interaction with – H^+, OH^- – Metal ions – Ligands (ligand exchange) **Thermodynamics of Surface Complex Formation** – K (mass law constants, corrected for electrostatic effects) – ΔG, ΔH **Kinetics of Surface Complex Formation** Rates of sorption and desorption **Structure of Surface Compounds** (Surface Speciation) – Inner-sphere versus outer-sphere – Monodentate versus binuclear – Monodentate versus bidentate **Establishment of Surface Charge** **Structure of Lattice** – Defect sites – Adatoms, kinks, steps, ledges – Lattice statistics **Microtopography**	**Binding of Reactive Elements to Aquatic Particles in Natural Systems** – Regulation of metals in soil, sediment, and water systems – Regulation of oxyanions of P, As, Se, Si in water and soil systems – Interaction with phenols carboxylates and humic acids – Transport of reactive elements including radio-nuclides in soils and aquifers **Binding of Cations, Anions and Weak Acids to Hydrous Oxides in Technical Systems** – Corrosion; passive films – Processing of ores, flotation – Coagulation, flocculation, filtration – Ceramics, cements – (Photo)electrochemistry (electrodes, oxide electrodes and semiconductors) **Surface Charge resulting from the Sorption of Solutes** – Particle-particle interaction; coagulation, filtration	**Natural Systems Dissolution of Oxides, Silicates and other minerals** – Weathering of minerals – Proton and ligand promoted dissolution – Reductive dissolution of Fe(III) and Mn(III, IV) oxides **Formation of Solid Phases** – Heterogeneous nucleation – Surface precipitation, crystal growth – Biomineralization **Surface Catalyzed Processes** – (Photo)redox processes – Hydrolysis of esters – Transformations of organic matter by Fe and Mn (photo)redox-cycles – Oxygenation of Fe(II), Mn(II), Cu(I) and V(IV) **Technical Systems** – Passive films (corrosion) – Photoredox processes with colloidal semiconductor particles as photo-catalyst, e.g. degradation of refractory organic substances – Photoelectrochemistry, e.g. photoredox processes at semiconductor electrodes

or Lewis acid sites. Of course, the properties of the surfaces are influenced with the properties and conditions of the bulk structure and the action of special surface structural entities will be influenced by the properties of both surface and bulk. Table 1.1 gives an overview of the major concepts and important applications. Emphasis is on surface chemistry of the oxide-water interface not only because the oxides are of great importance at the mineral-water (including the clay water) interface but because its coordination chemistry is much better understood than that of other surfaces. Experimental studies on the surface interactions of carbonates, sulfides, disulfides, phosphates and biological materials are only now emerging. The results of these studies show, that the concepts of surface coordination chemistry can also be applied to these interfaces.

Some emphasis is given in the first two chapters to show that complex formation equilibria permit to predict quantitatively the extent of adsorption of H^+, OH^-, of metal ions and ligands as a function of pH, solution variables and of surface characteristics. Although the surface chemistry of hydrous oxides is somewhat similar to that of reversible electrodes the charge development and sorption mechanism for oxides and other mineral surfaces are different. Charge development on hydrous oxides often results from coordinative interactions at the oxide surface. The surface coordinative model describes quantitatively how surface charge develops, and permits to incorporate the central features of the *Electric Double Layer* theory, above all the *Gouy-Chapman* diffuse double layer model.

The Hydrophobic Effect

The hydrophobic effect, due to the incomptability of the hydrophobic substance with water, plays an important role in the adsorption of non-polar organic substances (Tanford, 1980). The sorption of hydrophobic substances to solid materials (particles, soils, sediments) that contain organic carbon may be compared with the partitioning of a solute between two solvents – water and the organic phase. It is possible to characterize the sorption of a wide range of the organic compounds based on a single property of the compound, i.e., its octanol-water partition constant, K_{OW}, (Fig. 1.3) and the property of the sorbent, i.e., the fraction of the sorbent that is organic carbon (Westall, 1987). Many organic substances, such as fatty acids, detergents, contain a hydrophobic part and a hydrophilic polar or ionic group; they are amphipathic. Such substances may, depending on the configuration, become adsorbed either by hydrophobic effect or by coordinative interaction.

Colloids

Colloids will receive attention throughout this book. They are usually defined on the basis of size; they are entities having at least in one direction a dimension between 1 nm and 1 μm (Lyklema, 1991). Colloids are ubiquitous in seawater, in fresh surface waters, in soils and sediments and in groundwaters and are typically present at substantial concentrations (usually more than 10^6 colloids cm^{-3}). A renewed research interest concerns the *stability of colloids,* their genesis and dissolution, their coagulation and attachment and their role in the transport of reactive elements, of radionuclides and other pollutants. The presence of colloids causes major operational difficulties in distinguishing between dissolved and particulate matter. All what we learn about interfaces is applicable to the colloid surface; because of the small size of the colloids they have relatively large area per given volume.

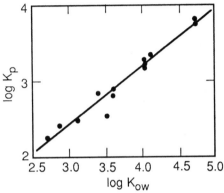

Figure 1.3

Partition constant for the distribution of various aromatic substances (mono-, di-, tri-, and tetramethyl, and chlorobenzenes) between water and an aquifer material (0.15 % organic carbon) as a function of the octanol-water partition coefficient, K_{ow}. The values of log K_p have been adjusted to be correct for a sorbent of 100 % organic carbon. K_{ow} is defined for the partition of a substance A between octanol water: $K_{ow} = [A_{oct}]/[A(aq)]$.
(Modified from Westall, 1987)

With a chapter on particle-particle interaction (coagulation) the characteristics of particles and colloids as chemical reactants are discussed. Since charge, and in turn the surface potential of the colloids is important in coagulation, it is illustrated how in simple cases the modelling of surface complex formation permits the calculation of surface charge and potential. The role of particle-particle interaction in natural water and soil systems and in water technology (coagulation, filtration, flotation) is exemplified.

Surface Structure and Surface Reactivity

Three applications in geochemistry, in soil science and sediment chemistry are of importance:
1) Dissolution (weathering) of minerals;
2) The formation of the solid phase (nucleation, precipitation, crystal growth, biomineralization);
3) Redox processes at the solid-water interface.

In analyzing the kinetics of surface reactions, it will be illustrated that many of these processes are rate-controlled at the surface (and not by transport). Thus, the surface structure (the surface speciation and its microtopography) determine the kinetics. Heterogeneous kinetics is often not more difficult than the kinetics in homogeneous systems; as will be shown, rate laws should be written in terms of concentrations of surface species.

Because surfaces can adsorb reductants and oxidants and modify redox intensity, the solid-solution interface can *catalyze many redox reactions*. The geochemical cycling of electrons is not only mediated by microorganisms but is of importance at particle-water interfaces (especially at the sediment-water interface due to strong redox gradients) and in surface waters due to *heterogeneous photo-chemical processes*. Many of the naturally occurring soid phases, such as Fe(III) oxides, TiO_2, CdS, have electronic structures with semiconductor properties. Light can induce – as in biological photosynthesis – transformations that are important in the cycling of elements; such light-catalyzed redox processes are also of importance in prebiotic geochemistry. Applications of heterogeneous photochemical redox processes include the catalytic degradation of toxic inorganic and organic substances in waters and wastes, and, of course, the exploration of the possibility of using semiconducting minerals in the splitting of water.

The Bonding between Solids and Solutes; The Need for a better Understanding

The structural identity of the surface species, the geometry of the coordinating shell of surface sites and of reactants at surfaces need to be known. The overlapping orbital of the *inner-sphere surface complex* interconnects the solid phase (metal, ionic or covalent solid, polymer) with the aqueous solution phase; it is a key to understanding of the reactivity of the solid-water interface (dissolution and formation of solids, heterogeneous catalysis). The mechanism of most surface controlled processes depend on the coordinative environment at the solid-water interface. We lack sufficient knowledge on the ways molecules, atoms and ions interact at solid-water interfaces, above all, on the electronic structure of the bonding between solids and solutes. The recent book by R. Hoffmann on *Solids and Surfaces; a Chemist's View of Bonding in Extended Structures* (1988) shows how chemistry and physics come together in the solid state and on surfaces and how the basic mechanistic steps in heterogeneous catalysis can be understood. Although *water* at the interface of the solid is not considered in Hoffmanns book, it gives us an idea in which direction we should go. A better understanding on the electronic structure of the bonding between solids and aquatic solutes would push the boundaries of aquatic surface chemistry.

Appendix

The International Units, some useful Conversion Factors, and numerical Constants

The "Système International" (SI) units, based on the metric system, were designed to achieve maximum internal consistency. The SI system is based on the following set of defined units:

Physical quantity	Unit	Symbol
Length	meter	m
Mass	kilogram	kg
Time	second	s
Electric current	ampère	A
Temperature	kelvin	K
Luminous intensity	candela	cd
Amount of material	mole	mol

The main derived units are: [1]

Force	newton	N	$= kg\,m\,s^{-2}$
Energy, work, heat	joule	J	$= N\,m$
Pressure	pascal	1 Pa	$= Nm^{-2}$
Power	watt	W	$= J\,s^{-1}$
Electric charge	coulomb	C	$= As$
Electric potential	volt	V	$= W\,A^{-1}$
Electric capacitance	farad	F	$= As\,V^{-1}$
Electric resistance	ohm	Ω	$= V\,A^{-1}$
Frequency	hertz	Hz	$= s^{-1}$
Conductance	siemens	S	$= AV^{-1}$

Useful Conversion Factors

Energy, Work, Heat 1 joule = 1 volt-coulomb = 1 newton meter
= 1 watt-second = 2.7778×10^{-7} kilowatt hours

---

[1] In this book we will continue to use the following traditional pressure and concentration units:
1 atm (= 1.013×10^5 Pa)
1 mol kg^{-1} (mass of solvent), molality
1 mol ℓ^{-1} (volume of solution) = M, molarity

$$= 10^7 \text{ erg}$$
$$= 9.9 \times 10^{-3} \text{ liter atmospheres}$$
$$= 0.239 \text{ calorie}$$
$$= 1.0365 \times 10^{-5} \text{ volt-faraday}$$
$$= 6.242 \times 10^{18} \text{ e V}$$
$$= 5.035 \times 10^{22} \text{ cm}^{-1} \text{ (wave number)}$$
$$= 9.484 \times 10^{-4} \text{ BTU (British thermal unit)}$$
$$\approx 3 \times 10^{-8} \text{ kg coal equivalent}$$

Power   1 watt $= 1 \text{ kg m}^2 \text{ s}^{-3}$
$= 2.39 \times 10^{-4} \text{ kcal s}^{-1} = 0.860 \text{ kcal h}^{-1}$

Entropy (S)   1 entropy unit, cal mol^{-1} K^{-1} $= 4.184$ J mol^{-1} K^{-1}

Pressure   1 atm $= 760 \text{ torr} = 760 \text{ mm Hg}$
$= 1.013 \times 10^5 \text{ N m}^{-2} = 1.013 \times 10^5 \text{ Pa (Pascal)}$
$= 1.013 \text{ bar}$

Coulombic Force   Coulomb's law of electrostatic force is written, in SI units, as

$$F = \frac{q_1 \times q_2}{4\pi\varepsilon\varepsilon_0 d^2} \tag{1}[1]$$

The charges q_1 and q_2 are expressed in coulombs (C), the distance in meters (m), and the force F in newtons (N). The dielectric constant ε is dimensionless. The permittivity in vacuum is $\varepsilon_0 = 8.854 \times 10^{-12}$ C^2 m^{-1} J^{-1}. Thus, to calculate a coulombic energy, E, we have

$$E\text{(joules)} = \frac{q_1 \times q_2}{4\pi\varepsilon\varepsilon_0 d} \tag{2}$$

Important Constants

   Avogadro's number (^{12}C = 12.000...) $N_A = 6.022 \times 10^{23}$ mol^{-1}
   Electron charge, e $= 4.803 \times 10^{-10}$ abs esu

---

[1] In the old cgs system of units, Eq. (1) was written as $F = q_1 \times q_y/\varepsilon d^2$ in which units were so defined that ε was dimensionless; with ε in vacuum, $\varepsilon = 1$.

(= charge of a proton)	$= 1.602 \times 10^{-19}$ C
1 Faraday	$= 96.490$ C mol^{-1} (= electric charge of 1 mol of electrons)
Electron mass, m	$= 9.1091 \times 10^{-31}$ kg
Permittivity of a vacuum, ε_0	$= 8.854 \times 10^{-12}$ C^2 m^{-1} J^{-1}
Speed of light in a vacuum, c	$= 2.998 \times 10^8$ m s^{-1}
Gas constant, R	$= 8.314$ J mol^{-1} K^{-1}
	$= 0.082057$ liter atm deg^{-1} mol^{-1}
	$= 1.987$ cal deg^{-1} mol^{-1}
Molar volume (ideal gas, 0° C, 1 atm)	$= 22.414 \times 10^3$ cm^3 mol^{-1}
Planck's constant, h	$= 6.626 \times 10^{-34}$ J s
Boltzmann's constant, k	$= 1.3805 \times 10^{-23}$ J K^{-1}
Ice point	$= 273.15$ K
R ln 10	$= 19.14$ J mol^{-1} K^{-1}
$RT_{298.15}$ ln χ	$= 5706.6 \log \chi$ J mol^{-1} or $1364.1 \log \chi$ cal mol^{-1}
RTF^{-1} ln 10	$= 59.16$ mV at 298.15 K
RTF^{-1} ln χ	$= 0.05916 \log \chi$, volt at 298.15 K

The Earth–Hydrosphere System

Earth area	5.1×10^{18} cm^2
Oceans area	3.6×10^{18} cm^2
Land area	1.5×10^{18} cm^2
Atmosphere mass	52×10^{17} kg
Ocean mass	13700×10^{17} kg
Pore waters in rocks	3200×10^{17} kg
Water locked in ice	165×10^{17} kg
Water in lakes, rivers	0.34×10^{17} kg
Water in atmosphere	0.105×10^{17} kg
Total stream discharge	0.32×10^{17} kg year^{-1}
Evaporation = precipitation	4.5×10^{17} kg year^{-1}

Reading Suggestions

Adamson, A. W. (1990), *Physical Chemistry of Surfaces. 5th Edition*, Wiley-Interscience, New York.

Davis, J. A., and D. B. Kent (1990), "Surface Complexation Modeling in Aqueous Geochemistry", in M. F. Jr. Hochella and A. F. White, Eds., *Mineral-Water Interface Geochemistry*, Mineralogical Society of America, pp. 177-260. (A very comprehensive, excellent review.)

Drever, J. I. (1988), *The Geochemistry of Natural Waters, 2nd Ed*, Prentice Hall, NJ, 437 pp. (A modern, easily understandable introduction.)

Dzombak, D. A., and F. M. M. Morel (1990), *Surface Complexation Modeling; Hydrous Ferric Oxide*, Wiley-Interscience, New York. (This book addresses general issues related to surface complexation and its modeling, using the results obtained for hydrous ferric oxide as a basis for discussion.

Lyklema, J. (1991) *Fundamentals of Interface and Colloid Science, Volume I: Fundamentals*, Academic Press, London. (This book treats the most important interfacial and colloidal phenomena starting from basic principles of physics and chemistry.)

Motschi, H. (1987), "Aspects of the Molecular Structure in Surface Complexes; Spectroscopic Investigations", in W. Stumm, Ed., *Aquatic Surface Chemistry*, Wiley-Interscience, New York, pp. 111-124.

Parks, G. A. (1990), "Surface Energy and Adsorption at Mineral/Water Interfaces: An Introduction", in M. F. Jr. Hochella and A. F. White, Eds., *Mineral-Water Interface Geochemistry*, Mineralogical Society of America, pp. 133-175.

Schindler, P. W., and W. Stumm (1987), "The Surface Chemistry of Oxides, Hydroxides, and Oxide Minerals", in W. Stumm, Ed., *Aquatic Surface Chemistry*, Wiley-Interscience, New York, pp. 83-110.

Somorjai, G. A. (1981), *Chemistry in Two Dimensions: Surfaces*, Cornell Univ. Press, Ithaca, N. Y. (An innovative treatment of modern inorganic surface chemistry.)

Sposito, G. (1984), *The Surface Chemistry of Soils*, Oxford University Press, New York. (This monograph gives a comprehensive and didactically valuable interpretation of surface phenomena in soils from the point of view of coordination chemistry.)

Stumm, W. (1987), *Aquatic Surface Chemistry; Chemical Processes at the Particle-Water Interface*, Wiley-Interscience, New York.

Westall, J. C. (1987), "Adsorption Mechanisms in Aquatic Surface Chemistry", in W. Stumm, Ed., *Aquatic Surface Chemistry*, Wiley-Interscience, New York, pp. 3-32.

Chapter 2

The Coordination Chemistry of the Hydrous Oxide-Water Interface

2.1 Introduction

Oxides, especially those of Si, Al and Fe, are abundant components of the earth's crust. Hence a large fraction of the solid phases in natural waters, sediments and soils contain such oxides or hydroxides. In the presence of water the surface of these oxides are generally covered with surface hydroxyl groups (Fig. 2.1).

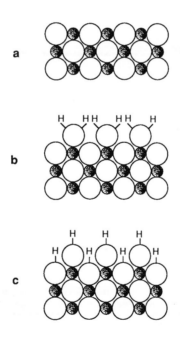

Figure 2.1

Schematic representation of the cross section of the surface layer of a metal oxide. ●, Metal ions; ○, oxide ions. The metal ions in the surface layer (a) have a reduced coordination number. They thus behave as Lewis acids. In the presence of water the surface metal ions may first tend to coordinate H_2O molecules (b). For most of the oxides dissociative chemisorption of water molecules (c) seems energetically favored.

(From P. Schindler, in *Adsorption of Inorganics at the Solid/Liquid Interface,* Anderson, N. and Rubin, A., Eds., Ann Arbor Science, Ann Arbor, 1981)

Geometrical considerations and chemical measurements indicate an average surface density of 5 (typical range 2 – 12) hydroxyls per square nanometer of an oxide mineral.

The various surface hydroxyls formed may structurally and chemically not be fully equivalent, but to facilitate the schematic representation of reactions and of equilibria, one usually considers the chemical reaction of "a" surface hydroxyl group, S–OH[1] (see the remarks on mean field statistics in Chapter 3.7).

These functional groups contain the same donor atoms as found in functional groups of soluble ligands; i.e. the surface hydroxyl group on a hydrous oxide has similar donor properties as the corresponding counterparts in dissolved solutes, such as hydroxides, carboxylates, e.g., (S–OH is a surface group)

$$R\text{–COOH} + Cu^{2+} = RCOOCu^+ + H^+ \tag{2.1}$$
$$S\text{–OH} \quad + Cu^{2+} = S\text{–OCu}^+ \quad + H^+ \tag{2.2}$$

i.e. deprotonated surface groups (S–O⁻) behave as Lewis bases and the sorption of metal ions (and protons) can be understood as competitive complex formation.

The adsorption of ligands (anions and weak acids) on metal oxide (and silicate) surfaces can also be compared with complex formation reactions in solution, e.g.,

$$Fe(OH)^{2+} + F^- = FeF^{2+} + OH^- \tag{2.3}$$
$$S\text{–OH} \quad + F^- = S\text{–F} \quad + OH^- \tag{2.4}$$

The central ion of a mineral surface (in this case we take for example the surface of a Fe(III) oxide and S–OH corresponds to ≡Fe–OH) acts as Lewis acid and exchanges its stuctural OH against other ligands (ligand exchange). Table 2.1 lists the most important adsorption (= surface complex formation) equilibria. The following criteria are characteristic for all surface complexation models: (Dzombak and Morel, 1990.)

i) Sorption takes place at specific surface coordination sites;
ii) Sorption reactions can be described by mass law equations;
iii) Surface charge results from the sorption (surface complex formation) reaction itself; and
iv) The effect of surface charge on sorption (extent of complex formation) can be taken into account by applying a correction factor derived from the electric double layer theory to the mass law constants for surface reactions.

---

[1] The following surface groups can be envisaged (Schindler, 1985):

$$\begin{array}{c} S \\ \diagdown \\ S \diagup \end{array} OH \, , \quad S\text{–}OH \quad S \diagup \!\!\!\!\! \begin{array}{c} OH_2 \\ \diagdown OH \end{array} \quad S \diagup \!\!\!\!\! \begin{array}{c} OH \\ \text{–} OH \\ \diagdown OH \end{array}$$

Table 2.1   Adsorption (Surface Complex Formation Equilibria)

Acid base equilibria

$\quad$ S–OH + H$^+$ $\qquad \rightleftharpoons \qquad$ S–OH$_2^+$

$\quad$ S–OH (+ OH$^-$) $\qquad \rightleftharpoons \qquad$ S–O$^-$ + (+ H$_2$O)

Metal binding

$\quad$ S–OH + M^{z+} $\qquad \rightleftharpoons \qquad$ S–OM$^{(z-1)+}$ $\quad$ + H$^+$

$\quad$ 2 S–OH + M^{z+} $\qquad \rightleftharpoons \qquad$ (S–O)$_2$M$^{(z-2)+}$ $\quad$ + 2 H$^+$

$\quad$ S–OH + M^{z+} + H$_2$O $\qquad \rightleftharpoons \qquad$ S–OMOH$^{(z-2)+}$ + 2 H$^+$

Ligand exchange (L$^-$ = ligand)

$\quad$ S–OH + L$^-$ $\qquad \rightleftharpoons \qquad$ S–L $\quad$ + OH$^-$

$\quad$ 2 S–OH + L$^-$ $\qquad \rightleftharpoons \qquad$ S$_2$–L$^+$ + +2 OH$^-$

Ternary surface complex formation

$\quad$ S–OH + L$^-$ + M^{z+} $\qquad \rightleftharpoons \qquad$ S–L–M^{z+} + OH$^-$

$\quad$ S–OH + L$^-$ + M^{z+} $\qquad \rightleftharpoons \qquad$ S–OM–L$^{(z-2)+}$ + H$^+$

From Schindler and Stumm, 1987 (modified)

2.2 The Acid – Base Chemistry of Oxides; pH of Zero Point of Charge

Uptake and release of protons can be described by the acidity constants (assuming a solution of constant ionic strength, we imply that the activity coefficients of the surface species are equal):

$$K_{a1}^s = \frac{\{SOH\}\,[H^+]}{\{SOH_2^+\}} \; mol/\ell \qquad\qquad (2.5)$$

$$K_{a2}^s = \frac{\{SO^-\}\,[H^+]}{\{SOH\}} \; mol/\ell \qquad\qquad (2.6)$$

where { } denotes the concentrations of surface species in moles per kilogram of adsorbing solid and [ ] denotes the concentrations of solutes [M][1]. In many cases, it is more desirable to express the concentrations of the surface species as surface densities (mol m^{-2}) or in the same units as the concentrations of dissolved species mol ℓ^{-1}]. Then conversion is easily accomplished with the equations.

---

[1] M means mol/ℓ. We will frequently use solute concentrations (rather than activities). Often the experiments are done in a constant ionic medium. If the concentrations of the solutes are smaller than the concentrations of the background electrolyte, it is justified setting the aqueous phase activity coefficients equal to 1 on the scale of the constant ionic medium reference state (Stumm and Morgan, 1981).

$$\langle SOH \rangle = s^{-1}\{SOH\} \ [\text{mol m}^{-2}] \tag{2.7a}$$

$$[SOH] = a\{SOH\} \ [\text{mol } \ell^{-1}] \tag{2.7b}$$

where $\langle SOH \rangle$ is the surface concentration in mol m^{-2} and s = specific surface area of the solid [m^2 kg^{-1}] and a is the quantity of oxide used [kg ℓ^{-1}].

Example 2.1: Evaluation of Surface Charge from Alkalimetric and Acidimetric Titration Curves and Determination of Surface Acidity Constants

We will demonstrate how the surface charge of a hydrous oxide (α-FeOOH) can be calculated from an experimental titration curve[1] (e.g., Fig. 2.2).

In titrating a suspension of α-FeOOH (6 g ℓ^{-1}, 120 m^2 g^{-1}; 2×10^{-4} mol g^{-1} surface functional groups ($\equiv$FeOHTOT)) in an inert electrolyte (10^{-1} M NaClO$_4$) with NaOH or HCl (C_B and C_A = concentration of base and acid, respectively, added per liter), we can write for any point on the titration curve

$$C_A - C_B + [OH^-] - [H^+] = [\equiv FeOH_2^+] - [\equiv FeO^-] \tag{i}$$

where [] indicates concentrations of solute and surface species per unit volume solution (M). Equation (i) can also be derived from a charge balance. The right-hand side gives the net number of moles per liter of H$^+$ ions bound to α-FeOOH. The mean surface charge (i.e., the portion of the charge due to OH$^-$ or H$^+$) can be calculated as a function of pH from the difference between total added base or acid and the equilibrium OH$^-$ and H$^+$ ion concentration for a given quantity a (kg liter^{-1}) of oxide used:

$$\frac{C_A - C_B + [OH^-] - [H^+]}{a} = \{\equiv FeOH_2^+\} - \{\equiv FeO^-\} = Q \tag{ii}$$

where { } indicates the concentration of surface species in mol kg^{-1} (e.g., [$\equiv$FeO$^-$] / a = {$\equiv$FeO$^-$}). If the specific surface area s (m^2 kg^{-1}) of the iron oxide used is known (in this case 1.2×10^5 m^2 kg^{-1}), the surface charge σ (C m^{-2}) can be calculated:

$$\sigma = QFs^{-1} = F(\Gamma_{H^+} - \Gamma_{OH^-}) \tag{iii}$$

---

[1] [H$^+$] is measured potentiometrically with a glass electrode. Briefly, the method involves the use of a glass electrode and a double-junction calomel reference electrode in the titration cell:

glass electrode	suspension of solid background electrolyte	background electrolyte solution	liquid junctions	calomel electrode

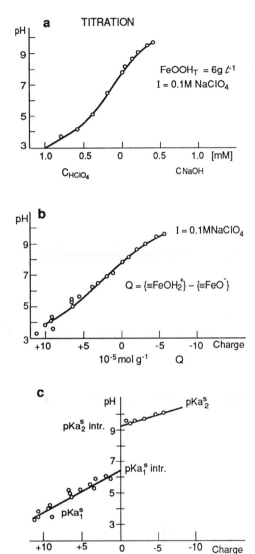

Figure 2.2

Titration of a suspension of α-FeOOH (goethite) in absence of specifically adsorbable ions.

a) Acidimetric-alkalimetric titration in the presence of an inert electrolyte

b) Charge calculated from the titration curve (charge balance)

c) Microscopic acidity constants calculated from a) and b). Extrapolation to charge zero gives intrinsic pK_{a1}^s and pK_{a2}^s.

(Data from Sigg and Stumm, 1981)

where F is the Faraday constant (96,490 C mol^{-1}) and Γ_{H^+} and Γ_{OH^-} are the "adsorption" densities of H^+ and OH^- (mol m^{-2}).

The *point of zero charge* pH$_{pzc}$ corresponds to the *zero proton condition* at the surface:

pH$_{pzc}$ (point of zero charge) (for definitions of points of zero charge see Chapter 3.2):

$$\{\equiv FeOH_2^+\} = \{\equiv FeO^-\}; \qquad \Gamma_{H^+} = \Gamma_{OH^-} \tag{iv}$$

In this case pH$_{pzc}$ = 7.9 (cf. Fig. 2.2b).

We can now calculate the surface acidity equilibrium constants (Eqs. 2.5, 2.6). There are 5 species: $\equiv FeOH_2^+$, $\equiv FeOH$, $\equiv FeO^-$, H^+, OH^-, that are interrelated by the two acidity mass law constants (Eqs. 2.5, 2.6), by the ion product of water ($K_W' = [H^+][OH^-]$, where $[H^+]$ and $[OH^-]$ are the activities of H^+ and OH^-, respectively and two mass balance equations:

$$C_A - C_B \quad = [\equiv FeOH_2^+] - [\equiv FeO^-] + [H^+] - [OH^-]$$

(compare Eq. i) and

$$[\equiv FeOHTOT] = [\equiv FeOH_2^+] + [\equiv FeOH] + [\equiv FeO^-] \tag{v}$$

The calculation is facilitated by the following assumptions:

$$Q \cong \{\equiv FeOH_2^+\} \text{ for pH} < pH_{pzc} \;\; ; \;\; Q \cong \{\equiv FeO^-\} \text{ for pH} > pH_{pzc}$$

Then, Eq. (2.5) becomes

$$K_{a1}^s = \frac{(\{\equiv FeOHTOT\} - Q) [H^+]}{Q} \qquad \text{for pH} < pH_{pzc}$$

and Eq. (2.6)

$$K_{a2}^s = \frac{Q [H^+]}{(\{\equiv FeOHTOT\} - Q)} \qquad \text{for pH} > pH_{pzc}$$

The acidity constants calculated from every point in the titration curve (Figure 2.2a and b) are microscopic acidity constants (Eqs. 2.5, 2.6). Each loss of a proton reduces the charge on the surface and thus affects the acidity of the neighboring groups. The free energy of deprotonation consists of the *dissociation* as measured by an intrinsic acidity constant, i.e., a constant valid for an uncharged surface, and the *removal* of the proton from the site of the dissociation in the bulk of the solution. As shown in Fig. 2.2, the intrinsic values for the acidity constants can be obtained by linear extrapolation of the log K^s vs charge, Q, curve to the zero charge condition; (as we shall show later, this somewhat empirical approach can be justified theoretically).

$$\log K_a^s = \log K_a^s(int) - \beta Q \tag{vi}$$

where Q is the surface charge in mol kg^{-1}, and β is a coefficient.

The pH$_{pzc}$ (zero proton condition, point of zero charge) is not affected by the concentration of the inert electrolyte. As Fig. 2.3 shows, there is a common intersection point of the titration curves obtained with different concentrations of inert electrolyte.

Obviously at the condition where $[\equiv FeOH_2^+] = [\equiv FeO^-]$

$$pH_{pzc} = 0.5\,[pK_{a1}(int) + pK_{a2}(int)] \tag{vii}$$

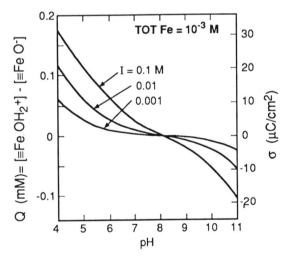

Figure 2.3

Surface charge as a function of pH and ionic strength (1 : 1 electrolyte) for a 90-mg/ℓ (TOTFe = 10^{-3} M) suspension of hydrous ferric oxide.

(From Dzombak and Morel, 1990)

Points of Zero Charge. Points of zero charge (pzc) are pH-values where the net surface charge is zero. We consider here above all the surface conditions where the charge is established by proton exchange at the surface. For a more detailed definition of points of zero charge see Chapter 3.1. If the surface charge is established solely by H^+ exchange (binding and dissociation of H^+) one may also refer to the point of zero net proton charge, pznpc. We can estimate the pH_{pznpc} of metal oxides from electrostatic considerations. Parks (1967) has shown that the pH_{zpnc} of a simple oxide is related to the appropriate cationic charge and radius of the central ion. As shown by Parks and illustrated by a few examples in Table 2.2, the points of zero charge of a composite oxide is approximately the weighted average of the values of its components. Predictable shifts in points of zero charge occur in response to state of hydration, cleavage habit, and crystallinity.

The points of zero charge of salt-type minerals depends, sometimes in a complicated way, upon pH and on the concentration (activities) of all potential-determining ions. Thus, in the case of calcite, possible potential-determining species, in addition to H^+ and OH^-, are HCO_3^-, CO_2, and Ca^{2+}; various mechanisms of charge development are possible. When referring to a point of zero charge of such non-oxides the solution composition should be specified. In the absence of complications such as those caused by structural or adsorbed impurities, the point of zero charge of the solid should correspond to the pH of charge balance (electroneutrality) of potential-determining ions.

Table 2.2   Point of Zero Charge caused by Binding or Dissociation of Protons [a]

Material	pH_{pznpc}	Material	pH_{pznpc}
α-Al_2O_3	9.1	δ-MnO_2	2.8
α-$Al(OH)_3$	5.0	β-MnO_2	7.2
γ-$AlOOH$	8.2	SiO_2	2.0
CuO	9.5	$ZrSiO_4$	5
Fe_3O_4	6.5	Feldspars	2 – 2.4
α-$FeOOH$	7.8	Kaolinite	4.6
α-Fe_2O_3	8.5	Montmorillonite	2.5
"$Fe(OH)_3$" (amorph)	8.5	Albite	2.0
MgO	12.4	Chrysotile	>10

[a]   The values are from different investigators who have used different methods and are not necessarily comparable. They are given here for illustration.

Table 1.1 summarizes some of the concepts of the coordination chemistry of the solid-water interface and illustrates some important applications in natural and technical systems. Some of these applications will be discussed in later chapters.

2.3 Surface Complex Formation with Metal Ions

Surface complex formation of cations by hydrous oxides involves the coordination of the metal ions with the oxygen donor atoms and the release of protons from the surface, e.g.,

$$S–OH + Cu^{2+} \rightleftharpoons S–OCu^+ + H^+ \tag{2.8}$$

There is also the possibility that bidentate surface complexes are formed:

$$2 \, S–OH + Cu^{+2} = (S–O)_2Cu + 2 \, H^+ \tag{2.9a}$$

$$
\begin{array}{l}
-S-OH \\
\quad | \qquad + Cu^{2+} \rightleftharpoons \\
-S-OH
\end{array}
\begin{array}{l}
-S-O \\
\quad | \qquad \diagdown \\
-S-O \diagup
\end{array} Cu + 2 \, H^+ \tag{2.9b}
$$

Eqs. (2.8) and (2.9) can also be formulated in terms of mass laws, e.g.,

$$K_{Cu}^s = \frac{\{S–OCu^+\} \, [H^+]}{\{S–OH\} \, [Cu^{2+}]} \tag{2.10a}$$

$$\beta_{2Cu}^s = \frac{\{(S–O)_2Cu\} \, [H^+]^2}{\{(S–OH)_2\} \, [Cu^{2+}]} \tag{2.10b}$$

These constants have the rank of conditional stability constants. For exact considerations we need to correct by a coulombic term for electrostatic interaction (see Chapter 4).

Inner-sphere and Outer-sphere Complexes

As Fig. 2.4 illustrates, a cation can associate with a surface as an inner sphere, or outer-sphere complex depending on whether a chemical, i.e., a largely covalent bond, between the metal and the electron donating oxygen ions, is formed (as in an inner-sphere type solute complex) or if a cation of opposite charge approaches the surface groups within a critical distance; as with solute ion pairs the cation and the base are separated by one (or more) water molecules. Furthermore, ions may be in the diffuse swarm of the double layer.

It is important to distinguish between outer-sphere and inner-sphere complexes. In inner-sphere complexes the surface hydroxyl groups act as σ-donor ligands which increase the electron density of the coordinated metal ion. Cu(II) bound inner-spherically is a different chemical entity than if it were bound outer-spherically or present in the diffuse part of the double layer; the inner-spheric Cu(II) has chemically different properties, e.g., a different redox potential (with regard to Cu(I)) and its equatorial water are expected to exchange faster than in Cu(II). As we shall see (Chapters 5, 9), the reactivity of a surface is affected by inner-sphere complexes.

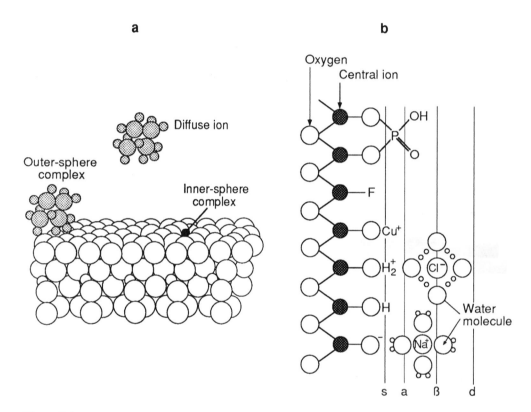

Figure 2.4

Surface complex formation of an ion (e.g., cation) on the hydrous oxide surface. The ion may form an inner-sphere complex ("chemical bond"), an outer-sphere complex (ion pair) or be in the diffuse swarm of the electric double layer. (From Sposito, 1989)

Fig. b shows a schematic portrayal of the hydrous oxide surface, showing planes associated with surface hydroxyl groups ("s"), inner-sphere complexes ("a"), outer-sphere complexes ("β") and the diffuse ion swarm ("d"). (Modified from Sposito, 1984)

Structural Identity

Direct evidence for inner-sphere complexes comes from spectroscopic methods; unfortunately, spectroscopic methods alone are seldom sufficiently sensitive to reveal the specific structure of surface complexes. Motschi (1987) used electron spin resonance spectroscopy to study Cu(II) surface complexes. Additional studies were carried out with electron nuclear double resonance spectroscopy (ENDOR) and electron spin echo envelope modulation (ESEEM) in order to elucidate structural aspects of surface-bound Cu(II), of ternary copper complexes (in which coordinated water is replaced by ligands), and of vanadyl ions on δ-Al$_2$O$_3$. Application of ENDOR spectroscopy allows the resolution of weak interactions between the unpaired electron and nuclei within a distance of about 5 Å. From these so-called hyperfine data, structural parameters can be derived, e.g., bond distances of the paramagnetic center to the coupling nuclei or ligands. In the ENDOR spectrum of adsorbed VO^{2+} on δ-Al$_2$O$_3$, signals caused by the coupling with the surface Lewis center (^{27}Al) are more strongly split than is calculated from molecular modeling. The existence of an inner-sphere coordination between the hydrated oxide and the metal is confirmed experimentally (Motschi, 1987).

Direct in situ X ray (from synchroton radiation) adsorption measurements (EXAFS) (Hayes et al., 1987, Brown et al., 1989) permit the determination of adsorbed species to neighboring ions and to central ions on oxide surfaces in the presence of water. Such investigations showed, for example, that selenite is inner-spherically and selenate is outer-spherically bound to the central Fe(III) ions of a goethite surface. It was also shown by this technique that Pb(II) is inner-spherically bound to δ-Al$_2$O$_3$ (Chisholm-Brause et al., 1989).

A simple method to distinguish between inner-sphere and outer-sphere complexes is to assess the effect of ionic strength on the surface complex formation equilibria. A strong dependence on ionic strength is typical for an outer-sphere complex. Furthermore, outer-sphere complexes involve electrostatic bonding mechanisms, and therefore are less stable than inner-sphere surface complexes, which necessarily involve largely covalent bonding or some combinations of covalent and ionic bonding.

As we shall see (Chapter 4), the kinetics of surface complex formation is often related to the rate of H$_2$O loss from the aquo cation. This is another (indirect) evidence for inner-sphere complex formation.

pH-Dependence of Surface Binding

As evidenced by the mass laws of Eqs. (2.8), (2.9), the binding of a metal ion by surface ligands – similar to the binding of a metal ion by a solute ligand – is strongly pH dependent (Fig. 2.5a). Complex formation is competitive (e.g., metal ion vs H$^+$ ion or vs another metal ion). Fig. 2.5b illustrates the sorption of various metal ions on hydrous ferric oxide. For each metal ion there is a narrow interval of 1 – 2 pH units where the extent of sorption rises from zero to almost 100 %.

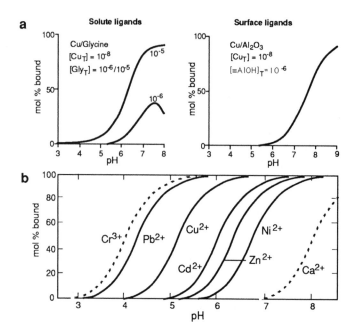

Figure 2.5

pH-dependence of the binding of metal ions by solute and surface ligands

a) Comparison of the complexation of Cu^{2+} by dissolved ligand (glycine) and by surface OH groups of Al_2O_3 as a function of pH. (The curves are calculated on the basis of experimentally determined equilibrium constants.)

b) Extent of surface complex formation as a function of pH (measured as mol % of the metal ions in the system adsorbed or surface bound).
[TOTFe] = 10^{-3} M (2×10^{-4} mol reactive sites ℓ^{-1}); Metal concentrations in solution = 5×10^{-7} M; I = 0.1 M $NaNO_3$. (The curves are based on data compiled by Dzombak and Morel, 1990.)

2.4 Ligand Exchange; Surface Complex Formation of Anions and Weak Acids

The main mechanism of ligand adsorption is ligand exchange; the surface hydroxyl is exchanged by another ligand. This surface complex formation is also competitive; OH^- ions and other ligands compete for the Lewis acid of the central ion[1] of the hydrous oxide (e.g., the Al(III) or the Fe(III) in aluminum or ferric (hydr)oxides). The extent of surface complex formation (adsorption) is, as with metal ions, strongly

---

[1] A Lewis acid site is a surface site capable of receiving a pair of electrons from the adsorbate. A Lewis base is a site having a free pair of electrons (like an oxygen donor atom in a surface OH-group) that can be transferred to the adsorbate.

dependent on pH. Since the adsorption of anions is coupled with a release of OH⁻ ions, adsorption is favored by lower pH values (Fig. 2.6) e.g.,

$$\equiv AlOH + F^- \rightleftharpoons \equiv AlF + OH^- \tag{2.11}$$

With bidentate ligands (mono nuclear or binuclear), surface chelates are formed.

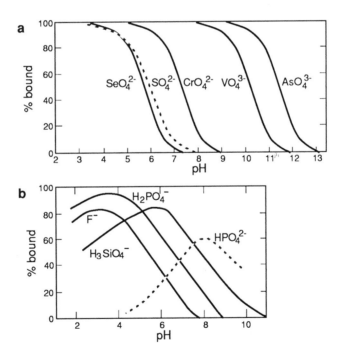

$$(2.12)$$

$$(2.13)$$

Figure 2.6

Surface complex formation with ligands (anions) as a function of pH

a) binding of anions from dilute solutions (5×10^{-7} M) to hydrous ferric oxide [TOTFe = 10^{-3} M]. Based on data from Dzombak and Morel, 1990. I = 0.1.

b) binding of phosphate, silicate and fluoride on goethite (α-FeOOH); the species shown are surface species. (6g FeOOH per liter, $P_T = 10^{-3}$ M, $Si_T = 8 \times 10^{-4}$ M.) (Sigg and Stumm, 1981).

(The curves are calculated with the help of experimentally determined equilibrium constants.)

The extent of surface coordination and its pH dependence can again be explained by considering the affinity of the surface sites for metal ion or ligand and the pH dependence of the activity of surface sites and ligands. The tendency to form surface complexes may be compared with the tendency to form corresponding (inner-sphere) solute complexes (Fig. 2.7), e.g.,

$$\equiv Fe\text{-}OH + F^- \quad \rightleftharpoons \quad \equiv Fe\text{–}F + OH^- \tag{2.14a}$$

$$Fe(OH)^{2+}(aq) + F^- \rightleftharpoons FeF^{2+} + OH^- \tag{2.14b}$$

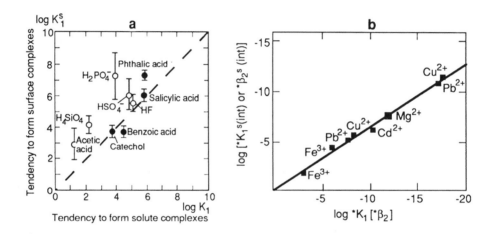

Figure 2.7

Linear free energy relations between the tendency to form solute complexes and corresponding surface complexes.

a) Comparison between intrinsic equilibrium constants of the reactions

$$\equiv MeOH \quad + H_2A \rightleftharpoons \equiv MeHA \quad + H_2O \quad ; \ K_1^s$$
$$MeOH^{2+} + H_2A \rightleftharpoons \equiv MeHA^{z+} + H_2O \quad ; \ K_1$$

where $\bigcirc$ Me = Fe^{3+}, and $\bullet$ Me = Al^{3+}.
(From Sigg and Stumm, 1981)

b) Correlation of stability constants of surface complexes of amorphous silica with metal ions

$$\equiv SiOH \quad + Me^{z+} \rightleftharpoons \equiv SiOMe^{(z-1)} \quad + H^+ \quad ^*K_1^s$$
$$2 \equiv SiOH \quad + Me^{z+} \rightleftharpoons (\equiv SiO)_2Me^{(z-2)} \quad + H^+ \quad ^*\beta_2^s$$

with corresponding hydrolysis reactions

$$H_2O \quad + Me^{z+} \rightleftharpoons MeOH^{(z-1)+} \quad + H^+ \quad ^*K^s$$
$$2 H_2O \quad + Me^{z+} \rightleftharpoons Me(OH)_2^{(z-2)+} \quad + 2 H^+ \quad ^*\beta_2^s$$

(From Schindler, 1985)

As with metal complexes, we can distinguish between outer-sphere and inner-sphere ligand complexes. Orbital overlap with the surface metal centers changes the electro density in inner-sphere ligand complexes and may reduce the activation energy for reduction of a higher valent surface metal center such as Fe(III) or Mn(III,IV). The orbital overlap in outer-sphere complexes is much smaller than in inner-sphere complexes.

Ternary Complexes

Since the coordination sphere of a complex of a metal on the surface of a hydrous oxide is only partially occupied by the surface ligands, further ligands may be acquired to form a ternary complex (type A) (Schindler, 1990).

$$S-OH + Me^{2+} + L^- \rightleftharpoons S-OMe-L + H^+ \qquad (2.15)$$

A (type B) ternary complex can also be formed when it is a polydentate ligand

$$S-OH + L^- + Me^{2+} \rightleftharpoons S-L-Me^{2+} + OH^- \qquad (2.16)$$

Hydrolysis and Adsorption. Some years ago, a theory was advanced, that hydrolyzed metal species, rather than free metal ions, are adsorbed to hydrous oxides. The pH-dependence of adsorption (the pH edge for adsorption is often close to the pH for hydrolysis) was involved to account for this hypothesis. As Figs. 2.7b and c illustrate, there is a correlation between adsorption and hydrolysis; but this correlation is caused by the tendency of metal ions to interact chemically with the oxygen donor atoms with OH^-, and with $S-OH$. The kinetic work of Hachiya et al. (1984) and spectroscopic information are in accord with the reaction of (free) metal ions with the surface.

There is however the possibility, especially with trivalent ions, and within given pH-ranges, that surface hydroxo species can be formed.

$$S-OH + Al^{3+} + H_2O \rightleftharpoons S-O-AlOH^+ + 2\,H^+$$

Example 2.2: pH-Dependence of Surface Complex Formation

In order to exemplify simple complex formation equilibria, we calculate the pH-dependence of the binding of a) a metal ion, Me^{2+}, and b) of a ligand, A^-, to a hydrous oxide. The metal oxide, $S-OH$, is characterized by two surface acidity constants $pK_1^s = 4$ and $pK_2^s = 9$. Its specific surface area is $10\ m^2\ g^{-1}$. The surface contains 10^{-4} mol surface sites per gram (~ 6 sites nm^{-2}) and we use $1\ g\ \ell^{-1}$. Thus, there are 10^{-4} mol surface sites per liter solution. The ligand A^- is characterized by a pK-value of its conjugate acid of $pK_{HA} = 5$. The total concentration of the adsorbates are a) TOTMe = 10^{-7} M and b) TOTHA = 10^{-7} M. We first make the calculation without correction for electrostatic interaction (we will come back in Chapter 3 (Example 3.3, Figs. 3.14 and 3.15) with the same problem and repeat the calculations using

Gouy Chapman theory). We use the following equilibrium constants:

$$S - OH_2^+ \rightleftharpoons S - OH + H^+ \quad ; \quad \log K_1^s = -4 \qquad \text{(i)}$$

$$S - OH \rightleftharpoons S - O^- + H^+ \quad ; \quad \log K_2^s = -9 \qquad \text{(ii)}$$

$$S - OH + Me^{2+} \rightleftharpoons S - OMe^+ + H^+ \quad ; \quad \log K_M^s = -1 \qquad \text{(iii)}$$

$$S - OH + HA \rightleftharpoons S - A + H_2O \quad ; \quad \log K_L^s = 5 \qquad \text{(iv)}$$

$$HA = H^+ + A^- \qquad\qquad \log K_{HA} = -5 \qquad \text{(v)}$$

For a) the following total concentrations can be defined:

$$\text{Tot SOH} = [SOH_2^+] + [SOH] + [SO^-] + [SOMe^+] = 10^{-4} \text{ M} \qquad \text{(vi a)}$$

$$\text{Tot Me} = [Me^{2+}] + [SOMe^+] \qquad\qquad = 10^{-7} \text{ M} \qquad \text{(vii a)}$$

For b):

$$\text{TOT SOH} = [SOH_2^+] + [SOH] + [SO^-] + [SA] \quad = 10^{-4} \text{ M} \qquad \text{(vi b)}$$

$$\text{TOT HA} = [HA] + [A^-] + [SA] \qquad\qquad = 10^{-7} \text{ M} \qquad \text{(vii b)}$$

We keep this example as simple as possible so that it can be calculated by hand.

There are various ways to approach the solving of the problem. One way is to use a trial and error approach expressing Eqs. (vi a) and (vii a) as functions of $[H^+]/[Me^{2+}]$ and $[SOH]$:

$$[\text{TOT SOH}] = 10^{-4} \text{ M} = [SOH] \left(\frac{[H^+]}{K_1^s} + 1 + \frac{K_2^s}{[H^+]} + \frac{K_M^s [Me^{2+}]}{[H^+]} \right) \qquad \text{(viii a)}$$

$$[\text{TOT Me}] = 10^{-7} \text{ M} = [Me^{2+}] \left(1 + \frac{K_M^s}{[H^+]} [SOH] \right) \qquad \text{(ix a)}$$

These two equations contain (for any preselected $[H^+]$ two unknowns ($[Me^{2+}]$ and $[SOH]$), and can readily be solved simultaneously by trial and error (systematic variations of assumed values of $[SOH]$ and $[Me^{2+}]$) until the left and right hand sight of these equations are equal; since $[SOH] \gg [Me^{2+}]$, one may start by assuming $[SOH] \approx 10^{-4}$ M.

Results are given in Figs. 2.8a and b. Of course such calculations are more conveniently carried out with a computer program. The solution techniques are described by Dzombak and Morel (1990). A program, to include adsorption on charged surfaces, developed by Westall (1979) (MICROQL, II) has been adapted to personal computers. The same approach can be used to calculate adsorption equilibrium of the ligand A⁻.

The two equations, expressing total concentrations, are:

$$[\text{TOT SOH}] = 10^{-4}\,M = [\text{SOH}]\left(\frac{[H^+]}{K_1^s} + 1 + \frac{K_2^s}{[H^+]} + K_L^s\,[\text{HA}]\right) \qquad \text{(viii b)}$$

$$[\text{TOT HA}] = 10^{-7}\,M = [\text{HA}]\left(1 + \frac{K_{HA}}{[H^+]} + K_L^s\,[\text{SOH}]\right) \qquad \text{(ix b)}$$

The final results are displayed in Figs. 2.9a and b. Note that so far no corrections have been made for electrostatic effects (cf. Chapter 3 (Fig. 3.15)).

Figs. 2.8 and 2.9 exemplify the typical pH dependence for cation and ligand adsorption. As Fig. 2.9 illustrates the adsorption of the ligand A (or HA) goes through a maximum at a pH value that is near the pK value of HA. All kind of explanations have been given for the pH-dependence of this maximum; it is important to realize that this maximum is a consequence of the mass law.

Surface Catalyzed Ester Hydrolysis

Mineral surfaces may accelerate the rate of ester hydrolysis (Stone, 1989; Hoffmann, 1990; Torrents and Stone, 1991). One plausible scheme for this heterogeneous catalysis assumes a nucleophilic addition of the ester to the surface functional group, e.g., in case of a carboxylic acid ester

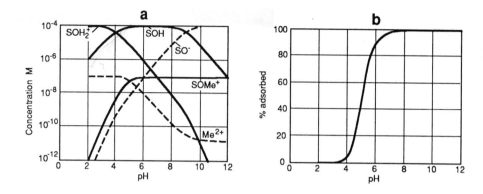

Figure 2.8

Metal binding by a hydrous oxide from a 10^{-7} M solution (SOH + Me^{2+} $\rightleftharpoons$ SOMe$^+$ + H$^+$) for a set of selected equilibrium constants (see Eqs. (i) – (iii)) and concentration conditions (see text). No corrections have been made for electrostatic interactions (compare Fig. 3.14). The pH edge reflects a narrow pH range in which the metal ion is adsorbed.

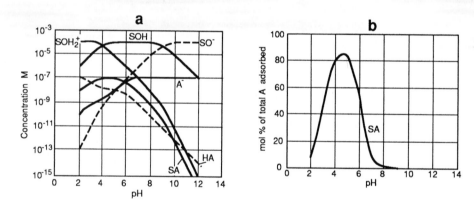

Figure 2.9

Ligand binding by a hydrous oxide from a 10^{-7} M solution (SOH + HA $\rightleftharpoons$ SA + H$_2$O) for a set of selected equilibrium constants (see Eqs. (i), (ii), (iv), (v)) and concentration conditions (see text). No corrections have been made for electrostatic interactions (compare Fig. 3.15).

Fig. b illustrates that a maximum of adsorption of the ligand occurs near the pK value of its conjugate acid.

2.5 Affinities of Cations and Anions for Surface Complex Formation with Oxides and Silicates

The relationship given in Fig. 2.7 illustrates the affinities of cations and anions to the Lewis bases and Lewis acids of oxide and silicate surfaces. For alkali and earth alkali cations the tendency to become sorbed increases with the ionic radius of the ion:

$$Cs^+ > Rb^+ > K^+ > Na^+ > Li^+$$
$$Ba^{2+} > Sr^{2+} > Ca^{2+} > Mg^{2+}$$

(2.17)

This selectivity sequence conform to what has often been observed in soil adsorption experiments. For transition elements, the electron configuration of the ions influences the adsorption affinity. The Irving-Williams order

$$Cu^{2+} > Ni^{2+} > Co^{2+} > Fe^{2+} > Mn^{2+}$$

(2.18)

is often observed (Fig. 2.10).

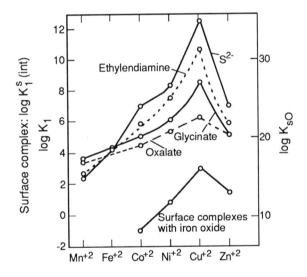

Figure 2.10

Stability constants (ethylendiamine, glycinate, oxalate), surface complex formation constants and solubility products (sulfides) of transition ions. The surface complex formation constant is for the binding of metal ions to hydrous ferric oxide: $\equiv Fe{-}OH + Me^{2+} \rightleftharpoons \equiv FeOMe^+ + H^+; K_1^s$.

The anions Cl^-, NO_3^-, ClO_4^-, for some oxides also SO_4^{2-} and SeO_4^{2-} are considered to adsorb mainly in outer-sphere complexes and as diffuse ion swarm.

Extensive tabulations on experimentally determined surface equilibrium constants (Schindler and Stumm, 1987; Dzombak and Morel, 1990) reflecting the acid-base characteristic of surface hydroxyl groups and the stability of surface metal complexes, of anion (ligand) complexes and of ternary surface complexes are now available and assist the application to practical problems. With the help of such constants one can describe how the concentration of reactive elements (metal ions, ligands) in soils, waters and sediments depend on the distribution between surfaces and the solution and on pH and other solution variables. The geochemical fate of most trace elements is controlled by the reaction of solutes with solid surfaces.

Example 2.3: Stoichiometry of H+ Release by Binding of Metal Ions;
                    The Kurbatov Plot

Metal ion binding to hydrous oxides can occur as monodentate or bidentate surface complexes (Eqs. 2.9a and 2.9b) where, respectively, one or two protons are released per mol of metal ion bound. Develop a simple graphical method to distinguish between monodentate and bidentate metal binding.

In principle, it appears possible to distinguish between the formation of monodentate and bidentate surface complexes if the stoichiometry of the H^+ release is know. A mean surface complex stoichiometry can be formulated

$$n \equiv S{-}OH + Me^{2+} \rightleftharpoons (\equiv S{-}O)_n\, Me^{(2-n)} + n\, H^+; \quad \beta_n^s \qquad (i)$$

Taking the log of the mass law of (i) (and setting $[Me_{ads}]$ for $[(\equiv S{-}O)_n\, Me^{(2-n)}]$) one obtains

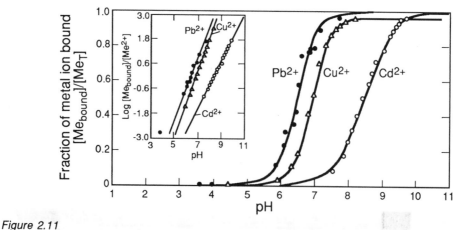

Figure 2.11

Binding of metal ions by amorphous silica. 300 mg SiO_2 dm^{-3}, $[Me_T \approx 10^{-5}$ M$]$. The fraction of metal ion bound was calculated from peak currents measured with differential pulse polarography.
In the inset, linearized plots of log $[Me_{bound}]/[Me^{2+}]$ are plotted as a function of pH. The slopes n = d log $[Me^{2+}]$ /d pH are 1.72, 1.77 and 1.12 for Pb^{2+}, Cu^{2+} and Cd^{2+}, respectively.
(From Wang and Stumm, 1987)

$$\log \frac{[Me_{ads}]}{[Me^{2+}]} = \log \beta_n^s + n \log [\equiv S\text{–}OH] + n \, pH \qquad \text{(ii)}$$

In a simplified form Eq. (ii) was used, decades ago, to assess metal ion adsorption to surfaces, by plotting $\log ([Me_{ads}] / [Me^{2+}])$ vs pH. (Kurbatov et al., 1951). The slope of this curve gives an idea on n. The model for this "Kurbatov-plot" assumes that the adsorbent $\equiv S$ is present in large excess and that the adsorption at constant pH is not affected by surface charge. Fig. 2.11 gives an example for the binding of metal ions to amorphous SiO_2.

The Kurbatov plot is a convenient tool to display in a simple way adsorption (surface complex formation) data. But care must be exercised in the interpretation of the data, because n varies with pH and may vary with the adsorption coverage. For an exact analysis of the proton release stoichiometry, see Hohl and Stumm (1976) or Honeyman and Leckie (1986).

Example 2.4: Shift in the Alkalimetric Titration Curve of an Oxide in the Presence of an Adsorbable Metal Ion or Ligand

We resume the Example 2.1, where we calculated the surface charge of FeOOH from alkalimetric (acidimetric) titration curves as a function of pH. We now make a similar calculation, using an Al_2O_3 suspension; but we add to this suspension a dilute $Pb(NO_3)_2$ solution from which the Al_2O_3 adsorbs some of the Pb^{2+} ions to form $\equiv AlOPb^+$. The adsorption of Pb^{2+} causes a shift in the titration curve (Fig. 2.12a) because of the reaction

$$\equiv AlOH + Pb^{2+} = \equiv AlOPb^+ + H^+ \qquad \text{(i)[1)]}$$

The charge balance (written for all species in mol/ℓ) in the absence of Pb^{2+} is as before (Example 2.1)

$$C_A - C_B + [OH^-] - [H^+] = [\equiv AlOH_2^+] - [\equiv AlO^-] \qquad \text{(ii)}$$

where C_A is the concentration [M] of a strong acid added (e.g., $[HClO_4]$ added = $[ClO_4^-]$) and C_B, the concentration of a strong base added (e.g., $[NaOH]_{added} = [Na^+]$). Alternatively, Eq. (ii) can be written as a charge Q_H (in mol kg^{-1}) or Γ_H (in mol m^{-2}) as

$$Q_H = \frac{C_A - C_B + [OH^-] - [H^+]}{a} = \{\equiv AlOH_2^+\} - \{\equiv AlO^-\} \qquad \text{(iia)}$$

[1)] In the corresponding experimental evaluation, a bidentate $(\equiv AlO)_2Pb$ complex was formed in addition to $\equiv AlOPb^+$. To keep our argumentation simple, we neglect the bidentate species (Hohl and Stumm, 1976).

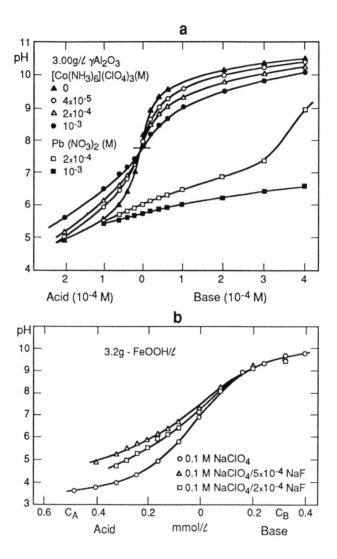

Figure 2.12

a) Comparison of the interaction of Al_2O_3 with $[CO(NH_3)_6]^{3+}$ with that of Pb^{2+}. The kinetically inert co-balt complex cannot form inner-sphere complexes and thus behaves like an inert electrolyte (common intersection at point of zero salt effect; cf. Eq. 3.6) whereas Pb^{2+} is specifically adsorbed to the positively charged aluminum oxide. The adsorption is accompanied by a displacement of the titration curve ($\equiv Al-OH + Pb^{2+} \rightleftharpoons \equiv AlOPb + H^+$). The shift in the titration curve at constant pH can be used to determine the extent of Pb^{2+} adsorption. The adsorption is accompanied by a decrease in surface protonation of the oxide surface (from Stumm, Hohl, Dalang, 1976).

b) The specific adsorption (surface concentration) of F^- causes a displacement of the titration curve for α-FeOOH from which the extent of adsorption and the resulting surface charge can be calculated. The adsorption is accompanied by an increase in surface protonation (from Sigg and Stumm, 1981) or

$$\sigma_H = Q_H F s^{-1} = F\left(\Gamma_H - \Gamma_{OH^-}\right) \qquad \text{(iib)}$$

where a = amount of oxide used kg ℓ^{-1}, and s = specific surface area m^2 kg^{-1} and F = Faraday = 96500 [coul mol^{-1}].

Upon addition of $Pb(NO_3)_2$ the charge condition (or the proton balance) is changed to

$$C_A^* - C_B^* - [OH^-] + [H^+] + [NO_3^-] = [\equiv AlOH_2^{+*}] + [\equiv AlOPb^+] - [\equiv AlO^{-*}] + 2\,[Pb^{2+}] \quad \text{(iii)}$$

Concentrations with an asterik are those in the presence of Pb^{2+}.

Because of the mass balance for the $Pb(NO_3)_2$

$$2\,[Pb_T] = [NO_3^-]$$
$$2\,[Pb^{2+}] + 2\,[\equiv AlOPb^+] = [NO_3^-] \qquad \text{(iv)}$$

Eq. (iii) can be rewritten as

$$C_A^* - C_B^* + [OH^-] - [H^+] + 2\,[Pb^{2+}] + 2\,[AlOPb^+] =$$
$$[\equiv AlOH_2^{+*}] + [\equiv AlOPb^+] - [\equiv AlO^{-*}] + 2\,[Pb^{2+}]$$

or

$$C_A^* - C_B^* + [OH^-] - [H^+] = [\equiv AlOH_2^{+*}] - [\equiv AlOPb^+] - [\equiv AlO^{-*}] \qquad \text{(v)[1]}$$

or the proton charge [mol kg^{-1}]

$$Q_H^* = \frac{C_A^* - C_B^* + [OH^-] - [H^+]}{a} = \{\equiv AlOH_2^{+*}\} - \{\equiv AlOPb^+\} - \{\equiv AlO^{-*}\} \qquad \text{(vi)}$$

while the net charge [mol kg^{-1}]

$$Q_{net} = \{\equiv AlOH_2^{+*}\} + \{\equiv AlOPb^+\} - \{\equiv AlO^-\} \qquad \text{(vii)}$$

or

$$Q_{net} = Q_H^* + 2\,\{\equiv AlOPb^{2+}\} \qquad \text{(viii)}$$

or, in terms of surface charge [coul m^{-2}]

---

[1] To facilitate understanding, Eq. (v) was derived on the basis of charge balance; it can be derived directly on the basis of the proton condition (using H_2O, and $\equiv AlOH$ as a reference).

$$Q_{net}\, Fs^{-1} = \sigma_P = F\left(\Gamma_H - \Gamma_{OH} + 2\,\Gamma_{Pb^{2+}}\right) \qquad \text{(ix)}$$

(cf. Eq. 3.1).

How is the shift in the titration curve related to the quantity of Pb(II) bound to Al_2O_3?

Considering that

$$\alpha_0 = \frac{[\equiv AlOH_2^+]}{[\equiv AlOHTOT]} = \left(1 + K_{a1}^s / [H^+] + K_{a1}^s\, K_{a2}^s / [H^+]^2\right)^{-1}$$

$$\alpha_1 = \frac{[\equiv AlOH]}{[\equiv AlOHTOT]} = \left([H^+] / K_{a1}^s + 1 + K_{a2}^s / [H^+]\right)^{-1}$$

$$\alpha_2 = \frac{[\equiv AlO^-]}{[\equiv AlOHTOT]} = \left([H^+]^2 / K_{a2}^s\, K_{a1}^s + [H^+] / K_{a2}^s + 1\right)^{-1}$$

and that in presence of adsorbed Pb^{2+}, $\alpha_0^* = [\equiv AlOH_2^{+*}] / [\equiv AlOHTOT] - [\equiv AlOPb^+]$, we can deduct Eq. (v) from Eq. (ii) under conditions of constant pH to obtain

$$\Delta C_B = (C_A - C_B) - (C_A^* - C_B^*) = [\equiv AlOHTOT]\,(\alpha_0 - \alpha_2) - \left([\equiv AlOHTOT] - [\equiv AlOPb^+]\right)\left(\alpha_0^* - \alpha_2^*\right) + [\equiv AlOPb^+] \qquad \text{(x)}$$

if we assume that the α^* values are the same as the α values (at least at low coverage) we can write

$$\Delta C_B = (1 + \alpha_0 - \alpha_2)\,[\equiv AlOPb^+] \qquad \text{(xi)}$$

Thus, the shift in the titration curve, ΔC_B, at constant pH, is directly related to the extent of Pb(II) binding to the oxide surface. The adsorption of a metal ion decreases the surface protonation.

In a similar way one can treat the shift (in the opposite direction) of an alkalimetric-acidimetric titration curve by the adsorption of a ligand (Sigg and Stumm, 1981), e.g.,

$$S-OH + A^{2-} = S-A^- + OH^- \qquad \text{(xii)}$$

For example the adsorption of F^- on goethite induces a shift in the titration curve (Fig. 2.10b) (Sigg and Stumm, 1981).

$$-\Delta C_B = \left(C_A^* - C_B^*\right) - \left(C_A - C_B\right) = \left(1 + \alpha_2 - \alpha_0\right) [\equiv FeF] \qquad \text{(xiii)}$$

The net charge [mol kg^{-1}] is given by

$$Q_{net} = \{\equiv FeOH_2^+\} - \{\equiv FeO^-\} \qquad \text{(xiv)}$$

while

$$Q_H^* = \{FeOH_2^+\} + \{\equiv FeF\} - \{\equiv FeO^-\} \qquad \text{(xv)}$$

2.6 The Coordinative Unsaturation of Non-Hydrous Oxide Surfaces

As we have seen in Fig. 2.1 the dissociative chemisorption of water to an oxide surface, schematically represented by

$$
\begin{array}{ccc}
O^{2-} & & \overline{OH}\ \ \overline{OH} \\
| & & |\ \ \ \ | \\
M^{n+}\ M^{n+} + H_2O & \rightleftharpoons & M^{n+}\ M^{n+}
\end{array}
\qquad (2.19)
$$

establishes the functional groups on hydrous oxide surfaces. The driving force for this water absorption is the coordinative unsaturation of the (non-hydrous) oxide.

In a similar way other molecules such as H_2, CO, O_2, NO, alkenes, alcohols become adsorbed, e.g.,

$$
\begin{array}{ccc}
 & & \overline{H}\ \ \ \ \ H^+ \\
 & & |\ \ \ \ \ | \\
-Zn^{2+} - O^{2-} + H^2 & \rightleftharpoons & -Zn^{2+} - O^{2-}
\end{array}
\qquad (2.20)
$$

$$
\begin{array}{ccc}
 & & RO\ \ \ \ H \\
 & & |\ \ \ \ \ | \\
Zn - O + ROH & \rightleftharpoons & Zn - O
\end{array}
\qquad (2.21)
$$

Transition metal oxides (in absence of water) are therefore essential catalysts for many chemical processes such as oxidation (e.g., oxidation of CO in emission control), dehydrogenation (e.g., production of aldehydes from alcohol), and selective reduction (e.g., reduction of NO). Usually, activation of an oxide by heating is a prerequisit for chemisorption and catalysis; one of the reasons is that H_2O "poisons" the surface and hydroxyl groups that are formed in a moist atmosphere usually inhibits molecular adsorption.

Coordinative unsaturation arises from the fact that because of steric and electronic reasons, only a limited number of ligands or nearest neighbors can be within bonding distance of a metal atom or ion. In most transition metal oxides, the oxygen anions in the bulk form closed-packed layers and the metal cations occupy holes among the anions as schematically depicted in Fig. 2.1. In this picture, the oxide ion ligands appear to have saturated the coordination sphere of the bulk cation.

There are two approaches to picture the formation of coordinative unsaturation depending on the way the surface is prepared:
1) The oxide surfaces prepared by condensation and polymerization of hydroxo metal ions (see Schneider, 1988 on iron(III) hydrolysis) usually have lower coordination numbers than bulk oxide ions of the surface hydroxyles. Often the coordinatively unsaturated M^{n+} site behaves like a Lewis acid and the coordinatively unsaturated O^{2-} ion is more basic than the bulk ions;
2) The second approach to picture the formation of surface coordinative unsaturation is by cleaving a single crystal which involves among many other phenomena the breaking of a small number of bonds. Different exposed surfaces possess ions of different degrees of coordinative unsaturation. Different crystalline faces have different degrees of coordinative unsaturation. In addition to these, ions or other types of coordinative unsaturation can be created by introducing defects in the surface. Although much information on the surface chemistry of non-hydrous oxides is of great interest in surface science we cannot cover this subject in this book. The reader is referred to an excellent introduction into transition metal oxide surfaces by Kung (1989).

Problems

1) a) Explain qualitatively the pH-dependence of cation- and of anion-adsorption, respectively.

   b) Why does surface complex formation with a weak acid lead to a relative maximum in the extent of surface complex formation (adsorption) at a pH which is usually near the value of –log acidity constant of the weak acid (pK$_{HA}$)?

2) A sample of goethite is characterized by the following reactions:

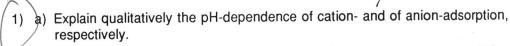

$$\equiv FeOH_2^+ \qquad = H^+ + \equiv FeOH \qquad pK_1^S = 6$$
$$\equiv FeOH \qquad = H^+ + \equiv FeO^- \qquad pK_2^S = 8.8$$
$$\equiv FeOH + Cu^{2+} = \equiv FeOCu^+ + H^+ \qquad pK^S = -8$$

Electrostatic effects are considered negligible.

   a) Calculate an adsorption isotherm for Cu^{2+} from a dilute $Cu(NO_3)_2$-solution at pH = 7 and $\equiv FeOHTOT = 10^{-6}$ M.

   b) What is the qualitative effect on the extent of adsorption of the following factors?:

   i)   presence of HCO_3^- in solution
   ii) increase in temperature of the solution
   iii) addition of 10^{-3} M Ca^{2+}

3) Discuss the binding of metal ions and of ligands in terms of the Lewis-acid-base theory.

4) Compare the alkalimetric titration of a polyprotic acid (e.g., polyaspartic acid) with that of an Al_2O_3 dispersion; show in either case the effect of the presence of a metal ion (e.g., Cu^{2+}) on the titration curve.

5) Describe semiquantitatively the effect of increased pH (at constant alkalinity) or of increased alkalinity (at constant pH) on the binding of Cu(II) to soil particles. (Consider that Cu^{2+} forms soluble carbonato complexes.)

6) Does the addition of small quantities of the following solutes to a suspension of α-Fe_2O_3 affect pH_{pznpc},
   i)   increase,
   ii) decrease, or
   iii) cause no effect?

   1) NaCl
   2) KF
   3) NaH_2PO_4

   4) $PbCl_2$
   5) $Na_2C_2O_4$ (oxalate)
   6) Humic acid

7) Compare the solubility of amorphous $Fe(OH)_3(s)$, as given in the figure below with the acid base properties of a solid hydrous ferric oxide (cf. Fig. 2.3). Is there a connection between the solubility minimum and the pH_{pznpc}?

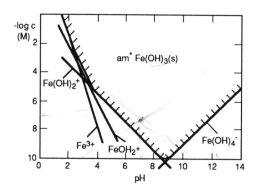

pH dependence of the solubility of amorphous $Fe(OH)_3$

8) Soil particles were found to have a capacity for ligand binding of 2×10^{-4} mol g^{-1}; these surface functional groups are characterized by an apparent "monoprotic" acidity constant

$$SOH \rightleftharpoons SO^- + H^+ \; ; \; K_a = 10^{-6}$$

The soil particles are characterized by a specific surface area of 30 m^2 g^{-1}; they are present at a concentration of 100 g ℓ^{-1}. What fraction of the surface, θ, is covered by organic matter if the organic matter is present (at equilibrium) at a concentration of 10^{-4} M? The organic matter is characterized by

$$HA \rightleftharpoons H^+ + A^- \; ; \; K = 4 \times 10^{-5}$$

and the adsorption is characterized by

$$SOH + HA \rightleftharpoons SA + H_2O \; ; \; K = 10^{-5}$$

9) The wall of a glass beaker contains $\equiv$SiOH groups. Why can dilute solutions of metal ions (pH > 7) not be stored in glass vessels?

10) Check the validity of the following statement: If a suspension of a hydrous oxide does not change its pH upon addition of $NaNO_3$ then this pH value is the pH_{pzc}.

Reading Suggestions

Dzombak, D. A., and F. M. M. Morel (1990), *Surface Complexation Modeling; Hydrous Ferric Oxide*, Wiley-Interscience, New York.

Kung, H. H. (1989), *Transition Metal Oxides: Surface Chemistry and Catalysis*, Elsevier, Amsterdam.

Parks, G. A. (1990), "Surface Energy and Adsorption at Mineral/Water Interfaces: An Introduction", in M. F. Hochella Jr. and A. F. White, Eds., *Mineral-Water Interface Geochemistry*, Mineralogical Society of America, pp. 133-175.

Schindler P.W., and W. Stumm (1987), "The Surface Chemistry of Oxides, Hydroxides, and Oxide Minerals", in W. Stumm, Ed., *Aquatic Surface Chemistry*, Wiley-Interscience, New York, pp. 83-110.

Sigg, L., and W. Stumm (1981), "The Interaction of Anions and Weak Acids with the Hydrous Goethite (α-FeOOH) Surface", *Colloids and Surfaces* **2,** 101-117.

Sposito, G. (1984), *The Surface Chemistry of Soils*, Oxford University Press, New York.

Westall, J. C. (1987), "Adsorption Mechanisms in Aquatic Surface Chemistry", in W. Stumm, Ed., *Aquatic Surface Chemistry*, Wiley-Interscience, New York, pp. 3-32.

Chapter 3

Surface Charge and the Electric Double Layer

3.1 Introduction
Acquiring Surface Charge

Fig. 3.1 shows that many suspended and colloidal solids encountered in waters sediments and soils have a surface charge and that this charge may be strongly affected by pH.

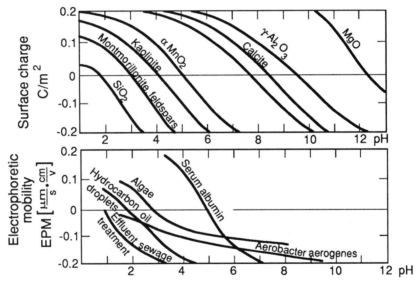

Figure 3.1

Effect of pH on charge and electrophoretic mobility. In the neutral pH range most suspended solids typically encountered in natural waters are negatively charged. These simplified curves are based on results by different investigators whose experimental procedures are not comparable and may depend upon solution variables other than pH. The curves are meant to exemplify trends and are meaningful in a semiquantative way only. The curve for calcite is for an equilibrium suspension of $CaCO_3$ with air ($p_{CO_2} = 10^{-3.5}$ atm).

Solid particle surfaces can develop electric charge in three principal ways:

1) The charge may arise from *chemical reactions* at the surface. Many solid surfaces contain ionizable functional groups: $-OH$, $-COOH$, $-OPO_3H_2$, $-SH$. The charge of these particles becomes dependent on the degree of ionization

43

(proton transfer) and consequently on the pH of the medium. As we have seen, the electric charge of a hydrous oxide can be explained by the acid base behavior of the surface hydroxyl groups S–OH

$$S-OH_2^+ \xrightleftharpoons{K_1^s} S-OH \xrightleftharpoons{K_2^s} S-O^-$$

Most oxides and hydroxides exhibit such amphoteric behavior; thus, the charge is strongly pH-dependent, being positive at low pH values (cf. Table 2.2). Similarly, for an organic surface, for example, that of a bacterium or of biological debris one may visualize the charge as resulting from protolysis of functional amino and carboxyl groups, for example,

$$R{\Big\langle}^{COOH}_{NH_3^+} \xrightleftharpoons{K_1} R{\Big\langle}^{COO^-}_{NH_3^+} \xrightleftharpoons{K_2} R{\Big\langle}^{COO^-}_{NH_2}$$

At low pH a positively charged surface prevails; at high pH, a negatively charged surface is formed. At some intermediate pH the net surface charge will be zero.

Charge can also originate by processes in which solutes become coordinatively bound to solid surfaces, for example,

$$\equiv Fe-OH + Cu^{2+} \rightleftharpoons \equiv FeOCu^+ + H^+$$

$$\equiv Fe-OH + HPO_4^{2-} \rightleftharpoons \equiv Fe-OPO_3^{2-} + H_2O$$

$$\equiv S + HS^- \rightleftharpoons \equiv S-SH^-$$

$$\equiv AgBr + Br^- \rightleftharpoons \equiv AgBr_2^-$$

$$\equiv RCOOH + Ca^{2+} \rightleftharpoons \equiv RCOOCa^+ + H^+$$

2) Surface charge at the phase boundary may be caused by lattice imperfections at the solid surface and by *isomorphous replacements* within the lattice. For example, if in any array of solid SiO_2 tetrahedra an Si atom is replaced by an Al atom (Al has one electron less than Si), a negatively charged framework is established:

$$\left[{}^{HO}_{HO}\!\!>\!\!Si\!\!<\!\!{}^O_O\!\!>\!\!Si\!\!<\!\!{}^O_O\!\!>\!\!Al\!\!<\!\!{}^O_O\!\!>\!\!Si\!\!<\!\!{}^{OH}_{OH} \right]^-$$

Similarly, isomorphous replacement of the Al atom by Mg atoms in networks of aluminum oxide octahedra leads to a negatively charged lattice. Clays are representative examples where such atomic substitution causes the charge at the phase boundary. Sparingly soluble salts also carry a surface charge because of lattice imperfections.

Thus, the net surface charge of a hydrous oxide is determined by the proton transfer and reactions with other cations or anions. In general, the net surface charge density of a hydrous oxide is given by

$$\sigma_P = F[\Gamma_H - \Gamma_{OH} + \Sigma(Z_M\Gamma_M) + \Sigma(Z_A\Gamma_A)] \tag{3.1}$$

where σ_P = net surface charge in Coulombs m^{-2}, F is the Faraday constant (96490 C mol^{-1}), Z the valency of the sorbing ion, Γ_H, Γ_{OH}, Γ_M and Γ_A, respectively are the sorption densities (mol m^{-2}) of H^+, OH^-, metal ions and anions.

The net proton charge (in Coulombs m^{-2}), the charge due to the binding of protons or H^+ ions – one also speaks of the surface protonation – is given by

$$\sigma_H = F(\Gamma_H - \Gamma_{OH})$$

where Γ_H and Γ_{OH} are the sorption densities of H^+ and OH^- expressed in mol m^{-2}, F is the Faraday constant 96490 [Coulomb mol^{-1}].

The surface charge in mol kg^{-1}, Q_H and Q_{OH}, is obtained as $Q_H = \Gamma_H s$ and $Q_{OH} = \Gamma_{OH} s$, where s is the specific surface area of the solid [m^2 kg^{-1}].

The net surface charge, Q_P, is experimentally accessible (by measuring cations and H^+ and OH^- and anions that have been bound to the surface), e.g., in case of adsorption of a metal, M^{2+}, or a ligand, A^{2-}:

$$Q_P = \{S\text{–}OH_2^+\} - \{S\text{–}O^-\} + \{S\text{–}OM^+\} \tag{3.2a}$$

$$Q_P = \{S\text{–}OH_2^+\} - \{S\text{–}O^-\} - \{SA^-\} \tag{3.2b}$$

where Q_P is the surface charge accumulated at the interface in mol kg^{-1}. Q_p can be converted into σ_P (Coulombs m^{-2}): $\sigma_P = Q_P F/s$ where s is the specific surface area in m^2 kg^{-1}.

Although aquatic particles bear electric charge, this charge is balanced by the charges in the diffuse swarm which move about freely in solution while remaining near enough to colloid surfaces to create the effective (counter) charge σ_D that balances σ_p

$$\sigma_P + \sigma_D = 0 \tag{3.3}$$

The following *points of zero charge* can be distinguished:

pzc: Point of zero charge: $\sigma_P = 0$ $\hspace{3cm}$ (3.4)

This is often referred to as isoelectric point. It is the condition where particles do not move in an applied electric field. If one wants to specify that the pzc is established solely due to binding of H^+ or OH^- one may specify:

pznpc: Point of zero net proton charge (or condition): $\sigma_H = 0$ $\hspace{1cm}$ (3.5)

pzse: Point of zero salt effect: $\delta\sigma_H/\delta I = 0$ $\hspace{2.5cm}$ (3.6)

where I = ionic strength. At pzse the surface charge is not effected by a change in concentration of an "inert" background electrolyte.

Each *diffuse swarm ion* contributes to σ_D the effective surface charge of an individual ion i can be apportioned according to

$$\sigma_{D_i} = \frac{Z_i}{m_s} \int_V [c_i(x) - c_{o\,i}]\ dV \tag{3.7}$$

where Z_i is the valence of the ion, $c_i(x)$ is its concentration at point x in the solution, and $c_{o\,i}$ is its concentration in the solution far enough from any particle surface to avoid adsorption in the diffuse ion swarm (Sposito, 1989). The integral in Eq. (3.7) is over the entire volume V of aqueous solution contacting the mass m_s of solid adsorbent. Thus, this equation represents the excess charge of ion i in aqueous solution: if $c_i(x) = c_{o\,i}$ uniformly, there could be no contribution of ion i to σ_D. Note that Eq. (3.7) applies to all ions in the solution including H^+ and OH^- and that σ_D is the sum of all σ_{D_i}.

3) A surface charge may be established by *adsorption* of a *hydrophobic species* or a surfactant ion. Preferential adsorption of a "surface active" ion can arise from so-called hydrophobic bonding (cf. Chapter 4.7), or from bonding via hydrogen bonds or from London-van der Waals interactions. The mechanism of sorption of some ions e.g., fulvates or humates, is not certain. Ionic species carrying a hydrophobic moiety may bind inner-spherically or outer-spherically depending on whether the surface-coordinative or the hydrophobic interaction prevails. (See Chapter 4.)

3.2 *The Net Total Particle Charge; Surface Potential*

Thus, different types of surface charge contribute to the *net total particle charge* on a colloid, denoted σ_P.

$$\sigma_P = \sigma_o + \sigma_H + \sigma_{IS} + \sigma_{OS}$$

where σ_P = total net surface charge

σ_o = permanent structural charge (usually for a mineral) caused by isomorphic substitutions in minerals. Significant charge is produced primarily in the 2 : 1 phyllosilicates;

σ_H = net proton charge, i.e., the charge due to the binding of protons or the binding of OH^- ions (equivalent to the dissociation of H^+). Protons in the diffuse layer are not included in σ_H;

σ_{IS} = inner-sphere complex charge;

σ_{OS} = outer-sphere complex charge.

The unit of σ is usually Coulomb m^{-2} (1 mol of charge units equals 1 Faraday or 96490 Coulombs).

As we have seen, the electric state of a surface depends on the spatial distribution of free (electronic or ionic) charges in its neighborhood. The distribution is usually idealized as an *electric double layer;* one layer is envisaged as a fixed charge or surface charge attached to the particle or solid surface while the other is distributed more or less diffusively in the liquid in contact (Gouy-Chapman diffuse model, Fig. 3.2). A balance between electrostatic and thermal forces is attained.

According to the Gouy-Chapman theory the surface charge density σ_P [C m^{-2}] is related to the potential at the surface ψ [volt] (Eq. (vi) in Fig. 3.2)

$$\sigma_P = (8\,RT\varepsilon\varepsilon_o\,c \times 10^3)^{1/2} \sinh (Z\psi F/2\,RT) \qquad (3.8a)$$

where R = molar gas constant (8.314 J mol^{-1} K^{-1}), T the absolute temperature (K), ε the dielectric constant of water (ε = 78.5 at 25° C), ε_o the permittivity of free space (8.854 × 10^{-12} C V^{-1} m^{-1} or 8.854 × 10^{-12} C^2 J^{-1} m^{-1}), c = molar electrolyte concentration [M]. Eq. (3.8a) is valid for a symmetrical electrolyte (Z = ionic charge). At low potential Eq. (3.8a) can be linearized as

$$\sigma_P = \varepsilon\varepsilon_o \kappa \psi \qquad (3.8b)$$

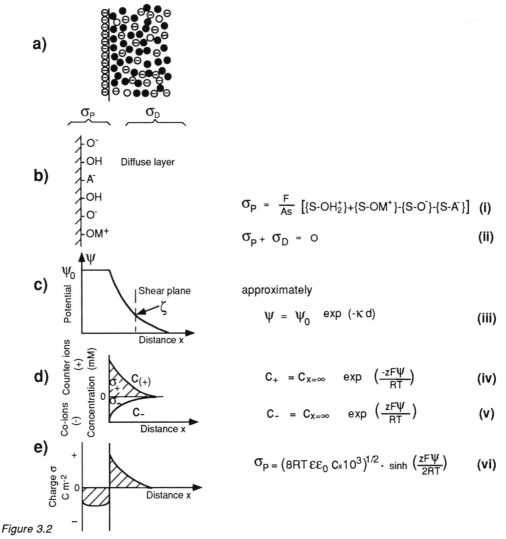

$$\sigma_P = \frac{F}{As}\left[\{S\text{-}OH_2^+\}+\{S\text{-}OM^+\}-\{S\text{-}O^-\}-\{S\text{-}A^-\}\right] \quad \text{(i)}$$

$$\sigma_P + \sigma_D = 0 \quad \text{(ii)}$$

approximately

$$\psi = \psi_0 \ \exp\ (-\kappa d) \quad \text{(iii)}$$

$$C_+ = C_{x=\infty} \ \exp\ \left(\frac{-zF\psi}{RT}\right) \quad \text{(iv)}$$

$$C_- = C_{x=\infty} \ \exp\ \left(\frac{zF\psi}{RT}\right) \quad \text{(v)}$$

$$\sigma_P = \left(8RT\,\varepsilon\varepsilon_0\, C_x 10^3\right)^{1/2}\cdot \ \sinh\ \left(\frac{zF\psi}{2RT}\right) \quad \text{(vi)}$$

Figure 3.2

The diffuse double layer

a) Diffuseness results from thermal motion in solution.
b) Schematic representation of ion binding on an oxide surface on the basis of the surface complexation model. s = specific surface area $m^2\ kg^{-1}$. Braces refer to concentrations in mol kg^{-1}.
c) The electric surface potential, ψ, falls off (simplified model) with distance from the surface. The decrease with distance is exponential when $\psi < 25$ mV. At a distance κ^{-1} the potential has dropped by a factor of $(1/e)$. This distance can be used as a measure of the extension (thickness) of the double layer (see Eq. 3.8c). At the plane of shear (moving particle) a zeta potential can be established with the help of electrophoretic mobility measurements.
d) Variation of charge distribution (concentration of positive and negative ions) with distance from the surface (Z = charge of the ion).
e) The net excess charge.

where the double layer thickness (compare Eq. (iii) in Fig. 3.2) κ^{-1} (in meters) is defined by

$$\kappa = \left(\frac{2\ F^2\ I \times 10^3}{\varepsilon\varepsilon_o\ RT}\right)^{1/2}$$ (3.8c)[1]

where I is the ionic strength [M].

Eq. (3.8a) and (3.8b) can be written for 25° C where ε, the dielectric constant of water, is 78.5 as

$$\sigma_P = 0.1174\ c^{1/2}\ \sinh (Z\psi \times 19.46)$$ (3.8d)

and

$$\sigma_P \cong 2.3\ I^{1/2}\ \psi$$ (3.8e)

σ_P = has the units $C\ m^{-2}$.

The Stern Layer. The Gouy-Chapman treatment runs into diffuculties at small κx values when the surface potential is high. The local concentrations of ions near the surface (Eqs. (3) and (4) in Table A.3.1 of the the Appendix) become far too large; this is because of the assumptions of point charges and neglect of ionic diameter. Stern suggested that the surface be divided into two parts, the first consisiting of a compact layer of ions adsorbed at the surface (Stern layer) and the second consisting of a diffuse double layer. In the Stern treatment H^+ and OH^- are incorporated in the solid and specifically adsorbed ions are placed at a plane separated from the solid by a short distance. In the surface complex formation model, specifically adsorbed (inner-spherically bound) ions belong like H^+ and OH^- ions to the solid phase.

The surface complexation approach is distinct from the Stern model in the primacy given the specific chemical interaction at the surface over electrostatic effects, and the assignment of the surface reaction to the sorption reactions themselves (Dzombak and Morel, 1990).

In the surface complex formation model the amount of surface charge that can be developed on an oxide surface is restricted by the number of surface sites. (This limitation is inherently not a part of the Gouy-Chapman theory.)

The Triple Layer Model. This model developped by Yates et al. (1974) and Davis et al. (1978) uses a similar idea as the Stern model: the specifically adsorbed ions are

---

[1] The simplified equation $\kappa = 3.29 \times 10^9\ I^{1/2}$ [m^{-1}] valid for 25° C is useful. Thus, κ^{-1} for a 10^{-3} M NaCl solution is ca. 10 nm; for seawater $\kappa^{-1} \approx 0.4$ nm.

placed as partially solvated ions at a plane of closest approach. From there on the Gouy-Chapman layer extends.

Zeta-Potential

The electrokinetic potential (zeta potential, ζ) is the potential drop across the mobile part of the double layer (Fig. 3.2c) that is responsible for electrokinetic phenomena, for example, elecrophoresis (= motion of colloidal particles in an electric field). It is assumed that the liquid adhering to the solid (particle) surface and the mobile liquid are separated by a shear plane (slipping plane). The electrokinetic charge is the charge on the shear plane.

The surface potential is not accessible by direct experimental measurement; it can be calculated from the experimentally determined surface charge (Eqs. 3.1 – 3.3) by Eqs. (3.3a) and (3.3b). The *zeta potential*, ζ, calculated from *electrophoretic* measurements is typically lower than the surface potential, ψ, calculated from diffuse double layer theory. The zeta potential reflects the potential difference between the plane of shear and the bulk phase. The distance between the surface and the shear plane cannot be defined rigorously.

Electrophoresis refers to the movement of charged particles relative to a stationary solution in an applied potential gradient, whereas in *electroosmosis* the migration of solvent with respect to a stationary charged surface is caused by an imposed electric field. The *streaming potential* is the opposite of electroosmosis and arises from an imposed movement of solvent through capillaries; conversely a *sedimentation potential* arises from an imposed movement of charged particles through a solution. Operationally the zeta potential can be computed from electrophoretic mobility and other electrokinetic measurements. For example, for nonconducting particles whose radii are large when compared with their double-layer ticknesses, the zeta potential, ζ, is related to the electrophoretic mobility, m_e (= velocity per unit of electric field). Frequently many corrections that are difficult to evaluate must be considered in the computation of ζ. The measurement of electrophoretic mobility is treated by Hunter (1989); the measurement in natural waters is discussed by Neihof and Loeb (1972).

Example 3.1

A hydrous iron(III) oxide suspension (10^{-3} moles per liter) at pH = 6.0 has adsorbed 20 % of Zn^{2+} from a solution that contained incipiently 10^{-4} mol/ℓ of Zn^{2+} and an inert electrolyte (I = 10^{-2} M) (25° C). The hydrous iron oxide has been characterized to have a specific surface of 600 m^2 g^{-1} and 0.2 moles of active sites per mol of $Fe(OH)_3$. From an alkalimetric acidimetric titration curve, we know that at pH = 6 the 10^{-3} M $Fe(OH)_3$ suspension contains $[\equiv Fe-OH_2^+] - [\equiv Fe-O^-] = 3 \times 10^{-5}$ M charge units.

Calculate the surface charge σ [C m^{-2}] and the surface potential and the "thickness" of the double layer, κ^{-1}.

i) The surface speciation in mol/ℓ is given by $[\equiv Fe-OZn^+] = 2 \times 10^{-5}$ M and by $([\equiv Fe-OH_2^+] - [\equiv Fe-O^-]) = 3 \times 10^{-5}$. (Since the active sites in the suspension is $[\equiv Fe-OH_{Tot}] = 2 \times 10^{-4}$ M and $[\equiv FeOH_2^+] > [\equiv FeO^-]$, the molar concentration of $[\equiv Fe-OH] \approx 1.5 \times 10^{-4}$ M.) Thus, the hydrous ferric oxide suspension carries on its surface a total of 5×10^{-5} moles of positive charge units per liter (cf. Eq. 3.2), i.e., ca. 25 % of the total sites are positively charged and nearly 75 % are uncharged. On a per surface area basis [m^2] we get

$$\sigma_P = \frac{5 \times 10^{-5} \text{ mol charge units}}{\ell} \frac{\ell}{10^{-3} \text{ mol Fe(OH)}_3(s)} \frac{1 \text{ g Fe(OH)}_3}{600 \text{ m}^2} \frac{\text{mol Fe(OH)}_3}{107 \text{ g}} \frac{96500 \text{ C}}{\text{mol}}$$

$$\sigma_P = 7.5 \times 10^{-2} \text{ Coulombs m}^{-2} \text{ or } 7.5 \text{ } \mu C \text{ cm}^{-2} \qquad \text{(i)}$$

ii) The surface potential is obtained from Eq. (3.8d):

$$\sinh (\psi \times 19.46) = \frac{7.5 \times 10^{-2}}{0.1174 \times 0.1} = 6.4$$

$$\psi \times 19.46 = 2.554$$

$$\psi = \frac{2.554}{19.46} = 0.13 \text{ V}$$

iii) κ can be calculated according to Eq. (3.8c) for I = 10^{-2} M;

$$\kappa = \left(\frac{2 \times (96490)^2 \frac{C^2}{mol^2} \times 10^{-2} \times 10^3 \text{ mol}}{8.854 \times 10^{-12} \frac{C}{mJ} \times 78.5 \times 2.478 \times 10^3 \frac{J}{mol}} \right)^{1/2} = 3.29 \times 10^8 \text{ m}^{-1}$$

Thus, the thickness of the double layer as characterized by $\kappa^{-1} = 3$ nm.

3.3 The Relation between pH, Surface Charge and Surface Potential

Dzombak and Morel, 1990, have illustratively and compactly summarized (Fig. 3.3) the interdependence of the Coulombic interaction energy with pH and surface charge density at various ionic strengths for hydrous ferric oxide suspensions in

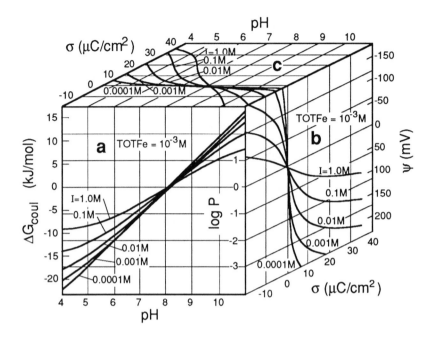

Figure 3.3

Relationship between pH, surface potential, ψ (or Coulombic term, log P, or Coulombic free energy, ΔG_{coul}), and surface charge density, σ (or surface protonation) for various ionic strengths of a 1:1 electrolyte for a hydrous ferric oxide surface (P = exp($-F\psi/RT$).
a) dependence of the coulombic term and surface potential on solution pH; note the near-Nernstian behavior at low ionic strength;
b) ψ versus σ; these curves correspond to the Gouy-Chapman theory;
c) σ versus pH; these are the curves obtained experimentally
(From Dzombak and Morel, 1990)

which H^+ is the only potential determining ion. In explaining this figure we follow largely their explanation. The influence of pH and ionic strength on the Coulombic interaction energy and on the Coulombic correction factor [exp($-F\psi/RT$)] is calculated according to the diffuse double layer model. The only experimental basis for the relationships depicted are the pH vs σ curves (panel C), as calculated from the surface protonation as measured from alkalimetric and/or acidimetric titration. The surface potential ψ cannot be measured. The ψ vs σ graphs (panel b) are obtained strictly from Gouy-Chapman theory. This relationship is predicted to be linear at low potentials, and exponential at higher potentials. Greater surface charges are developed at higher ionic strengths and greater surface potentials at lower ionic strengths. The Coulombic factor P [= exp($-F\psi/RT$)] varies by ca. seven orders of magnitude between pH = 4 and pH = 7. The Coulombic effect can be expressed in conventional energy units [kJ mol^{-1}] ($\Delta G = -F\psi$). As shown, Coulombic effects can contribute up to ca 20 kJ mol^{-1} (corresponding to 200 mV) to surface reactions.

Charge vs pH for Different Metal Oxides

Fig. 3.4a gives plots of charge resulting from surface protonation vs pH for various oxides. Dots represent experimental data from different authors (Table 3.1a) from titration curves at ionic strength $I = 0.1$ M (hematite = 0.2 M). It is interesting to note that the data "of different oxides" can be "normalised" i.e., made congruent, if we chose the master variable

$$\Delta pH = pH_{pzc} - pH \qquad\qquad (3.9)^{1)}$$

as the abscissa (Fig. 3.4b). A simple explanation is that the free energy of interaction for the surface protonation is composed of a chemical interaction ΔG_{int} and an electrostatic interaction term (ΔG^o_{coul})

$$\Delta G^o_{tot} = \Delta G^o_{intr} + \Delta G^o_{coul}$$

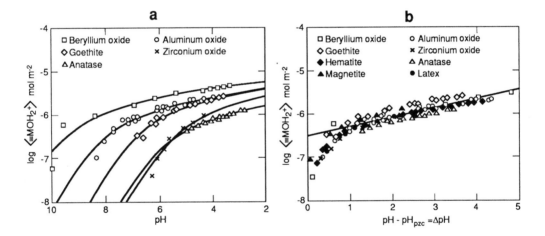

a   **b**

Figure 3.4

Surface protonation isotherms. Dots represent experimental data from titration curves at ionic strength $I = 0.1$ (Hematite, $I = 0.2$). References are indicated in Table 3.1. The concentration of protonated sites $\{=MOH_2^+\}$ is given in moles m^{-2}. BET surface data were used to calculate the surface concentration.
a) surface protonation vs pH;
b) surface concentration as a function of $pH_{pzc} - pH = \Delta pH$ (Eq. 3.5)
(From Wieland, Wehrli and Stumm, 1988)

---

[1] In this and many subsequent equations we often use the more general parameter pH_{pzc} rather than the more specific pH_{pznpc} despite the fact that in the cases discussed surface charge is due to protons only.

The chemical or intrinsic free energy term is reflected in the pH_{pzc}; this pH_{pzc} varies for every oxide depending on each oxide's proton affinity. The Coulombic term, however, is approximately the same for different oxides ($\Delta G^o_{coul} = ZF\psi$) at a given ΔpH and at a given ionic strength. This will be discussed further in Chapter 3.8.

Table 3.1   Protonation isotherm parameters of different oxides (see Fig. 3.4)

Surface		a) pH_{pzc}	b) $pK^s_{a_1}$ (int)	c) C $\mu F/cm^2$	d) a kJ/mol	e) n
TiO_2	f)	6.25	4.92	79	99	0.19
ZrO_2	g)	6.4	4.72	148	51	
$\delta\text{-}Al_2O_3$	h)	8.7	7.32	115	68	0.13
$\alpha\text{-}FeOOH$	i)	7.28	6.03	167	47	0.16
$\alpha\text{-}Fe_2O_3$	k)	8.67	7.47	94	83	0.16
Fe_3O_4	g)	6.8	5.63	151	52	
BeO	l)	10.2	8.71	134	58	0.085
Latex	m)	8.0	6.45	113	69	0.14

a) pH of the point of zero charge
b) Intrinsic protonation equilibrium constant (Eq. 2.5)
c) Integral double layer capacitance
d) Interaction energy parameter of the Frumkin Fowler Guggenheim FFG isotherm (see Chapter 4)
   The high site's density used in these calculations yields high values of a
e) Freundlich slope for $\Delta pH > 2$ (slope of Freundlich isotherm; see Chapter 4)
f) Wieland, Wehrli and Stumm (1988)
g) Regazzoni, Blesa and Maroto (1983)
h) Kummert and Stumm (1980)
i) Sigg and Stumm (1981)
k) Fokkink (1987)
l) Furrer and Stumm (1986); n was calculated for $\Delta pH > 4$
m) Harding and Healy (1985)

Effect of Metals and Ligands on Surface Charge

As we have seen, the net surface charge of a hydrous oxide surface is established by proton transfer reactions and the surface complexation (specific sorption) of metal ions and ligands. As Fig. 3.5 illustrates, the titration curve for a hydrous oxide dispersion in the presence of a coordinatable cation is shifted towards lower pH values (because protons are released as consequence of metal ion binding, S–OH + Me^{2+} ⇌ $SOMe^+$ + H^+) in such a way as to lower the pH of zero proton condition at the surface.

At this point (pH_{pznpc}) the portion of the charge due to H^+ and OH^- or their complexes[1] becomes zero. Because of the binding of M^{z+} to the surface ($\Gamma_{M^{z+}}$), the fixed surface charge increases or becomes less negative[2] and, at the pH where the fixed surface charge becomes zero, the point of zero charge, pzc, is shifted to higher pH values. Correspondingly, specifically adsorbable anions increase the pH of the zero proton condition but lower the pH of the pzc (Fig. 3.5).

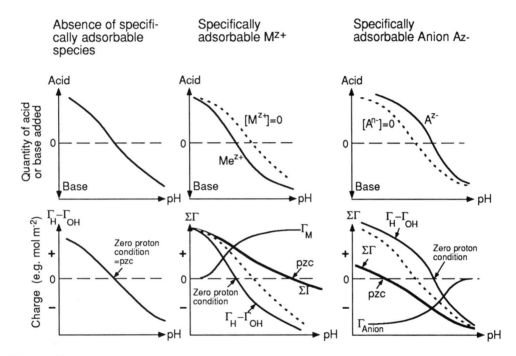

Figure 3.5

The net charge at the hydrous oxide surface is established by the proton balance (adsorption of H or OH^- and their complexes at the interface *and* specifically bound cations or anions. This charge can be determined from an alkalimetric-acidimetric titration curve and from a measurement of the extent of adsorption of specifically adsorbed ions. Specifically adsorbed cations (anions) increase (decrease) the pH of the point of zero charge (pzc) or the isoelectric point but lower (raise) the pH of the zero net proton condition (pznpc).

Addition of a ligand, at constant pH, increases surface protonation while the addition of a metal ion (that is specifically adsorbed) lowers surface protonation.

(Modified from Hohl, Sigg and Stumm,1980, and from Stumm and Morgan, 1981)

---

[1] If a hydrolyzed metal ion is adsorbed, its OH^- will be included in the proton balance; similarly, in case of adsorption of protonated anions, their H^+ will be included in the proton balance.

[2] Some colloid chemists often place these specifically bound cations and anions in the Stern layer (see Chapter 3.2). From a coordination chemistry point of view it does not appear very meaningful to assign a surface-coordinating ion to a layer different than H or OH in a $\equiv$MeOH group.

A Simplified Double Layer Model; (Constant Capacitance)

The simplest structure of the double layer is the surface charge in one plane and the counter charge in a similar parallel plane. Then, to a first approximation, the double layer may be visualized as a parallel plate condenser of distance d between the two plates and with its capacitance, C

$$C = \frac{\varepsilon \varepsilon_o}{d} \text{ or} \qquad\qquad (3.10a)$$

$$C = \varepsilon \varepsilon_o \kappa \qquad\qquad (3.10b)$$

where the distance d may be approximated by κ^{-1}. In this constant capacitance or Helmholtz model, the surface charge σ is related to the surface potential through a constant (for the conditions selected) capacitance

$$\sigma = C\psi \qquad\qquad (3.10c)$$

(Compare Eqs. (13), (14a) and (14b) in Table A.3.1 in the Appendix.)

This model is valid when the total surface charge is small in absolute magnitude or when the concentration of the inert electrolyte is large (compressed double layer).

3.4 Surface Charge on Carbonates, Silicates, Sulfides and Phosphates

Although we have used for exemplification largely the surfaces of hydrous oxides, the concepts given apply to all surfaces. As has been pointed out, most hydrous surfaces are characterized by functional groups that acquire charge by chemical interaction with H^+, OH^-, metal ions and by ligands. (For the moment we ignore redox reactions.)

There are various possibilities for functional groups on the surface of carbonates, sulfides, phosphates etc. Using a very simple approach similar to the one in Fig. 2.1 for hydrous oxides one could postulate surface groups for carbonates (e.g., $FeCO_3$) and sulfides (e.g., ZnS), as follows:

H	OH	H	OH	H	OH	water ↑	
CO_3	Fe	CO_3	Fe	CO_3	Fe	··········	
Fe	CO_3	Fe	CO_3	Fe	CO_3	**solid** ↓	(3.11a)
CO_3	Fe	CO_3	Fe	CO_3	Fe		

OH	H	OH	H	OH	H	water $\uparrow$	
Zn	S	Zn	S	Zn	S	solid $\downarrow$	(3.11b)
S	Zn	S	Zn	S	Zn		
Zn	S	Zn	S	Zn	S		

Stipp and Hochella (1991), on the basis X-ray photoelectron spectroscopy (XPS) and low energy electron diffraction (LEED), have shown that $CaCO_3$ exposed to water, contains at the surface $\equiv CO_3H$ and $\equiv CaOH$ functional groups and van Capellen (1991) has proposed a surface complex formation model for carbonates. Similarly, Rönngren et al. (1991) have proposed $\equiv SH$ and $\equiv ZnOH$ functional groups for the surface of hydrous ZnS(s).

The functional groups proposed can interact with the potential determining species H^+, OH^-, metal ions and in case of carbonate with H_2CO_3 and HCO_3^-; and in case of sulfides with H_2S and HS^-.

Carbonates

Many processes involving carbonates – ubiquitous minerals in natural systems – are controlled by their surface properties. In particular, flotation studies on calcite have revealed the presence of a pH-variable charge and of a point of zero charge (Somasundaran and Agar, 1967). Furthermore, electrokinetic measurements have shown that Ca^{2+} is a charge (potential) determining cation of calcite. (Thompson and Pownall, 1989).

It is reasonable to assume that H^+, OH^-, HCO_3^-, and $CO_2(aq)$ are able to interact as "potential determining" species (adsorption or desorption) with $CaCO_3(s)$ and affect its surface charge. In a system ($CaCO_3(s)$, CO_2, H_2O) where $CaCO_3(s)$ (calcite) is equilibrated with p_{CO_2} = constant, the equilibrium Eq. (3.12) is valid:

$$CaCO_3(s) + 2\,H^+ \rightleftharpoons Ca^{2+} + CO_2(g) + H_2O \;;\; K = 10^{9.7} \;(25° \,C, I = 0) \quad (3.12)\,[1]$$

---

[1] The equilibrium constant of (3.12) can be obtained by combining the following equilibria:

$CaCO_3$	$= Ca^{2+} + CO_3^{2-}$	K_{s0}	$= 10^{-8.3}$
$CO_3^{2-} + H^+$	$= HCO_3^-$	K_2^{-1}	$= 10^{10.2}$
$HCO_3^- + H^+$	$= CO_2(aq) + H_2O$	K_1^{-1}	$= 10^{6.3}$
$CO_2(aq)$	$= CO_2(g)$	K_H^{-1}	$= 10^{+1.5}$

$CaCO_3(s) + 2\,H^+ = Ca^{2+} + CO_2(g) + H_2O\;;\quad K_{s0}(K_1 K_2 K_H)^{-1} = 10^{9.7}$

The equilibrium concentration (activity) of the species H^+, OH^-, Ca^{2+}, HCO_3^-, CO_3^{2-} and $CO_2(aq)$ are known for a given p_{CO_2}. In addition to the four equilibria (Eq. 3.12 and footnote) the electroneutrality (charge balance) condition

$$2\,[Ca^{2+}] + [H^+] = [HCO_3^-] + 2\,[CO_3^{2-}] + [OH^-] \tag{3.13a}$$

or approximately,

$$2\,[Ca^{2+}] \approx [HCO_3^-] \tag{3.13b}$$

is needed to compute the composition. Fig. 3.6 plots these equilibria in a double logarithmic plot. The pH of a $CaCO_3$ (calcite) suspension in equilibrium with $p_{CO_2} = 10^{-3.5}$ atm and 1 atm is (25° C, I = 0) pH = 8.2 and pH = 6.45, respectively.

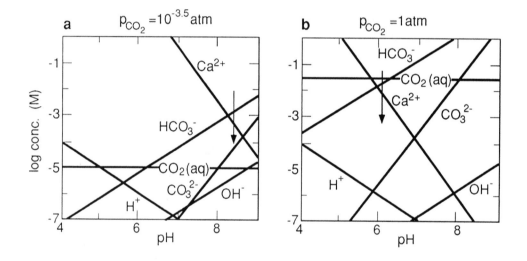

Figure 3.6

Equilibrium diagram of $CaCO_3$ as a function of pH at constant partial pressure of CO_2
a) $p_{CO_2} = 3.2 \times 10^{-4}$ atm;
b) $p_{CO_2} = 1$ atm.
The equilibrium composition reflected by the charge balance $2\,[Ca^{2+}] \cong [HCO_3^-]$ (log $[Ca^{2+}] + 0.3 = $ log $[HCO_3^-]$) corresponding to the vertical arrow. The point of zero charge is at pH = 8.2 (a) and 6.45 (b).

The pH value in accordance with Eqs. (3.13a) or (3.13b), at least in a first approximation, corresponds to *the point of zero charge*, because at this pH_{pzc} the surface species do not contribute to the overall electroneutrality of the aqueous suspension. Thus, pH_{pzc} depends on p_{CO_2}; the higher p_{CO_2} the lower is the pH of the suspension and the pH_{pzc}. The pH_{pzc} of different metal carbonates decreases with decreasing solubility (Stumm and Morgan, 1970).

The addition of acid (C_A) or base (C_B) to a $CaCO_3$ system (while p_{CO_2} = constant) will change the alkalinity in solution and produce (i) a shift in the HCO_3^-, CO_3^{2-}, Ca^{2+} equilibrium (and in pH), (ii) an adsorption of potential determining ions on the $CaCO_3$ surface, and (iii) a dissolution or precipitation of $CaCO_3$.

$$C_B + 2[Ca^{2+}] + [H^+] = [HCO_3^-] + 2[CO_3^{2-}] + [OH^-] + C_A \qquad (3.14)$$

Upon addition of acid the charge determining positively charged species will increase at the expense of the negatively charged charge determining species

$$2[Ca^{2+}] + [H^+] > [HCO_3^-] + 2[CO_3^{2-}] + [OH^-] \qquad (3.15a)$$

and upon addition of base

$$2[Ca^{2+}] + [H^+] < [HCO_3^-] + 2[CO_3^{2-}] + [OH^-] \qquad (3.15b)$$

Thus, the surface charge of the $CaCO_3$ will, respectively, increase or decrease with addition of acid or base.

In principle, the surface charge on solid carbonates can be determined – as with hydrous oxides from alkalimetric or acidimetric titration curves; but the procedure is more involved because in addition to the sorption of charge determining ions (the extent of which can be assessed from alkalimetric or acidimetric titration) there is also the additional effect of dissolution or precipitation of the carbonates. It was shown in case of $MnCO_3(s)$ (rhodochrosite) and $FeCO_3(s)$ (siderite) that the precipitation or dissolution of the carbonates was slow in comparison to the adsorption of the charge determining ions (Charlet et al., 1990) and thus, a rapid titration procedure permits a surface charge determination (see Fig. 3.7). In the case of $CaCO_3$ precipitation and dissolution was too fast, relatively, and thus, an experimental surface charge determination was not possible. But the same concept should also be valid for $CaCO_3$ and other carbonates.

We do not know the details at the atomic scale how the surface charge development on a carbonate mineral is established, but formally and schematically one could visualize the following type of charging reactions for a hydrous surface carbonate group of metal carbonate ($MeCO_3$) such as $\equiv CO_3H^\circ$ or $\left| C \xleftrightarrow{\text{OH}^\circ}_{O}^{O} \right.$

(Stipp and Hochella, 1991; van Capellen, 1991).

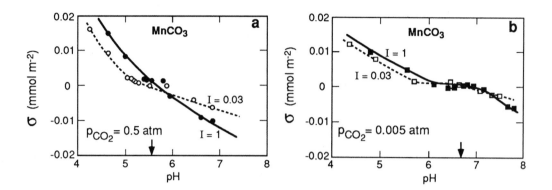

Figure 3.7

Surface charge of $MnCO_3$ (rhodochrosite) as a function of pH and p_{CO_2} as determined from surface titration curves. The values of pH_{pzc} (point of zero charge as calculated from equilibrium (cf. Eq. 3.12)) are given by arrows.

(From Charlet, Wersin and Stumm, 1990)

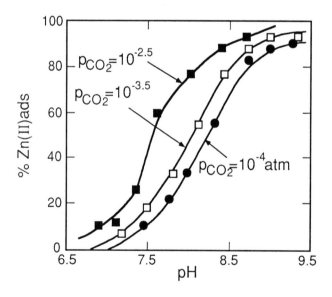

Figure 3.8

A plot of percent adsorbed Zn on calcite as a function of pH and different p_{CO_2} (data from Zachara, Kittrick and Harsh, 1988). Total $[Zn] = 10^{-6}$ M.

$$\equiv CO_3^- \xrightleftharpoons[\;]{H^+} \boxed{\equiv CO_3H^o} \xrightleftharpoons[\;]{Me^{2+}} \equiv CO_3Me^+$$

$$\equiv MeO^- \xrightleftharpoons[\;]{H^+} \boxed{\equiv MeOH^o} \xrightleftharpoons[\;]{H^+} \equiv MeOH_2^+ \qquad (3.16)$$

$$\Big\downarrow HCO_3^- \qquad \Big\updownarrow H_2CO_3$$

$$\equiv MeCO_3^- \qquad \equiv MeHCO_3^o$$

The charges assigned to the surface species in (3.16) indicate *relative* values. Surface equilibrium constants need to be established in order to estimate the species distribution as a function of $[H^+]$ and other potential determining species. An aqueous suspension of $CaCO_3$ crystals in the presence of CO_2 (p_{CO_2} = constant) at the point of zero charge (pH_{pzc}) at equilibrium – in line with the scheme of (3.16) – contains in addition to $\equiv CaOH$ and $\equiv CO_3H$, an equivalent number of positively and negatively charged surface species. If the pH is adjusted, e.g., with a base (at constant p_{CO_2}) the surface charge of the $CaCO_3$ becomes negative. It is thus not surprising that under these conditions $CaCO_3$ surfaces can adsorb specifically metal cations as a function of pH. Fig. 3.8 gives data from Zachara et al. on the adsorption of Zn(II) in dependence of pH.

Surface Charge on Silicates

In addition to the inorganic hydroxyl groups which are exposed on many mineral surfaces (metal oxides, phyllo-silicates and amorphous silicate minerals) we need to consider the particular features relating to charge on the silica surfaces of *layer silicates*.

The tetrahedral silica sheet in a layer silicate (see for example the schematic representation of the kaolinite structure in Fig. 3.9 and Figs. 3.10c, d) is called a *siloxane surface*. Sposito (1984) has described the nature of the *siloxane ditrigonal cavity* (see Fig. 3.10) in the following way: "The silica plane is characterized by a distorted hexagonal (i.e., trigonal) symmetry among its constituent oxygen atoms, that is produced when the underlying tetrahedra rotate to fit their apexes to contact points on the octahedral sheet. Further accomodation of the tetrahedra to the octahedral sheet is achieved throuth the tilting of their bases so that the silicon-oxygen bonds are directed towards the contact points instead of laying normal to the basal plane of the mineral. As a result of this adjustment one of the basal oxygen atoms in each tetrahedron is raised about 0.02 nm above the other two and the siloxane surface becomes corrugated."

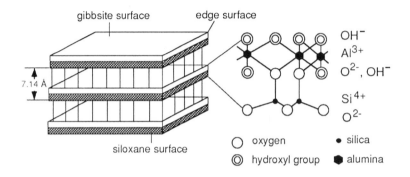

Figure 3.9

Schematic representation of the kaolinite structure. It reveals the 1:1 structure due to the alternation of silica-type (black) and gibbsite-type layers (white). Furthermore, the edge surface exposes aluminol and silanol groups.

The ditrigonal cavity formed by six corner sharing silica tetrahedra (Fig. 3.10) has a diameter of 0.26 nm and is bordered by six sets of lone-pair electron orbitals emanating from the surrounding ring of oxygen atoms. These structural features – as is pointed out by Sposito (1984) – qualifies the ditrigonal cavity as a soft Lewis base capable to complex water molecules (and possibly other neutral dipolar molecules).

Two cases of isomorphic substitution can be distinguished: In the tetrahedral sheet, or in the octahedral sheet (Sposito, 1984).
1) If isomorphic substitution of Si(IV) by Al(III) occurs in the tetrahedral sheet, the resulting negative charge can distribute itself over the three oxygen atoms of the tetrahedron (in which the Si has been substituted); the charge is localized and relatively strong inner-sphere surface complexes (Fig. 3.10a) can be formed.
2) If, on the other hand, isomorphic substitution occurs in the octahedral sheet (substitution of Al(III) by Fe(II) or Mg(II)), the resulting negative charge distributes itself over the ten surface oxygen atoms of the four silicon tetrahedra, that are associated through their apexes with a single octahedron in the layer. This distribution of negative charge enhances the Lewis base character of the ditrigo-

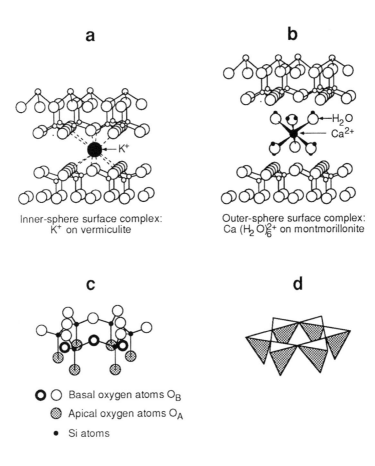

Figure 3.10

Surface complexes between metal cations and siloxane ditrigonal cavities on 2:1 phyllosilicates, shown in exploded view.

a) and b) linked SiO_4 groups in one siloxane Si_6O_6 ring (from Sposito, 1984);
c)   as a "ball and spike" model; and
d)   as linked tetrahedra

nal cavity and makes it possible to form complexes with cations as well as with dipolar molecules. An outer-sphere surface complex of this type of a Ca^{2+} cation is illustrated (Sposito, 1984) in Fig. 3.10b.

Example: Kaolinite

Three morphological planes of different chemical composition exist at the Kaolinite surface (Fig. 3.9): a gibbsite layer, a siloxane layer and an edge surface which is a complex oxide of the two constituents $Al(OH)_3$ and SiO_2. We can distinguish the following type of surface ligands at the gibbsite plane and edge surface:

i) ≡MOH groups are involved in proton (acid-base) equilibria, as discussed earlier; different surface species belong to this ligand type such as Al–OH–Al groups at the gibbsite surface, =AlOH and/or =SiOH groups at the edge surface and Si–O–Si groups at the siloxane layer;

ii) XO⁻ groups on the surface of the siloxane layer are induced by isomorphic substitution of Si by Al (Al–O–Si)⁻; these groups react by ion exchange.

The charge distribution on kaolinite platelets can be shown by particle association, in kaolinite suspensions and electron microscopic studies; such studies have shown convincingly a pH-dependent charge at the edges of the kaolinite mineral (Van Olphen, 1977). The protonation of the surface hydroxyl groups, displayed in Fig. 3.11, have been derived (Wieland, 1988; Wieland and Stumm, 1991) from alkalimetric-acidimetric titration curves. The surface proton concentration, Γ_H, denotes the proton density at =MOH groups with respect to the point of zero proton charge ($pH_{pzc} = 7.5$) of the hydroxyl groups at the edge face. At pH = 7.5 the proton density, Γ_H, is zero and at pH < 7.5 it equals the surface charge, $\Gamma_H = \sigma$. The total

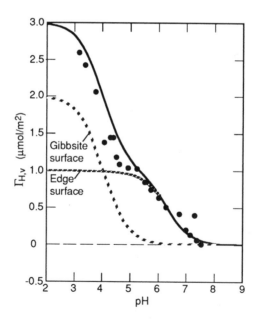

Figure 3.11

Surface protonation at the kaolinite surfaces. The excess proton density, $\Gamma_{H,V}$, at the surface hydroxyl group is displayed as a function of pH. Surface protonation is interpreted as a successive protonation of two distinct types of OH groups localized at the gibbsite and edge surfaces. The pH_{pzc} of the edge surface is about 7.5.

(From Wieland, 1988; Wieland and Stumm, 1991)

proton density (solid line) is computed from the superposition of two successive protonation equilibria at the kaolinite surface (broken lines). It is postulated that the edge surface is protonated in the near neutral and weakly acidic pH range whereas the gibbsite plane (Al–OH–Al groups) is protonated at pH < 5. The intrinsic acidity constant of =MOH groups at the edge surface (pK_{a1}^s = 6.3 and pK_{a2}^s = 8.7, cf. Wieland, 1988) are comparable to the values determined for γ-Al_2O_3 (pK_{a1}^s = 7.4, pK_{a2}^s = 10, cf. Kummert and Stumm, 1980). The intrinsic acidity constant of the more acidic =MOH groups localized at the gibbsite surface ($pK_{a1}^s \cong 4$). Furthermore, there is evidence for the existence of permanent negatively charged surface sites XO^- at the siloxane layer. The XOH groups at the siloxane layer are the main surface species at low pH and may undergo ion exchange reactions with Na^+ and Al(III). At pH < 4.3 Al^{3+} is the dominant Al(III) species and accessible to ion exchange reactions.

In summary, the model proposed on the basis of acid-base characteristics of kaolinite platelets explains the pH-dependent charge primarily to the protonation of the hydroxyl groups at the basal gibbsite and the edge surface. We will later illustrate how this charge characteristics (surface protonation) influences the reactivity (dissolution characteristics) of kaolinite.

Sulfides

Fig. 3.12 gives a recent Scanning Tunnelling Microscope (STM) image of a galena (PbS) {100} surface. STM imaging was accomplished on fresh fractured surfaces.

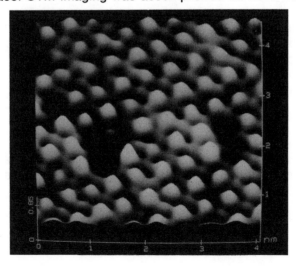

Figure 3.12

Scanning Tunnelling Microscope (STM) image of a galena {100} surface taken under oil. Two different surface sites are visible. The authors suggest that the larger peaks correspond to S sites. For specifications under which STM imaging was done see original publication.

(From Eggleston and Hochella, 1990)

The Pb and S sites are distinguishable and as Fig. 3.12 suggests the S sites appear to be imaged preferentially.

It is reasonable to assume that the surface of metal sulfides show amphoteric behavior and it has been shown that uptake of bivalent cations is pH-dependent. Metal sulfide precipitates are efficient scavengers for heavy metal ions.

Much information on the functional surface groups on sulfides is not yet available. Recent studies by Rönngren et al. (1991) gave the following acidity constants for ZnS(s):

$$\equiv ZnSH^+ \quad \rightleftharpoons \quad \equiv ZnS + H^+ \qquad K_1^s = 6.91 \qquad (3.17a)$$

$$\equiv SZn + H_2O \rightleftharpoons \equiv SZnOH^- + H^+ \qquad K_2^s = 10.28 \qquad (3.17b)$$

Thus, according to this interpretation the zero proton condition is at pH = 8.6. Furthermore, an ion exchange reaction

$$\equiv SZn + 2 H^+ \quad \rightleftharpoons \quad \equiv SH_2 + Zn^{2+} \qquad \log K^s = 9.59 \qquad (3.17c)$$

has to be considered.

Phosphates; Apatite

Hydroxyapatite and fluoroapatite surfaces differ from oxide surfaces in as far as they are expected to carry two different classes of surface groups (Wu et al., 1991). From a simple pictorial presentation

$$\equiv Ca - OH_2^+ \quad \text{and} \quad \begin{matrix} \equiv Ca - O \diagdown & \diagup O^- \\ & P \\ \equiv Ca - O \diagup & \diagdown O \end{matrix}$$

Wu et al. (1991) propose for fluoroapatite $\equiv Ca-OH_2^+$ and $\equiv P-O^-$ as dominating surface groups, characterized by the following equilibrium constants (0.1 M NaCl, 25° C).

$$\equiv P{-}O^- + H^+ = \equiv P{-}OH \qquad ; \log K^s = 6.6$$

$$\equiv Ca{-}OH_2^+ = \equiv Ca{-}OH + H^+ \quad ; \log K^s = -9.7$$

According to these equilibrium constants a pH_{pzc} of 8.15 would be expected for hydrous fluoroapatite.

3.5 *Correcting Surface Complex Formation Constants for Surface Charge*

We return to the complex formation equilibria described in Chapter 2 (Eqs. 2.1 – 2.10). The equilibrium constants as given in these equations are essentially intrinsic constants valid for a (hypothetically) uncharged surface. In many cases we can use these constants as apparent constants (in a similar way as non-activity corrected constants are being used) to illustrate some of the principal features of the interdependent variables that affect adsorption. Although it is impossible to separate the chemical and electrical contribution to the total energy of interaction with a surface without making non-thermodynamic assumptions, it is useful to operationally break down the interaction energy into a chemical and a Coulombic part:

$$\Delta G_{tot} = \Delta G^o_{chem} + \Delta G^o_{coul}$$

where ΔG^o_{chem} is the "intrinsic" free energy term and ΔG^o_{coul} is the variable electrostatic or Coulombic term

$$\Delta G^o_{coul} = \Delta ZF\psi \tag{3.18}$$

Theoretically this term reflects the electrostatic work in transporting ions through the interfacial potential gradient (e.g., Stumm et al., 1970; Stumm and Morgan, 1981; Morel, 1983). Since

$$\Delta G_{tot} = -RT \ln K$$

we can write, e.g., for the proton transfer of $\equiv$FeOH to $\equiv$FeO$^-$,

$$\frac{\{\equiv FeO^-\}[H^+]}{\{\equiv FeOH\}} = K^s_{a2}(app) = K^s_{a2}(int) \exp\left(-\frac{F\psi}{RT}\right) \tag{3.19}$$

i.e., the proton transfer is interpreted as a chemical dissociation if the proton from the $\equiv$FeOH (given by K^s(int)) and by the transfer of the proton from the surface to bulk solution given by exp($-F\psi/RT$)

Most generally for a surface complex formation reaction:

$$K^s(app) = K^s(int) \exp\left(-\frac{\Delta ZF\psi}{RT}\right) \tag{3.20}$$

In these equations K(int) and K(app) are the intrinsic and apparent equilibrium constants, respectively (cf. Fig. 2.2), F is the Faraday, ψ is the surface potential and ΔZ

is the *change* in the charge of the surface species for the reaction under consideration (as written for the equilibrium reaction for which K is defined).

The intrinsic equilibrium constants are postulated to be independent of the composition of the solid phase They remain conditional in the sense of the constant ionic medium reference state if interacting ions such as H^+ are expressed as concentrations.

There is no experimental way to measure ψ. (As we mentioned before, the zeta potential – as obtained, for example, from electrophoretic measurments – is in a not readily definable way – smaller than ψ.) But as discussed in section 3.3 we can obtain the surface charge (Eq. 3.2) and then compute the surface potential ψ on the basis of the diffuse double layer model with Eq. (3.8a); Eq. (3.8a) in simplified form for 25° C is

$$\sigma = 0.1174 \, c^{1/2} \sinh (Z\psi \times 19.46)$$

where Z is the valence of ions in the symmetrical background electrolyte.

Example 3.2

From alkalimetric-acidimetric titration curves on hydrous ferric oxide the following intrinsic acidity constants have been obtained: (I = 0.1 M, 25° C)

$$\log K^s_{a1}(\text{int}) = -7.18$$

$$\log K^s_{a2}(\text{int}) = -8.82$$

Furthermore the surface complex formation with Zn(II) has been determined from adsorption studies in 10^{-3} M suspensions of hydrous ferric oxide with dilute (10^{-7} M) Zn(II) solution. It can be described by the reaction

$$\equiv FeOH + Zn^{2+} \rightleftharpoons \equiv FeOZn^+ + H^+$$

for which the intrinsic constant is

$$\log K^s_1 = 0.66$$

The hydrous iron oxide has been characterized to have a specific surface area of 600 m^2 g^{-1} and 0.2 moles of active sites per mol of FeOOH. Then the concentration of the active sites is

$$\text{TOT}(\equiv FeOH) = \frac{10^{-3} \text{ mol Fe(OH)}_3}{\text{liter}} \frac{0.2 \text{ mol sites}}{\text{mol Fe(OH)}_3} = \frac{2 \times 10^{-4} \text{ mol sites}}{\text{liter}}$$

$$TOT(\equiv FeOH) = [\equiv FeOH] + [\equiv FeOH_2^+] + [\equiv FeO^-] + [\equiv FeOZn^+] = 10^{-3.7} M$$

Calculate the binding (adsorption of Zn(II)) as a function of pH.

The species are, H^+, OH^-, Zn^{2+}, $\equiv FeOH_2^+$, $\equiv FeOH$, $\equiv FeO^-$, $\equiv FeOZn^+$

For a given pH six equations are needed to calculate the speciation. They are four mass laws; [in the subsequent equations $P = \exp(-F\psi/RT)$]

$$[OH^-] \quad = [H^+]^{-1} \qquad\qquad 10^{-13.8} \tag{i}$$

$$[\equiv FeOH_2^+] \quad = [H^+][\equiv FeOH] \qquad P\ 10^{7.18} \tag{ii}$$

$$[\equiv FeO^-] \quad = [H^+]^{-1}[\equiv FeOH] \qquad P^{-1}\ 10^{-8.82} \tag{iii}$$

$$[\equiv FeOZn^+] \quad = [H^+]^{-1}[Zn^{2+}][\equiv FeOH] \quad P\ 10^{0.66} \tag{iv}$$

and three mol balance equations

$$(\equiv FeOHTOT) = [\equiv FeOH_2^+] + [\equiv FeOH] + [\equiv FeO^-] + [\equiv FeOZn^+] \tag{v}$$

$$TotZn \quad = [Zn^{2+}] + [\equiv FeOZn^+] \tag{vi}$$

Furthermore, the surface charge σ is given by

$$\sigma = (F/as)\left([\equiv FeOH_2^+] - [\equiv FeO^-] + [\equiv FeOZn^+]\right) \tag{vii}$$

and the charge potential relationship is

$$\sigma = 0.1174\ c^{\frac{1}{2}}\ \sinh(Z\psi \times 19.46) \tag{viii}$$

(As before a = conc. of solids [kg ℓ^{-1}] and s = specific area [m^2 kg^{-1}].)

It is of course convenient to make this calculation with the help of a computer program (such a program is described in Dzombak and Morel, 1990). But a manual solution is possible; it is less cumbersome if we realize that $[\equiv FeOZn^+]$ is negligible in Eqs. (v) and (vii). Thus, the surface species, the surface charge and the Coulombic interation depend on pH and ionic strength. The result is given in Fig. 3.13 and in Fig. 2.3 (or Fig. 3.3 (Panel c)).

Now the interaction of Zn(II) with the hydrous iron oxide surface can be readily computed because the surface charge is hardly affected by $[\equiv FeOZn^+]$, because this is negligible in comparison to $[\equiv FeOH_2^+] - [\equiv FeO^-]$ and is known for every pH. Thus, for example, at pH = 6, the surface charge $[\equiv FeOH_2^+] - [\equiv FeO^-]$ = 0.06 µM, or σ = 11 µC cm^{-2}. Correspondingly exp(−Fψ/RT) = 2.7 × 10^{-2} and log P = −1.57. (This corresponds to a surface potential of +92 mV or to a Coulombic energy Fψ ≅ 9 kJ mol^{-1} which the Zn^{2+} has to overcome to be adsorbed at the positively charged hydrous ferric oxide.

Now we can compute the concentration of surface bound (adsorbed) Zn(II), i.e., $\equiv FeOZn^+$ by considering Eqs. (iv) and (vi). Considering that $[\equiv FeOH]$ = 5 × 10^{-5} M, one calculates from Eq. (iv) that

$$\frac{[\equiv FeOZn^+]}{[Zn^{2+}]} = \frac{[\equiv FeOH]}{[H^+]} \; P \; 10^{-0.66} = 0.29$$

Since $[\equiv FeOZn] + [Zn^{2+}]$ = 10^{-7} M, $[\equiv FeOZn^+]$ = 2.9 × 10^{-8} M, or 29 % of the Zn(II) is adsorbed.

The speciation as a function of pH is given in Fig. 3.13.

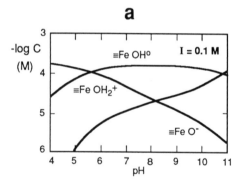

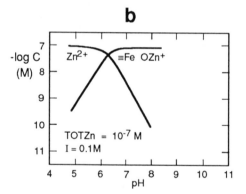

Figure 3.13

a) Calculated surface speciation as a function of pH at ionic strength 0.1 (1:1 electrolyte(for a 10^{-3} M hydrous ferric oxide suspension).

b) Calculated equilibrium speciation as a function of pH for zinc in a 10^{-3} M suspension of hydrous ferric oxide: TOTZn = 10^{-3} M, I = 0.1 M.

(From Dzombak and Morel, 1990)

Example 3.3:   pH-Dependence of Surface Complex Formation

We resume the problem discussed in Example 2.2 and solve the same problem, but now we correct for electrostatic effects[1]. Sumarizing the problem: Calculate the pH dependence of the binding of a) a metal ion Me^{2+}, and b) of a ligand A^- to a hydrous oxide, SOH, and compare the effect of a charged surface at an ionic strength I = 0.1. A specific surface area of 10 g m^{-2}; 10^{-4} mol surface sites per gram (~ 6 sites nm^{-2}); concentration used 1 g ℓ^{-1} (10^{-4} mol surface sites per liter solution). As before (Example 2.2) the surface complex formation constants are log K_M^s = –1 and log K_L^s = 5, respectively.

The diffuse double layer model is used to correct for Coulombic effects. The constant capacitance model depends on the input of a capacitance; but the result obtained is not very different.

The results are given in Figs. 3.14 and 3.15, respectively. They should be compared with Figs. 2.8 and 2.9.

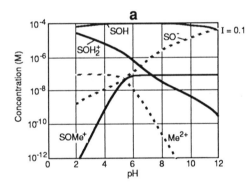 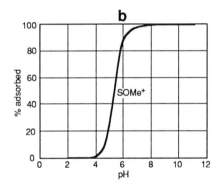

Figure 3.14

Metal binding by a hydrous oxide from a 10^{-7} M solution (SOH + Me^{2+} ⇌ $SOMe^+$ + H^+) for a set of equilibrium constants (see Eqs. (i) – (iii) from Example 2.3) and concentration conditions (see text). Corrected for electrostatic interactions by the diffuse double layer model (Gouy Chapman) for I = 0.1. The hydrolysis of Me^{2+} is neglected.

---

[1] The set of equilibrium constants are (as in Example 2.2):

$S-OH_2^+$	⇌	$SOH + H^+$	; log K_1^s	= –4	(i)
$S-OH$	⇌	$SO^- + H^+$	; log K_2^s	= –9	(ii)
$S-OH + Me^{2+}$	⇌	$SOMe^+ + H^+$	; log K_M^s	= –1	(iii)
$S-OH + HA$	⇌	$SA + H_2O$	; log K_L^s	= 5	(iv)
HA	⇌	$H^+ + A^-$	; log K_{HA}	= –5	(v)

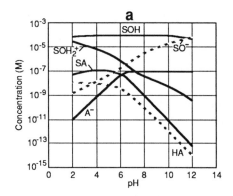
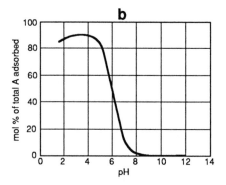

Figure 3.15

Ligand binding by a hydrous oxide from a 10^{-7} M solution (SOH + HA $\rightleftharpoons$ SA + H$_2$O) for a set of equilibrium constants (see Eqs. (i), (ii), (iv), and (v)) and concentration conditions (see text) corrected for electrostatic interactions by the diffuse double layer model (Gouy Chapman) for I = 0.1.

A comparison of Figs. 3.14 and 3.15 with those of Figs. 2.8 and 2.9 shows that the Coulombic effects change the relative concentrations of SOH_2^+ and of SO^- in comparison to the charge-uncorrected conditions. But it is of interest to note that the influence of the charge correction is not very significant for the extent of adsorption of Me^{2+} or A^- as a function of pH. In the case of the ligand there is a less pronounced maximum but the location of the pH-edge is similar.

Example 3.4:   Sorption of SO_4^{2-} to Aluminum Oxide

The (ad)sorption of SO_4^{2-} to aluminum oxide is of considerable interest in the infiltration of SO_4^{2-} bearing waters into soils.

The following data are available:

$$\equiv AlOH_2^+ \rightleftharpoons \equiv AlOH + H^+ \quad ; \log K = -7.4 \quad \text{(i)}$$
$$\equiv AlOH \rightleftharpoons \equiv AlO^- + H^+ \quad ; \log K = -10.0 \quad \text{(ii)}$$
$$\equiv AlOH + H^+ + SO_4^{2-} \rightleftharpoons \equiv AlSO_4^- + H_2O \quad ; \log K = 8.0 \quad \text{(iii)}$$
$$\equiv AlOH + SO_4^{2-} \rightleftharpoons \equiv AlOHSO_4^{2-} \quad ; \log K = 0.7 \quad \text{(iv)}[1]$$

Reasonable assumptions on specific surface area for aluminum oxides are 10 m^2 g^{-1}; we may further assume 10^{-4} surface sites g^{-1} and a concentration of 1 g/ℓ. Thus,

---

[1] The complex $\equiv AlOHSO_4^{2-}$ could be looked at as $\equiv Al\begin{smallmatrix} OH \\ OSO_3^{2-} \end{smallmatrix}$

if we look at the problem in terms of a batch process we have the following mass balance of surface Al-species on a per liter basis:

$$10^{-4} \text{ M} = [\equiv\text{AlOHTOT}] = [\equiv\text{AlOH}_2^+] + [\equiv\text{AlOH}] + [\equiv\text{AlO}^-] + [\equiv\text{AlOHSO}_4^{2-}] + [\equiv\text{AlSO}_4^-] \qquad \text{(v)}$$

A reasonable concentration of initial SO_4^{2-} is 2×10^{-4} M; (20 mg SO_4^{2-} per liter). Thus,

$$2 \times 10^{-4} \text{ M} = [\text{TOT SO}_4^{2-}] = [SO_4^{2-}] + [\equiv\text{AlSO}_4^-] + [\equiv\text{AlOHSO}_4^{2-}] \qquad \text{(vi)}$$

For charge correction we use the diffuse double layer model and assume I = 0.01.

We have to solve simultaneously the Eqs. (i) to (vi) and consider iteratively the correction for surface charge.

The results of the calculation on a MICROQL based computer program adapted to a personal computer are given in Fig. 3.16.

The calculation illustrates that SO_4^{2-} is bound strongly to aluminum (hydr)oxide surfaces under acid or slightly acid conditions. At pH above 7 adsorption of SO_4^{2-} occurs only to a very small extent.

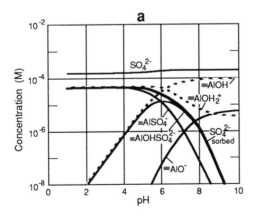

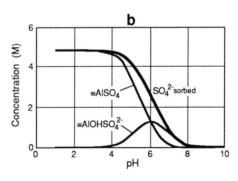

Figure 3.16

Binding of SO_4^{2-} to aluminum (hydr)oxide
Equilibrium distribution of surface and solute species for [TOT $\equiv$AlOH] = 10^{-4} M and [TOTSO$_4^{2-}$] = 2×10^4 M, I = 0.01.

Constant Capacitance

As we have seen, when surface potentials are small or the solution side of the electric double layer is compressed (high ionic strength), the surface potential is proportional to the surface charge (as given before in Eqs. (3.10a) and (3.10c)).

$$\psi = \frac{\sigma}{C} \qquad (3.21)$$

where C is the integral capacitance [Farad m^{-2}] and ψ and σ are, as before, the potential [V] and the surface charge [C m^{-2}]. Reconsidering Eq. (3.19) we can now replace the ψ by Eq. (3.21) to obtain

$$K^s(app) = K^s(int) \exp\left(-\frac{\Delta ZF\sigma}{RTC}\right) \qquad (3.22)$$

In case of an acid-base equilibrium of SOH_2^+–SOH, the surface charge is proportional to $\{SOH_2^+\}$ and

$$\log K_{a1}^s(app) = \log K_{a1}^s(int) - \frac{F^2}{RT\,2.3\,Cs}\,\{SOH_2^+\} \qquad (3.23)$$

where s is the specific surface area in m^2 kg^{-1}. This is equivalent to the empirical Eq. (vi) given in Example 2.1. The difference in the Coulombic correction for the diffuse double layer (Eq. 3.20) or the constant capacitance double layer is in most instances quite small. They both fit the experimental data equally well (Westall and Hohl, 1980). Schindler and Stumm (1987) discuss the surface chemistry of oxide minerals in terms of the constant capacitance model.

Surface Complexation Models and Mean Field Statistics

We pointed out in 2.1 that the surface of a hydrous oxide is not uniform because the various hydroxyl groups are structurally and chemically not equivalent (see for example Fig. 5.6). It has been shown by Sposito (1984) that the equations given for surface complexation are in accord with a statistical mechanical interpretation.

The surface complexation models used are only qualitatively correct at the molecular level, even though good quantitative description of titration data and adsorption isotherms and surface charge can be obtained by curve fitting techniques. Titration and adsorption experiments are not sensitive to the detailed structure of the interfacial region (Sposito, 1984); but the equilibrium constants given reflect – in a mean field statistical sense – quantitatively the extent of interaction.

In addition to the diffuse double layer and the constant capacitance model discussed here, other models such as the triple layer model (Davies and Leckie, 1978), and the Stern model (cf. Chapter 3.3). have been proposed. These models differ essentially in how the adsorption energy is separated into electrostatic and chemical contributions. As Westall and Hohl (1980) have shown, all models may be viewed as beeing of the correct mathematical forms to represent the data. The problem of separating the chemical and electrostatic energy is experimentally indeterminant for oxide-water interfaces.

Hiemstra et al. (1989) have elaborated on a multisite proton adsorption model taking into account the various types of surface groups; intrinsic log K values for the protonation of various types of surface groups can be estimated with this model.

3.6 Some Thermodynamic Aspects of Interactions on Oxide Surfaces

In Fig. 3.4a the surface charge resulting from surface protonation was plotted for various oxides. In Fig. 3.4b it was shown that the curves for surface charge become congruent if charge is plotted vs ΔpH (= $pH_{pzc} - pH$) because the free energy of interaction is composed of a chemical interaction (which depends on the type of oxide) and an electrostatic interaction. Lyklema (1987) distinguishes between *specific* and *generic* aspects. A specific feature is a phenomenon different for each object, whereas a generic feature is something that is independent of the nature of the object.

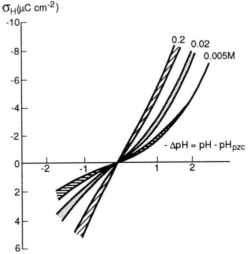

Figure 3.17

Uniformity of the electrical double layer on oxides: plot of $-\sigma_H$ vs $-\Delta pH$ master curves for rutile, ruthenium dioxide and hematite. The concentration of KNO_3 is indicated.

(From Lyklema, 1987)

Similar information as in Fig. 3.4 is given in Fig. 3.17 which plots surface charge (due to protonation or deprotonation vs $-\Delta pH$ (= $pH - pH_{pzc}$) for three oxides. Within a certain band width these curves coincide.

This *pH-congruence* can be explained (Lyklema, 1987) by inferring that the process of charge formation is independent of the nature of the oxide; hence it must be determined by the solution side of the double layer. On the other hand, the specific part, determined by the surface side, is determined by the chemical nature and is reflected only by the pH_{pzc}.

Temperature Congruence. On rutile and hematite, (the two oxides for which temperature dependence has been studied), the surface charge is not only pH-congruent but also temperature-congruent (Fig. 3.18). Thermodynamically it is possible to obtain the entropy and enthalpy of the forming of the double layer from the temperature dependence of the surface charge (Fokking et al., 1989) (Table 3.2). Proton binding is partially enthalpically and partially entropically driven. As Lyklema

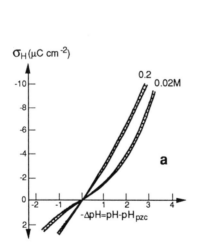

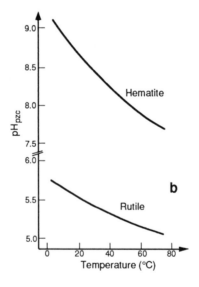

Figure 3.18a

Temperature congruence for the double layer on rutile in the presence of the indicated concentrations of KNO_3. Between 5 and 50° C all data coincide within the band width.

(From Lyklema, 1987)

Figure 3.18b

Temperature dependence of the points of zero charge on hematite and rutile.

(From Lyklema, 1987)

(1987) pointed out the affinity of protons as given by the enthalpy change, $-\Delta_{ads}$ H^o_{H+}, for hematite is much larger than for rutile because hematite has a larger affinity for H^+ than rutile. On the other hand $T\Delta_{ads} S^o_{H+}$ is the same for rutile and hematite and probably for the other oxides; it is indicative of the solution side of the process, i.e., the building up of surface charge. Interestingly the entropy production for the reaction of H^+ with a surface hydroxyl group is about half that for the protonation of a solute OH^-.

As Lyklema (1987) points out, the congruence observation implies the applicability of Nernst's law for the oxides[1] under consideration (Figs. 3.4 and 3.17):

$$\psi_0 = \frac{-2.3\ RT}{F}\ \Delta pH \tag{3.24}$$

(at 25° C: $\psi_0 = -0.059\ \Delta pH$)

Table 3.2   Thermodynamic data for proton adsorption T = 20° C
All quantities in kJ mol⁻¹ (from Lyklema, 1987)

	$\Delta_{ads}G^o_{H+}$	$\Delta_{ads}H^o_{H+}$		$T\Delta_{ads}S^o_{H+}$
		temperature dependence	microcal	
Rutile	−31.0	−17.6	−21	+13.4
Hematite	−48.8	−36.6	−36	+12.5
OH⁻ in solution	−79.5	−57.5		+22.5

Enthalpy changes per mol of proton are given as calculated from temperature dependence or as measured microcalorimetrically.

Anion Binding. This discussion illustrates how valuable information on enthalpy changes of surface reactions (either from temperature dependence or from direct calorimetric measurements) are. Zeltner et al. (1986) have studied calorimetrically the surface complex formation of phosphate and salicylate on goethite. They show that these reactions are exothermic (at pH = 4) with ΔH_{ads} values at low coverage (~ 10 %) of ca. −24 kJ mol⁻¹, they argue tentatively that these values indicate bidentate surface complex formation. They also show that $-\Delta H$ decreases with increasing surface coverage.

---

[1] The data of SiO_2 cannot be included in data related to pH congruence.

From the little data available one may conclude
  i) adsorption reactions where coordinative bonds are formed, are exothermic;
  ii) a large portion of the calculated entropy change results from reordering of water structure and from the protonation of the oxide surface which accompanies ligand binding even at constant pH;
  iii) adsorption due to hydrophobic effect (cf. Chapter 4.7) is accompanied by very small release of heat of adsorption ΔH (+ or −).

Appendix

A.3.1 The Gouy-Chapman Theory

Table A.3.1 gives a brief derivation of the equations. The theory is based on the validity of the Poisson equation for distances measured over molecular dimensions. Furthermore, the theory depends on the following assumptions (Sposito, 1984):

  i) The surface from which x is measured is a uniform infinite plane of charge;
 ii) the charged species in the solution are point ions; these ions interact with themselves and with the surface only by Coulomb force;
iii) the water in the solution is a uniform continuum characterized by the dielectric constant;
 iv) the (inner) potential, ψ, at a distance x is proportional to the average energy, $W_i(x)$, required to bring a ion i from infinity to the point x in the solution.

$$W_i(x) = Z_i F \psi(x) \tag{i}$$

The limitations imposed on this theory, have been discussed (see for example Sposito, 1984).

Table A.3.1   Gouy-Chapman Theory of Singlet Flat Double Layer [a]

I. Variation of Charge Density in Solution

Equality of electrochemical potential, $\bar{\mu}$ $(= \mu + zF\Psi)$ of every ion, regardless of position

Electrochemical Potential: $\bar{\mu}_{+(x)} = \bar{\mu}_{+(x=\infty)}, \qquad \bar{\mu}_{-(x)} = \bar{\mu}_{-(x=\infty)}$ (1)

$$zF(\Psi_{(x)} - \Psi_{(x=\infty)}) = -RT \ln \frac{n_{+(x)}}{n_{+(x=\infty)}} = RT \ln \frac{n_{-(x)}}{n_{-(x=\infty)}} \tag{2}$$

Cations: $$n_+ = n_{+(x=\infty)} \exp\left(\frac{-zF\Psi_{(x)}}{RT}\right) \tag{3}$$

Anions: $$n_- = n_{-(x=\infty)} \exp\left(\frac{zF\Psi_{(x)}}{RT}\right) \tag{4}$$

Space Charge Density: $$q = zF(n_+ - n_-) \tag{5}$$

(if $z_+ = z_-$)

Table A.3.1 (continued)

II. Local Charge Density and Local Potential

Ψ and q are related by Poisson's equation:

Poisson's Equation
$$\frac{d^2\Psi}{dx^2} = -\frac{q}{\varepsilon\varepsilon_0} \tag{6}$$

Combining (3), (4), and (5) with (6) and considering that $\sinh x = (e^x - e^{-x})/2$ gives the

Double-Layer Equation:
$$\frac{d^2\Psi}{dx^2} = \frac{\kappa^2 \sinh(zF\Psi/RT)}{(2\,F/RT)} \tag{7}$$

where κ is the reciprocal thickness of the double layer (the reciprocal Debye length)

$$\kappa = \left(\frac{e^2 \sum_i n_i z_i^2}{\varepsilon\varepsilon_0 kT}\right)^{1/2} \quad (\text{cm}^{-1}) \tag{8}$$

For convenience the following substitutions can be made:

$$y = \frac{zF\Psi}{RT}; \qquad \bar{z} = \frac{zF\Psi_d}{RT}; \qquad \zeta = \kappa x \tag{9}$$

Considering (9), (8) becomes the

Substituted Double-Layer Equation:
$$\frac{d^2y}{d\zeta^2} = \sinh y \tag{10}$$

For boundary conditions, if $\zeta = \infty$, $dy/d\zeta = 0$ and $y = 0$

first integration:
$$dy/d\zeta = -2\sinh(y/2), \text{ or} \tag{11}$$

$$d\Psi/dx = -\frac{RT}{zF} 2\kappa \sinh(y/2) \tag{11a}$$

and for boundary conditions, if $\zeta = 0$, $\Psi = \Psi_d$ or $y = \bar{z}$

second integration:
$$e^{y/2} = \frac{e^{\bar{z}/2} + 1 + (e^{\bar{z}/2} - 1)e^{-\zeta}}{e^{\bar{z}/2} + 1 - (e^{\bar{z}/2} - 1)e^{-\zeta}} \tag{12}$$

Table A.3.1 (continued)

Simplified Equations for $\Psi_d \ll 25$ mV:

Instead of (7)
$$\frac{d^2\Psi}{dx^2} = \kappa^2\Psi \tag{7a}$$

Instead of (12)
$$\Psi = \Psi_d \exp(-\kappa x) \tag{12a}$$

III. Diffuse Double-Layer Charge and Ψ_d

Surface Charge Density: $\quad \sigma = -\int_0^\infty q\,dx = \varepsilon\varepsilon_0 \int_0^\infty \left(\frac{d^2\Psi}{dx^2}\right)dx$

$$= -\varepsilon\varepsilon_0\left[\frac{d\Psi}{dx}\right]_{x=0} \tag{13}$$

Inserting (11a), $\qquad \sigma = (8\varepsilon\varepsilon_0 n_s kT)^{\frac{1}{2}} \sinh\left(\frac{zF\Psi_d}{2\,RT}\right) \tag{14}$

$$= 0.1174\, c_s^{\frac{1}{2}} \sinh\left(\frac{zF\Psi_d}{2\,RT}\right)(C\ m^{-2})\ \ (\text{at } 25°\ C) \tag{14a}$$

If $\Psi_0 \ll 25$ mV, $\sigma \approx \varepsilon\varepsilon_0\,\kappa\Psi_d$, or $\sigma = 2.3\,I^{\frac{1}{2}}\,\Psi_d$ \hfill (14b)

---

a)
μ	= chemical potential;		I	= ionic strength (mol liter^{-1});
$\bar\mu$	= electrochemical potential;		z	= valence of ion;
Ψ	= local potential (V);		κ	= reciprocal thickness of double layer (cm^{-1});
Ψ_d	= diffuse double-layer potential (V);			
q	= (volumetric) charge density (C cm^{-3});		e	= charge of electron (elementary charge), 1.6×10^{-19} C;
σ	= diffuse surface charge (C cm^{-2});			
x	= distance from surface (cm);		k	= Boltzmann constant, 1.38×10^{-23} J K^{-1};
n_+	= local cation concentration (mol cm^{-3});		kT	= Boltzmann constant times absolute temperature, 0.41×10^{-20} V C at 20° C;
n_-	= local anion concentration (mol cm^{-3});			
$n_{(x=\infty)}$	= bulk ion concentration (mol cm^{-3});		RT	= $N \times kT = 2.46 \times 10^3$ V C mol^{-1} at 20° C;
n_i	= number of ions i per cm^3 [$Nn_{(x=\infty)}$] (cm^{-3});		ε	= relative dielectric permittivity (dimensionless) ($\varepsilon = 78.5$ for water at 25° C);
n_s	= number of ion pairs (cm^{-3});		ε_0	= permittivity in vacuum, 8.854×10^{-14} C V^{-1} cm^{-1};
c_s	= salt concentration (mol liter^{-1} or M);			
N	= Avagadro's number, 6.03×10^{23} mol^{-1};		F	= Faraday = $6 \times 10^{23} \times e = 96.490$ C eq^{-1};
			$F\Psi/RT$	= $e\Psi/kT = 1$ for $\Psi = 25$ m V at 20° C

A.3.2 The solid Phase; References

Introductory Remarks. Since the surfaces can be looked at as extending structures of the solid phase, the reader should familiarize himself with the structure chemistry of the solid phase. Since it is not the objective of this book to cover this subject, the reader should consult some of the books on crystal structure and solid state chemistry such as for example Cox (1987), Greenland and Hayes (1987), Newman (1987), Stucki, Goodman and Schwertmann (1985), Wells (1984), West (1984).

To assist the reader, some references on some of the solid phases of importance are given..

1) *Iron Oxides and Hyxdroxides*
   General reference: U. Schwermann and R.M. Cornell, *Iron Oxides in the Laboratory*, VCH Weinheim (1991).

2) *Iron(III) Hydrolysis*
   General reference: W. Schneider, *Iron Hydrolysis and the Biochemistry of Iron – The Interplay of Hydroxide and Biogenic Ligands*, Chimia **42**, 9–20 (1988).

3) *Aluminum Oxides and Hydroxides*
   General reference: J.A. Davis and J.D. Hem, *The Surface Chemistry of Aluminum Oxides and Hydroxides*, in: G. Sposito, The Environmental Chemistry of Aluminum, CRC Press, Boca Raton (1989).

4) *Silicates and Clays*
   General references: J.E. Gieseking, *Soil Components*, Vol. 2, Springer, New York (1975); R.E. Grim, *Clay Mineralogy*, 2nd ed., McGraw Hill, New York (1968)

Problems

1) The surface potential, ψ, cannot be measured directly. It can be estimated how-ever, e.g., with the help of Eq. (3.8a) from the surface charge density. Discuss the assumptions involved in applying such a calculation.

2) Consider the date given in Fig. A on surface charge of river sediments. How can these data be interpreted in terms of possible sediment constituents, presence of organic matter?

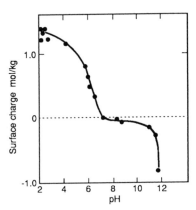

Figure A

Variation of sediment charge with pH in the River Meuse at T = 25° C and I = 0.01 M. (From Mouvet and Bourg, 1983)

3) Some researchers suggest to estimate ψ from the zeta potential ζ by assuming $\psi \approx \zeta$. Discuss the validity of this approximation.

4) Consider the results given in Fig. B on the zeta potential of Al_2O_3 (corundum) in solutions in various electrolytes by Modi and Fuerstenau (1957). Explain the various potential increasing and decreasing effects; identify the ions that are specifically adsorbed.

5) Recalculate question 2 of the problem set in Chapter 2, but consider coulombic interaction.
   i) diffuse double layer model, I = 10^{-1} M, and
   ii) constant capacitance model.

6) Explain the Lewis-base properties of a siloxane ditrigonal cavity.

7) Discuss some of the consequences of Fig. 3.4. Why does the congruence of sur-face charge vs ΔpH imply the applicability of Nernst's law for oxides?

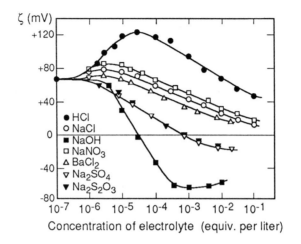

Figure B

Zeta potential of Al_2O_3 (corundum) in solutions of various electrolytes. The concentration unit is equivalents per liter.

(From Modi and Fuerstenau, 1957)

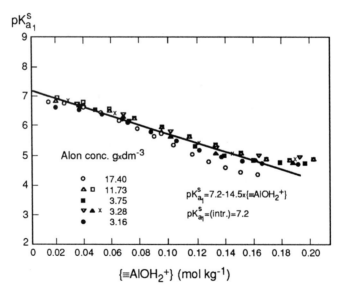

Figure C

The microscopic acidity constants $K_{a_1}^s$ as a function of $\langle \equiv AlOH_2^+ \rangle$. Extrapolation to zero charge gives intrinsic acidity constants.

(From Stumm et al., 1976)

8) In Fig. C microscopic acidity constants of the reaction $\equiv AlOH_2^+ \rightleftharpoons \equiv AlOH + H^+$ for γ-Al_2O_3 are plotted as a function of $\{\equiv AlOH_2^+\}$. The data are for 0.1 M $NaClO_4$. This figure illustrates (within experimental precision) the conformity of the proton titration data to the constant capacitance model. Calculate the capacitance.

Reading Suggestions

Brown Jr., G. E. (1990), "Spectroscopic Studies of Chemisorption Reaction Mechanisms at Oxide-Water Interfaces", in M. F. Hochella Jr. and A. F. White, Eds., *Mineral-Water Interface Geochemistry*, pp. 309-363.

Davis, J. A., and D. B. Kent (1990), "Surface Complexation Modeling in Aqueous Geochemistry", in M. F. Hochella Jr. and A. F. White, Eds., *Mineral-Water Interface Geochemistry*, Mineralogical Society of America, pp. 177-260.

Dzombak, D. A., and F. M. M. Morel (1990), *Surface Complexation Modeling; Hydrous Ferric Oxide*, Wiley-Interscience, New York.

Hohl, H., L. Sigg, and W. Stumm (1980), "Characterization of Surface Chemical Properties of Oxides in Natural Waters; The Role of Specific Adsorption Determining the Specific Charge," in M. C. Kavanaugh and J. O. Leckie, Eds., *Particulates in Water, Advances in Chemistry Series, ACS* **189**, 1-31.

Lyklema, J. (1987), "Electrical Double Layers on Oxides: Disparate Observations and Unifying Principles", *Chemistry and Industry*, 741-747.

Sposito, G. (1989), "Surface Reactions in Natural Aqueous Colloidal Systems", *Chimia* **43**, 169-176.

Wieland, E., B. Wehrli, and W. Stumm (1988), "The Coordination Chemistry of Weathering: III. A Generalization on the Dissolution Rates of Minerals", *Geochim. Cosmochim. Acta* **52**, 1969-1981.

Chapter 4

Adsorption

4.1 Introduction
Intermolecular Interaction between Solute and Solid Phase

The geochemical fate of most trace elements is controlled by the reaction of solutes with solid surfaces. Adsorption reactions affect the surface charge of suspended particles and colloids which influence their aggregation and transport; they are also important in water and soil technology. Furthermore, the rates of processes such as dissolution of mineral phases (of importance in the weathering of rocks, in the formation of soils and sediments and in the corrosion of metals) and precipitation (heterogeneous nucleation and crystal growth) and ion exchange depend on the reactivity of surfaces and their molecular (atomic) surface structures which in turn are influenced by adsorption to these surfaces.

A full understanding of adsorption requires that the interaction of a solute with a surface be characterized in terms of the fundamental physical and chemical properties of the solute, the sorbent and the solvent (water) (Westall, 1987). The adsorption reactions of importance in waters, sediments and soils are listed in terms of intermolecular reactions in Table. 4.1. The fundamental chemical interactions of solutes with the surfaces by formation of coordinative bonds were already discussed in Chapter 2. The electrostatic interactions and the electric double layer were considered in Chapter 3.

Table 4.1   Intermolecular Interactions at the Solid-water Interface
(Modified from Westall, 1987)

A	*Chemical Reactions with Surfaces*	
	Surface hydrolysis	
	Surface complexation	
	Surface ligand exchange	
	Hydrogen bond formation	
B	*Electrical Interactions at Surfaces*	
	Electrostatic interactions	
	Polarization interactions	
C	*Interactions with Solvent*	
	Hydrophobic expulsion	

4.2 Gibbs Equation on the Relationship between Interfacial Tension and Adsorption

Molecules in the surface or interfacial region are subject to attractive forces from adjacent molecules, which result in an attraction into the bulk phase. The attraction tends to reduce the number of molecules in the surface region (increase in inter-molecular distance). Hence work must be done to bring molecules from the interior to the interface. The minimum work required to create a differential increment in surface $d\overline{A}$ is $\gamma d\overline{A}$, where $\overline{A}$ is the interfacial area and $\overline{\gamma}$ is the surface tension[1] or interfacial tension. One also refers to γ as the interfacial Gibbs free energy for the condition of constant temperature, T, pression, P, and composition (n = number of moles)

$$\gamma = \left(\frac{\delta G}{\delta \overline{A}}\right)_{T,p,n} \tag{4.1}$$

γ is usually expressed in $J\ m^{-2}$ or in $N\ m^{-1}$

In water the intermolecular interactions which produce surface tension are essentially composed of
114

    a) London-Van der Waals dispersion interactions, $\gamma_{H_2O(L)}$, and
    b) hydrogen bonds, $\gamma_{H_2O(H)}$:

$$\gamma = \gamma_{H_2O(L)} + \gamma_{H_2O(H)} \tag{4.2}$$

It is estimated that about one third of the interfacial tension is due to Van der Waals attraction, and the remainder is due to hydrogen bonding.

The interfacial tension between two phases is subject to the resultant force field made up of components arising from attractive forces in the bulk of each phase and the forces, usually the London dispersion forces, operating accross the interface itself (Fowkes, 1965; Adamson 1990) (see Appendix, this Chapter).

The Gibbs-Equation. Thermodynamically from Eq. (4.1) the Gibbs Equation

---

[1] Strictly speaking, the surface tension of water is the tension of water with respect to vacuum; but one usually refers to the interfacial tension between water and air. As will be discussed in Chapter 6, the interfacial tension between water and solid minerals is of importance in the kinetics of nucleation and precipitation.

$$\Gamma_i = -\frac{1}{RT}\left(\frac{\delta\gamma}{\delta \ln a_i}\right)_{T,p} \tag{4.3}$$

can be derived. Here Γ_i is the surface concentration [mol m^{-2}] or more specifically the surface excess[1] with regard to a reference condition (with pure water the adsorption density of $H_2O = 0$).

R = gas constant
T = absolute temperature
γ = surface tension or interfacial tension [J m^{-2}]
a_i = activity (or concentration) of species i

The Gibbs equation relates the extent of adsorption at an interface (reversible equilibrium) to the change in interfacial tension; qualitatively, Eq. (4.3) predicts that a substance which reduces the surface (interfacial) tension [$(\delta\gamma/\delta \ln a_i) < 0$] will be adsorbed at the surface (interface). Electrolytes have the tendency to increase (slightly) γ, but most organic molecules, especially surface active substances (long chain fatty acids, detergents, surfactants) decrease the surface tension (Fig. 4.1). *Amphipathic* molecules (which contain hydrophobic and hydrophilic groups) become oriented at the interface.

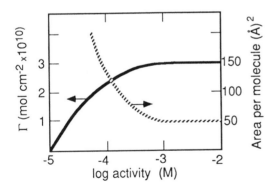
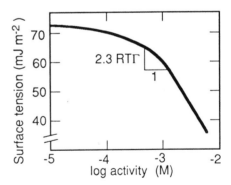

Figure 4.1

Gibbs adsorption equation
The surface excess Γ can be obtained – cf. Eq. (4.3) – from a plot of surface tension $\bar{\gamma}$ vs log activity (concentration) of adsorbate. The area occupied per molecule or ion adsorbed can be calculated.

---

[1] To define a surface excess concentration rigorously, we must decide whether or not to recognize the finite thickness of surfaces. In view of the difficulty of defining surface thickness, Gibbs defined the surface (for thermodynamic purposes) as a mathematical plane or dividing surface of zero thickness near the physical surface, and surface properties as the *net* positive or negative excess in the vicinity of the surface over the magnitude of the same property in the bulk (Adamson, 1990).

At the water-air interface hydrophilic groups are oriented toward the water, hydrophobic groups are oriented toward air. At solid-water interfaces, the orientation depends on the relative affinities for water and for the solid surface. The hydrophilic groups of amphipathic molecules may – if the hydrophobic tendency is relatively small – interact coordinatively with the functional groups of the solid surface (Ulrich et al., 1988) (see Fig. 4.10).

Although the Gibbs equation applies also to the solid-liquid interface, direct measurements of the interfacial tension are difficult. In a qualitative sense, it is important to realize that adsorption at a solid surface – of both, molecules and ions – reduce interfacial tension. For example, the interfacial tension of an oxide surface, e.g., Quartz, is reduced upon adsorption or desorption of (charge determining) ions. Such an oxide has a relative maximum value at its point of zero charge (corresponding to the capillary maximum of a Hg electrode in water); its interfacial energy decreases at pH values above and below pH_{pzc}. Parks (1984) has elaborated this concept by discussing the surface and interfacial free energies of quartz.

4.3 Adsorption Isotherms

Adsorption is often described in terms of isotherms which show the relationship between the bulk activity (concentration) of adsorbate[1] and the amount adsorbed at constant temperature.

The Langmuir Isotherm. The most simple assumption in adsorption is that the adsorption sites, S, on the surface of a solid (adsorbent) become occupied by an adsorbate from the solution, A. Implying a 1:1 stoichiometry

$$S + A \rightleftharpoons SA \qquad (4.4)$$

where

   S   = surface site of adsorbens
   A   = adsorbate in solution
   SA = adsorbate on surface sites

S and SA can be expressed in mol/ℓ or in mol m^{-2}. For simplicity, we use first mol/ℓ. Applying the mass law to Eq. (4.4)

---

[1] Adsorbate is the substance in solution to become adsorbed at the adsorbent. If the mechanism of accumulation at the solid surface (e.g., adsorption or precipitation) is not known one may refer to *sorption* (and sorbent, and sorbate).

$$\frac{[SA]}{[S]\,[A]} = K_{ads} = \exp\left(-\frac{\Delta G^{o}_{ads}}{RT}\right) \tag{4.5}$$

The maximum concentration of surface sites, S_T, is given by

$$[S_T] = [S] + [SA] \tag{4.6}$$

Thus,

$$[SA] = [S_T] \frac{K_{ads}\,[A]}{1 + K_{ads}\,[A]} \tag{4.7}$$

If we define the surface concentration

$$\Gamma = [SA] / \text{mass adsorbent} \tag{4.8}$$

and

$$\Gamma_{max} = [S_T] / \text{mass adsorbent} \tag{4.9}$$

we obtain

$$\Gamma = \Gamma_{max} \frac{K_{ads}\,[A]}{1 + K_{ads}\,[A]} \tag{4.10a}$$

Usually the Langmuir equation is known in the form of (4.10a)[1]. Frequently one can also write it as

$$\frac{\theta}{1 - \theta} = K_{ads}\,[A] \tag{4.10b}$$

where

$$\theta = \frac{[SA]}{[S_T]}$$

The conditions for the validity of a Langmuir type adsorption equilibrium are i) thermal equilibrium up to the formation of a monolayer, $\theta = 1$; ii) the energy of adsorption is independent of θ, (i.e., equal activity of all surface sites). There is no difference between a surface complex formation constant and a Langmuir adsorption

---

[1] The Langmuir equation is derived here from application of the mass law, in a similar way as the surface complex formation equilibria were derived in Chapter 2. In principle at a constant pH there is no difference between a Langmuir constant and a surface complex formation constant.

constant. If there are two adsorbates, A and B, one can formulate

$$\Gamma_A = \frac{\Gamma_{max} \, K_A \, [A]}{1 + K_A \, [A] + K_B \, [B]} \qquad (4.11)$$

where K_A and K_B are the adsorption constants for A and B, respectively. There are further assumptions implied in Eq. (4.11), e.g., the non-interaction of A and B on the surface.

Alternatively a two term series Langmuir equation can be written for two adsorbents (or for an adsorbent with two sites of different affinities)

$$\Gamma_A = \frac{\Gamma_1 \, K_1 \, [A]}{1 + K_1 \, [A]} + \frac{\Gamma_2 \, K_2 \, [A]}{1 + K_2 \, [A]} \qquad (4.12)$$

A Langmuir type adsorption isotherm is given in Fig. 4.2a. As is shown in Fig. 4.2b the evaluation of the equilibrium adsorption constant and of Γ_{max} is readily obtained from experimental data by plotting Eq. (4.10a) in the reciprocal form

$$\Gamma^{-1} = \Gamma_{max}^{-1} + K_{ads}^{-1} \, \Gamma_{max}^{-1} \, [A]^{-1} \qquad (4.10c)$$

It is often convenient to plot the Langmuir isotherm in a double logarithmic plot (Fig. 4.3).

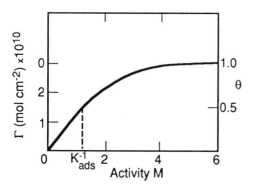 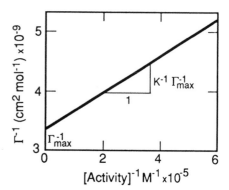

Figure 4.2

Langmuir adsorption isotherm
From the adsorption isotherm [plotted in accordance with (4.10a)], the equilibrium constant K and the adsorption capacity, Γ_{max}, is obtained by plotting Γ^{-1} versus the reciprocal activity of the sorbate (Eq. 4.10c).
(From Stumm and Morgan, 1981)

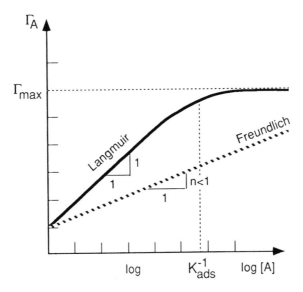

Figure 4.3

Plot of adsorption data in a double logarithmic plot. In a Langmuir isotherm the initial slope is unity. A Freundlich isotherm shows in a double log plot a slope of n < 1. Such a Freundlich isotherm is obtained if the adsorbent is heterogeneous (decreasing tendency for adsorption with increasing θ).
(Modified from Morel, 1983)

The fits of experimental data to a Langmuir (or another) adsorption isotherm does not constitute evidence that adsorption is the actual mechanism that accounts for the loss of the sorbate from the solution. Very frequently adsorption to a surface is followed by additional interactions at the surface, e.g., a surfactant undergoes two-dimensional association subsequent of becoming adsorbed; or charged ions tend to repel each other within the adsorbed layer.

The Frumkin Equation (also referred to as the Frumkin-Fowler-Guggenheim, FFG, equation) has been specifically developped to take lateral interactions at the surface into account. In the FFG equation, the term $\theta / (1-\theta)$ in (4.10b) is multiplied by the factor $\exp(-2\,a\theta)$ which reflects the extent of lateral interactions

$$\frac{\theta}{1 - \theta} \; \exp(-2\;a\theta) \; = \; B[A] \tag{4.13}$$

This may be compared with the Langmuir Equation

$$\frac{\theta}{1 - \theta} \; = \; B[A]$$

where B is the adsorption constant and [A] is the equilibrium (bulk) concentration (activity) of the adsorbate, a is the interaction coefficient. If a = 0 Eq. (4.13) reduces to the Langmuir equation (4.10b); a > 0 indicates attraction, a < 0 means repulsion. The value for the adsorption constant B and for the interaction coefficient a can be determined from the intercept and the slope of the resulting straight line in a plot of log [θ / (1 − θ)] vs θ. Fig. 4.4 plots data on the adsorption of caprylic acid at pH = 4 on a hydrophobic surface (liquid mercury). At pH = 4, the caprylic acid is uncharged and lateral interaction (between the hydrocarbon moiety of the fatty acids) occurs (a > 0). At higher pH, the species getting into the adsorption layer are anions which repel each other; under such conditions, repulsive lateral interaction is observed (a < 0). As shown in Example 4.1, surface complex formation equilibria using a correction for electrostatic interaction in terms of the constant capacitance model (see Eqs. (3.20) and (3.23)) are equivalent to the FFG equation.

As every surface complex formation equilibrium constant can be converted into an equivalent Langmuir adsorption constant (Stumm et al., 1970), every FFG equation reflects the surface complex formation constant corrected with the interaction coeffi-

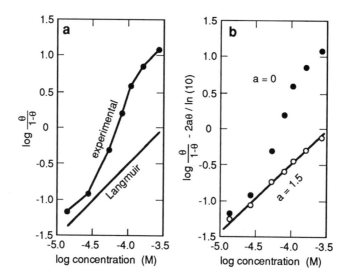

Figure 4.4

Data on the adsorption of caprylic acid on a hydrophobic (mercury) surface in terms of a double logarithmic plot of Eq. (4.13) Panel a) compares the experimental values with a theoretical Langmuir isotherm, using the same values for the adsorption constant B for both curves. Panel b) shows that the adsorption process can be described by introducing the parameter a, which accounts for lateral interaction in the adsorption layer. Eq. (4.13) postulates a linear relation between the ordinate [= log [θ / 1 − θ)] − 2a θ / (ln 10)] and the abscissa (log c). If the correct value for a is inserted, a straight line results. For caprylic acid at pH 4, a value of a = 1.5 gives the best fit.
(From Ulrich, Ćosović and Stumm, 1988)

cient due to electrostatic interaction (constant capacitance model, Schindler and Stumm, 1977).

The Freundlich Isotherm

This equation is very convenient to plot adsorption data empirically in a log Γ vs log [A] plot (Fig. 4.3)

$$\Gamma = m[A]^n \tag{4.14}$$

As pointed out by Sposito (1984) this equation initiated the surface chemistry of naturally occurring solids. Maarten van Bemmelen published this equation (now referred to as the Freundlich isotherm) more than 100 years ago and distilled from his results, that the adsorptive power of ordinary soils depends on the colloidal silicates, humus, silica, and iron oxides they contain.

The equation applies very well for solids with heterogeneous surface properties and generally for heterogeneous solid surfaces. As has been shown by Sposito (1984), Eq. (4.14) can be derived by generalizing Eq. (4.12) to an integral over a continuum of Langmuir equations. As pointed out by Sposito (1984), the van Bemmelen-Freundlich isotherm can be thought of as the result of a log-normal distribution of Langmuir parameters K (i.e., a normal distribution of ln K) in a soil (or on a natural aquatic particle surface).

Example 4.1:   Surface Complex Formation, Langmuir Equation and Frumkin Equation

Consider a reaction such as

$$SH + Cu^{2+} \rightleftharpoons SCu^+ + H^+ \; ; \; K_{Cu}^s \tag{i}$$

and illustrate that the mass law of Eq. (i) can be converted into a Langmuir-type equation and that Eq. (i), after correction for electrostatic effects, corresponds to a Frumkin (FFG) equation.

The mass law of (i) is

$$\frac{[SCu^+]}{[SH][Cu^{2+}]} = \frac{K_{Cu}^s}{[H^+]} \tag{ii}$$

Furthermore,

$$[SCu^+] + [SH] + [S^-] = S_T \tag{iii}$$

Depending on the pH-range [SH] or [S⁻] predominates. For most oxides in the pH-range of interest $[SH] \gg [S^-]$; thus,

$$[SCu^+] + [SH] \approx S_T \tag{iii a}$$

Combination of Eqs. (ii) and (iii a) gives

$$[SCu^+] = \frac{S_T \, K_{Cu}^s \, [H^+]^{-1} \, [Cu^{2+}]}{1 + K_{Cu}^s \, [H^+]^{-1} \, [Cu^{2+}]} \tag{iv}$$

If we define

$$\Gamma_{Cu} = [SCu^+] / \text{mass adsorbent, and}$$
$$\Gamma_{max} = S_T / \text{mass adsorbent}$$

we obtain

$$\Gamma_{Cu} = \frac{\Gamma_{max} \, K_{Cu}^s \, [H^+]^{-1} \, [Cu^{2+}]}{1 + K_{Cu}^s \, [H^+]^{-1} \, [Cu^{2+}]} \tag{v}$$

where K_{Cu}^s is an *intrinsic* surface complex formation constant and $K_{Cu}^s / [H^+]$ is an intrinsic adsorption constant at a given [H⁺]; i.e.,

$$K_{Cu}^s / [H^+] = K_{ads(pH)} \tag{vi}$$

Eq. (vi) can also be written as

$$\Gamma_{Cu} = \frac{\Gamma_{max} \, K_{Cu}^s \, [Cu^{2+}]}{[H^+] + K_{Cu}^s \, [Cu^{2+}]} \tag{vii}$$

Constant Capacitance

In line with Eq. (3.20) we have

$$\frac{[SCu^+]}{[Cu^{2+}] \, [SH]} = K_{Cu}^s(\text{app}) \, [H^+]^{-1} = K_{Cu}^s \, [H^+]^{-1} \exp\left(-\frac{F\psi}{RT}\right) \tag{viii}$$

where K_{Cu}^s, as before, is the intrinsic constant.

Because in the constant capacitance model

$$\psi = \frac{\sigma}{C} \tag{ix}$$

where C is the capacitance (cf. Eq. 3.12);

where the surface charge, σ [C m^{-2}], is

$$\sigma = \frac{F \, [SCu^+]}{\bar{a} \times s} \tag{x}$$

where F is the Faraday (96490 C mol^{-1}), $\bar{a}$ is the surface concentration [kg ℓ^{-1}] and s is the specific surface area [m^2 kg^{-1}]. The surface charge is due to $[SCu^+]$

$$[SCu^+] = \theta \, S_T \tag{xi}$$

We can rewrite Eq. (viii)

$$\frac{[SCu^+]}{[Cu^{2+}] \, [SH]} = K^s_{Cu}(app) \, [H^+]^{-1} = K^s_{Cu} \, [H^+]^{-1} \exp\left(-\frac{F^2}{CRT \, s\bar{a}} \theta \, [S_T]\right) \tag{xii}$$

or as in the form of Eq. (4.13)

$$\frac{\theta}{1 - \theta} \exp\left(\frac{F^2}{CRT \, s\bar{a}} \theta \, [S_T]\right) = K^s_{Cu} \, [H^+]^{-1} \, [Cu^{2+}] \tag{xiii}$$

The interaction coefficient a in Eq. (4.13) corresponds in the case considered to

$$a = -\tfrac{1}{2} \frac{F^2}{CRT \, s\bar{a}} \, [S_T]$$

4.4 Adsorption Kinetics

Various possible steps are involved in the transfer of an adsorbate to the adsorption layer. Transport to the surface by convection or molecular diffusion, attachment to the surface, surface diffusion, dehydration, formation of a bond with the surface constituents.

We consider here first, the kinetics of the adsorption of metal ions on a hydrous oxide surface.

Adsorption of Metal Ions to a Hydrous Oxide Surface

We have argued that (inner-sphere) surface complex formation of a metal ion to the oxygen donor atoms of the functional groups of a hydrous oxide is in principle similar to complex formation in homogeneous solution, and we have used the same type of equilibrium constants. How far can we apply similar concepts in kinetics?

The Homogeneous Case. Margerum (1978) and Hering and Morel (1990) have elaborated on mechanisms and rates of metal complexation reactions in solution. In the Eigen mechanism, formation of an outer-sphere complex between a metal and a ligand is followed by a rate limiting loss of water from the inner coordination sphere of the metal, Thus, for a bivalent hexaaqua metal ion

$$Me(H_2O)_6^{2+} + L^{n-} \underset{k_{-1}}{\overset{k_1}{\rightleftharpoons}} Me(H_2O)_6L^{(2-n)+} \tag{4.15}$$

$$Me(H_2O)_6L^{(2-n)+} \underset{slow}{\overset{k_{-w}}{\longrightarrow}} Me(H_2O)_5L^{(2-n)+} + H_2O \tag{4.16}$$

The ligand (as shown here schematically without H_2O) is also hydrated and has to loose some of its water to form a complex. The water loss from the ligand is not included here explicitly because it is considered to be fast relative to the water loss from the metal.

The stability constant for the outer sphere complex depends on the charge of the reacting species and the ionic strength of the medium, and can be readily calculated from electrostatic considerations.

Formally the rate for reaction (4.16) is

$$\frac{d[Me(H_2O)_5L^{(2-n)+}]}{dt} = k_{-w}[Me(H_2O)_6L^{(2-n)+}] \tag{4.17}$$

and applying the steady state principle to the formation of $Me(H_2O)_6L^{(2-n)+}$ we have

$$\frac{d[Me(H_2O)_6L^{(2-n)+}]}{dt} = k_1[Me(H_2O)_6]^{2+}[L^{n-}] - (k_{-1} + k_{-w})[Me(H_2O)_6L^{(2-n)+}] = 0 \tag{4.18}$$

We can simplify by considering that $k_{-1} \gg k_{-w}$, and by setting $k_1/k_{-1} = K_{os}$. K_{os} is the equilibrium constant of the outer sphere complex. For the rate of the formation of $MeL^{(2-n)+}$ inner-sphere complex (now written without water), we have

$$\frac{d\,[MeL^{(2-n)+}]}{dt} = k_f\,[Me^{2+}]\,[L^{n-}] = K_{os}\,k_{-w}\,[Me^{2+}]\,[L^{n-}] \tag{4.19}$$

In other words, the complex formation rate constant k_f $[M^{-1}\,s^{-1}]$ depends on the outer sphere electrostatic encounter and on the water exchange rate of the aquo metal ion k_{-w}

$$k_f = K_{os}\,k_{-w} \tag{4.20}$$

Fig. 4.5 gives data on the rate of water loss of aquo metal ions. It is well known that hydrolysis of the metal ion enhances the rate of water loss from the metal ion. Usually a monohydroxo complex has a k_{-w} that is two to three orders of magnitude larger than that of a non-hydrolyzed species.

The Heterogeneous Case. Hachiya et al. (1984) and Hayes and Leckie (1986) used the pressure-jump relaxation method to study the adsorption kinetics of metal ions to oxide minerals. Their results support in essence the same adsorption mechanism as that given for homogeneous complex formation.

$$\equiv Al - OH + Me(H_2O)_n^{2+} \overset{K_{os}}{\rightleftharpoons} \equiv Al - OH \cdots Me(H_2O)_n^{2+} \tag{4.21}$$

$$\equiv Al - OH \;\text{----}Me(H_2O)_n^{2+} \overset{k_{ads}}{\longrightarrow} \equiv Al - O\begin{smallmatrix} \nearrow H \\ \searrow Me(H_2O)_{n-1}^{2+} \end{smallmatrix} + H_2O \tag{4.22}$$

$$\equiv Al - OH \begin{smallmatrix} \nearrow H \\ \searrow Me(H_2O)_{n-1}^{2+} \end{smallmatrix} \overset{k}{\underset{fast}{\longrightarrow}} \equiv Al - O - Me(H_2O)_{n-1}^{+} + H^{+} \tag{4.23}$$

The dashed line in the complex in (4.21) and (4.22) indicates an outer-sphere (o.s.) surface complex, K_{os} stands for the outer-sphere complex formation constant and k_{ads} $[M^{-1}\,s^{-1}]$ refers to the intrinsic adsorption rate constant at zero surface charge (Wehrli et al., 1990). K_{os} can be calculated with the help of a relation from Gouy Chapman theory (Appendix Chapter 3).

$$K_{os} = \exp\left(-\frac{Z\psi}{RT}\right) \cong \exp\left(-z\,\frac{F^2}{CsRT}\,Q\right) \tag{4.24}$$

where Z = charge of adsorbate, F is the Faraday constant, C is the integral capacitance of a flat double layer [farad m^{-2}], s is the specific surface area [m^2 kg^{-1}] and Q refers to the experimentally accessible surface charge [mol kg^{-1}]. The overall rate for the formation of the surface complex is given in analogy to the homogeneous case (Eq. 4.19) by

$$\frac{d\,[\equiv AlOMe^+]}{dt} = K_{os}\,k_{ads}\,[M^{2+}]\,[\equiv Al - OH] \tag{4.25}$$

The term $F^2/CsRT$ is obtained from the constant capacitance model (Chapter 3.7). Fig. 4.6 gives a plot of the linear free energy relation between the rate constants for water exchange and the intrinsic adsorption rate constant, k_{ads}.

The mechanism given is in support of the existence of inner-sphere surface complexes; it illustrates that one of the water molecules coordinated to the metal ion has to dissociate in order to form an inner-sphere complex; if this H$_2$O-loss is slow, then the adsorption, i.e., the binding of the metal ion to the surface ligands, is slow.

The circles refer to the pressure jump experiments of Hachiya et al. The relationship is expressed by an almost linear correlation (slope 0.92):

$$\log k_{ads} = -4.16 + 0.92 \log k_{-w} \tag{4.26a}$$

where as before k_{ads} refers to the intrinsic adsorption rate constant [M^{-1} s^{-1}] and k_{-w} refers to the constant for water exchange [s^{-1}]. (This constant can be converted into k'_{H_2O} [M^{-1} s^{-1}] by dividing k_{-w} by the concentration of water (~ 55.6 M).) Then Eq. (4.33) becomes

$$\log k_{ads} = -2.42 + 0.92 \log \left(\frac{k_{-w}}{55.6}\right) \tag{4.26b}$$

The heterogeneous adsorption process is slower by a factor of ca. 250 than an equivalent homogeneous process.

Most metal ions have water exchange rates larger than 10^5 and then their adsorption rates are characterized by $k_{ads} > 10^{0.5}$ M^{-1} s^{-1}. Thus, with 10^{-3} M surface ligands the half-time for adsorption, $t'_{1/2}$, to an uncharged surface would be $t'_{1/2} < 4$ min. However, Cr^{3+}, VO^{2+}, Co^{3+}, Fe^{3+} and Al^{3+} have lower exchange rates (see Fig. 4.6; the water exchange rates for Co^{3+} and Cr^{3+}, which are not given in the figure, are $k_{-w} = 10^{-1}$ and 10$^{-5.6}$ [s^{-1}], respectively). These ions, if non-hydrolyzed, are probably adsorbed inner-spherically rather slowly. But we might recall that hydroxo species can exchange water by two or three order of magnitude faster than "free" aquo ions. Experiments by Wehrli et al. (1990) have shown that adsorption rates of V(IV) increase with pH (in the pH range 3 – 5). This increase can be accounted for by as-

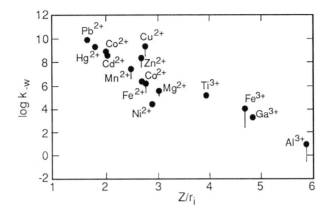

Figure 4.5

Rate of water loss from metal cations as a function of the ratio of the charge to the radius of the metal ion. Rate constants from Margerum et al. (1978). Z/r_i ratios calculated from ionic radii tabulated in the CRC Handbook of Chemistry and Physics.

(Adapted from Crumbliss and Garrison, 1988)

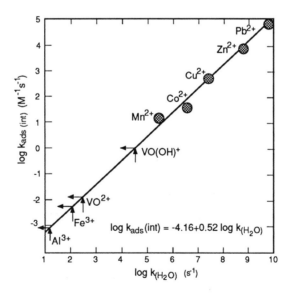

Figure 4.6

Linear-free energy relation between the rate constants for water exchange k_{-w} [s^{-1}] and the intrinsic adsorption rate constants $k_{ads(int)}$ [$M^{-1} s^{-1}$] from the pressure jump experiments of Hachiya et al. The intrinsic constants refer to an uncharged surface. The linear-free energy relations based on the experimental points are extended to some ions with lower H_2O exchange rate in order to predict adsorption rates.

(Modified from Wehrli et al., 1990)

suming that $VO(OH)^+$ is the reacting adsorbate. The half time for the adsorption of $VO(OH)^+$ onto 10^{-3} M surface ligands obtained from experimental data is in the order of 4 h. This is in good agreement with the rate predicted on the basis of the dissociative mechanism. Cr(III) was observed to adsorb to $Al_2O_3(s)$ above pH = 5. Under these conditions, dihydroxo and possibly multimeric hydroxo species are formed that are adsorbed faster.

Example 4.2: Adsorption Rate of Co^{2+}on Al_2O_3

Calculate the half-time for the adsorption of Co^{2+} on Al_2O_3 at pH = 4. We use for the Al_2O_3 the same equilibrium constants as those given in Example 3.4 and I = 0.1

$$\equiv AlOH_2^+ \rightleftharpoons \equiv AlOH + H^+ \qquad ; \log K = -7.4 \qquad (i)$$
$$\equiv AlOH \rightleftharpoons \equiv AlO^- + H^+ \qquad ; \log K = -10.0 \qquad (ii)$$

and assume a specific surface area for the aluminum oxide of $s = 10^4$ m^2 kg^{-1} and 10^{-1} mol surface sites kg^{-1} and use a suspension $\bar{a} = 10^{-3}$ kg/ℓ; i.e., there are 10^{-4} mol surface sites per liter. As Fig. 3.16 shows, at pH = 7, the surface charge is ca. $[\equiv AlOH_2^+] = 2 \times 10^{-5}$ M; k_{ads} for $Co^{2+} = 3.2 \times 10^1$ M^{-1} s^{-1} according to Fig. 4.6.

We can now calculate the outer-sphere complex formation constant according to Eq. (4.24):

$$K_{OS} = \exp\left(-\frac{ZF\Psi}{RT}\right) \qquad (iii)$$

we calculate Ψ for I = 0.1 according to the diffuse double layer theory (Eq. 3.8c). σ_P is calculated from $[\equiv AlOH_2^+]$ as follows:

$$\sigma_P = \frac{[\equiv AlOH_2^+]\ F}{\bar{a} \times s} = \frac{1 \times 10^{-5} \text{ mol } \ell^{-1} \times 96490 \text{ C mol}^{-1}}{10^{-3} \text{ kg } \ell^{-1}\ 10^4 \text{ m}^2 \text{ kg}^{-1}} = 0.095 \text{ C m}^{-2} \quad (iv)$$

From this value we compute Ψ by Eq. (3.8c)

$$\sinh (z\Psi \times 19.46) = \frac{\sigma_P}{0.1174 \sqrt{I}} = \frac{0.095}{0.1174 \times 3.16 \times 10^{-1}} = 2.56$$

$$z\Psi \times 19.46 \qquad = 1.67$$
$$\Psi \qquad\qquad = 0.086 \text{ V} \qquad (v)$$

K_{OS} can then be calculated according to (iii), where RT = 2.48×10^3 VC mol^{-1}

$$K_{OS} = \exp\left(-\frac{2 \times 96490 \text{ C mol}^{-1} \times 0.086 \text{ V}}{2480 \text{ VC mol}^{-1}}\right) = 1.25 \times 10^{-3}$$

Accordingly in line with Eq. (4.25) the rate is given by

$$-\frac{d[Co^{2+}]}{dt} = K_{OS} k_{ads} [Co^{2+}] [\equiv Al-OH]$$

$$= 1.25 \times 10^{-3} \times 3.2 \times 10^1 \times 10^{-4} [Co^{2+}] \tag{vi}$$

$$-\frac{d[Co^{2+}]}{[Co^{2+}] dt} = 4.03 \times 10^{-6} \text{ s}^{-1}$$

$$t_{1/2} = \ln 2 / 4.03 \times 10^{-6} \text{ s}^{-4} = 1.7 \times 10^5 \text{ s}$$
$$= 48 \text{ h} \tag{vii}$$

This result illustrates that the rate of adsorption is decreased (by nearly three orders of magnitude in comparison to an uncharged surface) by the opposite charge of adsorbate and adsorbent.

Kinetics of Anion Adsorption. A similar mechanism can be postulated for the kinetics of ligand exchange. Since the ligand exchanges with the central Me ion (Lewis acid) of the metal oxide, its water exchange rate may be rate determining. But the central ion exists already as an oxo, hydroxo species and thus the water exchange rate can be expected to be significantly larger than that of a non-hydolyzed metal ion. But in this context it is interesting to note that the rate of OH-exchange on α-Cr_2O_3 is known to be orders of magnitudes lower than that on goethite (α-FeOOH) (Yates and Healy, 1965). The extent and rate of NO_3^- adsorption by these two minerals are the same with equilibrium attained in minutes (outer-sphere adsorption). By contrast, the rate of phosphate adsorption is much lower with goethite (inner-sphere ligand exchange retarded because of slow H_2O exchange) than with α-Cr_2O_3, although the extent of adsorption is larger on goethite than on α-Cr_2O_3.

Semi-infinite linear Diffusion

For many adsorbates, especially organic substances, the concept of semi-infinite linear diffusion can give us some ideas on the time necessary for an adsorbate to be adsorbed. The number of mols adsorbate, n, diffusing to a unit area of a surface per second, is proportional to the bulk concentration of adsorbate, c:

$$\frac{dn}{A \, dt} = c\sqrt{\frac{D}{\pi \, t}} \tag{4.27}$$

It is assumed i) that the concentration c remains constant and ii) that transport by diffusion is rate controlling, i.e., the adsorbate arriving at the interface is adsorbed fast (intrinsic adsorption). This intrinsic adsorption, i.e., the transfer from the solution to the adsorption layer is not rate determining or in other words, the concentration of the adsorbate at the interface is zero; iii) furthermore, the radius of the adsorbing particle is relatively large (no spherical diffusion).

Integration of (4.27) gives

$$\Gamma = \frac{n}{A} = 2\,c\sqrt{\frac{Dt}{\pi}} \tag{4.28}$$

since $\Gamma_t = (n/A)_t$ is the surface concentration (mol m^{-2}) at a specific time, the time δ, necessary to cover the surface with a monomolecular layer is given by

$$\delta = \frac{\pi}{4}\frac{\Gamma_{max}^2}{D\,c^2} \tag{4.29}$$

Eq. (4.29) can be extended to obtain a fractional surface coverage, $\theta = \dfrac{\Gamma}{\Gamma_{max}}$ at time t

$$t = \frac{\pi}{4}\frac{\theta^2\,\Gamma_{max}^2}{D\,c^2} \tag{4.30}$$

Thus, the time that is necessary to attain a certain coverage, θ, or the time necessary to cover the surface completely ($\theta = 1$) is inversely proportional to the square of the bulk concentration (cf. Fig. 4.10b). Assuming molecular diffusion only, δ is of the order of 2 minutes for a concentration of 10^{-5} M adsorbate when the diffusion coefficient D is 10^{-5} cm^2 s^{-1} and $\Gamma_{max} = 4 \times 10^{-10}$ mol cm^{-2} [1]. Considering that transport to the surface is usually by turbulent diffusion, such a calculation illustrates that the formation of an adsorption layer is relatively rapid at concentrations above 10^{-6} M. But it can become slow at concentrations lower than 10^{-6} M.

Subsequent to the adsorption onto a surface, surfactants, especially long chain fatty acids and alcohols tend to undergo alterations such as two-dimensional associations in the adsorbed layer, presumably at rates kinetically independent of preliminary steps. These intra-layer reactions have been shown to be very slow.

---

[1] cf. Eq. (4.29)

$$\delta = \frac{\pi}{4}\frac{(4\times10^{-10})^2}{10^{-5}\ cm^2\ s^{-1}}\ \frac{[mol^2]}{cm^4}\ \frac{[\ell^2]\times10^6\,[cm^6]}{10^{-10}\,[mol^2]\,[\ell]^2} = 125\ s$$

The applicability of Eqs. (4.27) – (4.30) is somewhat restricted because the bulk concentration is assumed to be constant because either its depletion is negligible (adsorbed quantity << quantity present in the system) or because it is kept constant by a steady state mechanism. Analytical expressions of Γ as a function of time for situations where the approach to adsorption equilibrium is accompanied by a corresponding adjustment of c are available only for a few relatively simple cases.

A rigorous treatment of diffusion to or from a flat surface has been given by Lyklema (1991). Van Leeuwen (1991) has pointed out that in analyzing experimental adsorption data, that are always confined to a certain time window, it is tempting to fit the data to the sum of two or three exponential functions with different arguments. Although such fits are often apparently sucessful, the merit of the fit is purely mathematical; a mechanistic interpretation in terms of a first order dependence is usually not justified. With porous materials, diffusion into the pores renders the adsorption process very slow; often one gains the impression that the process is irreversible (e.g., Fig. 4.18).

The Elovich Equation

This equation has been used, especially in soil science, to describe the kinetics of adsorption and desorption on soils and soil minerals.

$$\frac{d\Gamma}{dt} = k_1 \exp(-k_2 \Gamma) \tag{4.31}$$

The solution of Eq. (4.31) (Sposito, 1984) is

$$\Gamma(t) = \frac{1}{k_2} \ln(k_1 k_2 t_0) + \frac{1}{k_2} \ln\left(1 + \frac{t}{t_0}\right) \qquad (t \geq t_c) \tag{4.32}$$

where $t_0 = \dfrac{\exp(k_2 \Gamma_c)}{k_1 k_2} - t_c \qquad (t_c \geq 0)$

Γ_c is the value of Γ at time t_c, the time at which the rate of sorption begins to be described by Eq. (4.31).

Although Eq. (4.32) can be derived as a general rate law expression under the assumption of an exponential decrease in number of available sorption sites with Γ and/or a linear increase in activation energy of sorption with Γ, the Elovich Equation is perhaps best regarded as an empirical one for the characterization of rate data (Sposito, 1984).

As an example, Eq. (4.32) appears to describe phosphate sorption by soils quite well (see Fig. 4.7).

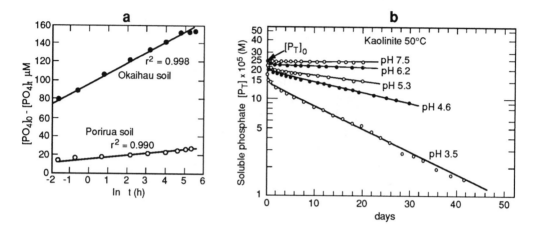

Figure 4.7

Adsorption kinetics of phosphate on soils or soil minerals

a) Plot of Elovich Equation
   (From Chien and Clapton, 1980)
b) Adsorption of phosphate on kaolinite. Semilogarithmic plot (Eq. 4.33) on kinetics of phosphate reaction. In the latter part of the phosphate sorption, a multilayer adsorption mechanism, i.e., the formation of an Al-phosphate coating at the surface of the kaolinite, occurs.

(From Chen, Butler and Stumm, 1973)

Phosphate sorption has also been shown to be described by the rate law

$$-\frac{d[PO_4]_T}{dt} = k[PO_4]_T \tag{4.33}$$

where $[PO_4]_T$ is the molar concentration of phosphate in the aqueous solution phase. Eq. (4.33) has been shown to apply for times of reactions longer than ca. 100 hours and to be associated with the appearance of discrete crystallites of phosphate solids (Chen et al., 1973).

4.5 Fatty Acids and Surfactants

Fig. 4.8 compares data on the adsorption of lauric acid (C_{12}) and caprylic acid (C_8) at a hydrophobic surface (mercury) as a function of the total bulk concentration[1] for different pH-values. As is to be expected the molecular species becomes adsorbed at much lower concentrations than the carboxylate anions. The latter cannot penetrate into the adsorption layer without being accompanied by positively charged counterions (Na^+). As was shown in Fig. 4.4, the adsorption data of pH = 4 can be plotted in the form of a Frumkin (FFG) equation. Fig. 4.9 compares the adsorption of fatty acids on a hydrophobic model surface (Hg) with that of the adsorption on $\gamma\text{-}Al_2O_3$.

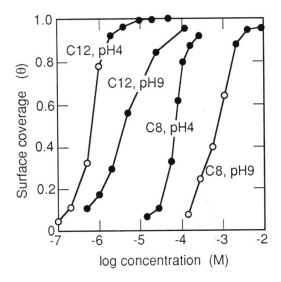

Figure 4.8

Comparison of the adsorption of lauric and caprylic acid as a function of the total bulk concentration. At pH 4, this concentration is made up predominantly of fatty acid molecules and at pH 9 predominantly by their conjugate bases. (●) Points at equilibrium ($d\theta/dt = 0$). (○) Points at which the equilibrium is not yet attained.

(From Ulrich, Ćosović and Stumm, 1987)

---

[1] Using a liquid Hg surface as a model for a hydrophobic solid surface here and in subsequent representations of data may be looked at as being far away from natural systems. On the other hand, the Hg surface represents an excellent tool to measure extent and rate of adsorption. Adsorption of organic substances displaces water from the surface, thus, causing a change in the electric capacity of the souble layer at the Hg-water interface. Using the Hg surface as an electrode (whose surface can readily and conveniently be renewed, e.g., on a dropping Hg-electrode and whose surface potential can be adjusted by a potentiometer) the capacity can be readily measured using an alternating current technique. The measurements reported have been carried out at an incipiently uncharged Hg surface. (Ulrich et al., 1991; Ćosović, 1990)

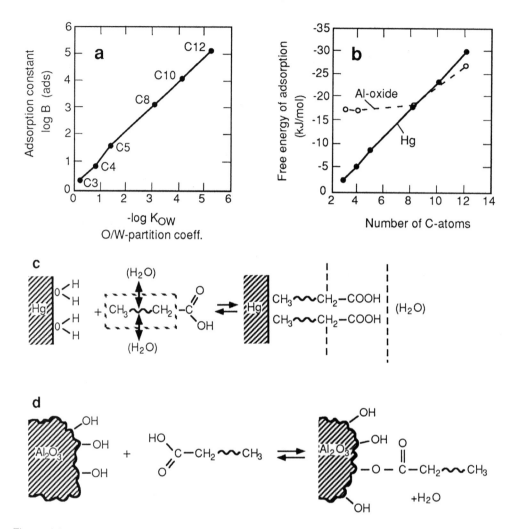

Figure 4.9

a) Adsorption on the mercury electrode vs hydrophobic properties (as measured by the octanol/water partition coefficient) of adsorbing species at pH 4. B_{ads} is the adsorption constant (M^{-1}) (Eq. 4.13).

b) The free energy of adsorption as a function of the number of carbon atoms (n_C) in the fatty acid, on the mercury electrode and on the alumina.

Adsorption of fatty acids at two model surfaces.

c) At the nonpolar Hg electrode, the hydrophobic interaction between the hydrocarbon chain and the water phase leads to a displacement of the first-row H_2O molecules and to adsorption at the Hg surface.

d) At the polar aluminum oxide surface, the fatty acid molecules become adsorbed because the functional carboxylic groups exchange for the surface OH groups (ligand exchange). At sufficiently high chain length, adsorption due to the hydrophobic expulsion outweighs adsorption by coordinative interaction; the orientation of the fatty acid molecules may revert.

(From Ulrich, Ćosović and Stumm, 1988)

The adsorption of fatty acids on the non-polar hydrophobic surface (Hg) is dominated by their hydrophobic properties. The extent of adsorption increases with increasing chain length. The following relationship of the free energy of adsorption, ΔG_{ads}, and the number of C atoms, n_c, of the fatty acids can be established:

$$-RT \ln B = \Delta G_{ads} = 0.7 - 3.1\, n_c \ (\text{kJ mol}^{-1}) \tag{4.34}$$

As seen in Fig. 4.8, the adsorption of lauric acid (C_{12}) is slow because of slow transport (diffusion) at concentrations smaller than 10^{-6} M. In case of Na^+-caprylate (C_8) the attainment of equilibrium is delayed most probably by structural rearrangement at the surface. In case of anions, such association reactions are slower than with free acids.

In the case of γ-Al_2O_3 the adsorption of fatty acids of low molecular weight results, at least in part from coordinative interactions of the carboxylic groups with the Al ions in the surface layer (surface complex formation) (Fig. 4.9c,d). While $-\Delta G_{ads}$ for the hydrophobic surface increases linearly with chain length (Eq. 4.34), this is not the case for $-\Delta G_{ads}$ on the alumina surface. Plausibly, the adsorption energy due to coordinative interaction, among short-chain fatty acids is not dependent on chain length; with molecules such as caprylic acid (C_8) and larger ones a free energy contribution due to the hydrophobic effect of the larger hydrocarbon moiety becomes preponderant. More than one layer may become adsorbed, presumably because of the formation of a hemicelle (two-dimensional micelle) at the interface.[1]

In dilute solutions of surfactants adsorption processes are controlled by transport of the surfactant from the bulk solution towards the surface as a result of the concentration gradient formed in the diffusion layer; the inherent rate of adsorption usually is rapid. For non-equilibrium adsorption the apparent (non-equilibrium) isotherm can be constructed for different time periods that are shifted with respect to the true adsorption isotherm in the direction of higher concentration (Ćosović, 1990) (see Fig. 4.10).

Mixtures of Adsorbates. The adsorption behavior of mixtures of adsorbates is more complicated. In the most simple case, concentrations C_1 and C_2 are very low and surface coverage is $\theta_1 + \theta_2 < 1$. Then , when interaction between the adsorbed species is negligible, the adsorption of both types of adsorbates occurs independently, and we can write Langmuir isotherms

$$\frac{\theta_1}{1 - \theta_1 - \theta_2} = B_1 C_1 \tag{4.35}$$

---

[1] Micelles are organized aggregates of surfactants, often surfactant ions in which the hydrophobic hydrocarbon chains are oriented towards the interior of the micelle leaving the hydrophilic groups in contact with the aqueous medium. The concentration above which micelle formation becomes appreciable is termed the *critical micelle concentration,* (c.m.c).

$$\frac{\theta_2}{1 - \theta_1 - \theta_2} = B_2 C_2 \qquad\qquad (4.36)$$

If this is not the case, the adsorption behavior can be described by competitive adsorption. The extent of adsorption depends not only on the individual adsorption constants but also on the intrinsic kinetics of adsorption of the individual adsorbates and thus, the adsorption layer formed is influenced by the qualitative and quantitative composition of the complex mixture of adsorbable solutes; full equilibrium is often not attained (Ćosović, 1990).

Formation of Hemicelles at the Solid-Water Interface

The evidence accumulated in the literature suggests that the structure of surfactant adsorbed layers is, in some respects, analogous to that of surfactant micelles. Fluorescence probing techniques – e.g., pyrene and dinaphtylpropane (DNP) fluorescence probes are used to investigate the structure of adsorbed layer of a surfactant – give information on the polarity of the microenvironment in the adsorbed

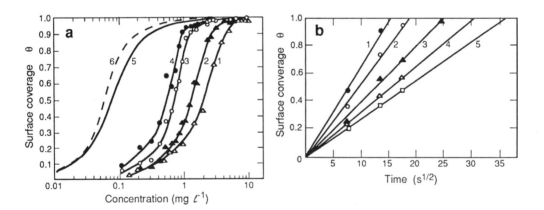

Figure 4.10

Adsorption on Hg surface

a) Apparent adsorption isotherms of Triton-X-100 in 0.55 mol ℓ^{-1} NaCl obtained with various accumulation times: (1) 30, (2) 60, (3) 180, (4) 300 s; and threoretical Frumkin adsorption isotherms of Triton-X-100[1] with interaction factors a = 1.0 (curve 5) and 1.25 (curve 6).
b) Surface coverage vs. square root of accumulation time for adsorption of Triton-X-100 at Hg surface. Concentration of Triton-X-100 in 0.55 mol ℓ^{-1} NaCl: (1) 1.25; (2) 0.94; (3) 0.73; (4) 0.63; (5) 0.52 mg ℓ^{-1}.

(From Batina, Ruzić and Ćosović 1985)

---

[1] tert. Octylphenol ethoxilate with 9 – 10 ethoxy groups

layer. The results of such investigations by Chandar et al. (1987) on the structure of the adsorbed layer of sodium dodecyl sulfate at the alumina-water interface indicate the presence of highly organized surfactant aggregates at the solid liquid interface, formed by the association of hydrocarbon chains.

The adsorption isotherm of sodium dodecyl sulfate (SDS) on alumina at pH = 6.5 in 0.1 M NaCl (Fig. 4.11a) is characteristic of anionic surfactant adsorption onto a positively charged oxide. As shown by Somasundaran and Fuerstenau (1966) and by Chandar et al. (1987), the isotherm can be divided into four regions. These authors give the following explanation for the adsorption mechanism:
1) The slope of unity in region I indicates that the anionic surfactant adsorbs as individual ion through electrostatic interaction with the positively charged surface.
2) The sharp increase in adsorption in region II marks the onset of surfactant association at the surface through lateral interaction of the hydrocarbon chains.
3) The decreasing slope in region III can be attributed to an increasing electrostatic hindrance to the surfactant association process following interfacial charge reversal.
4) Plateau adsorption could correspond to either complete surface coverage or a value limited by constant surfactant monomer activity in solution as a result of bulk micellization.

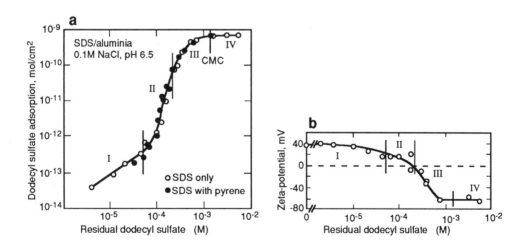

Figure 4.11

a) Adsorption isotherm of sodium dodecyl sulfate (SDS) on alumina at pH 6.5 in 10^{-1} M NaCl.

b) Zeta potential of alumina as a function of equilibrium concentration of SDS (designation of regions based on isotherm shape).

From Chandar et al. (1987) based on the data of Somasundaran and Fuerstenau (1966). Chandar et al. (1987) have shown with the aid of fluorescent probe studies that in region II and above adsorption occurs through the formation of surfactant aggregates of limited size.

The corresponding zeta potential behavior of alumina particles is found to correlate with the amount of adsorbed SDS as shown in Fig. 4.11b. In region I, the potential is relatively constant and equal to that of alumina particles at pH 6.5 in the absence of SDS (+40 mV). A significant decrease in the positive potential is observed beyond the transition from region I to II which correlates with the marked increase in anionic surfactant adsorption. The isoelectric point corresponding to the neutralization of the positively charged surface by the adsorbed anionic surfactant appears to coincide with the transition from region II to III. Region III is therefore characterized by SDS adsorption resulting in a net negative potential at the surface. Constant negative potentials (−65 mV) are measured when the adsorption reaches its plateau level in region IV.

The fluorescence probe studies (Chandar et al., 1987) have shown that in region II and above adsorption occurs through the formation of surfactant aggregates of limited size. As these authors point out, in region II, where the surface is essentially bare and sufficient number of positive sites are available (zeta potential is positive), increase in adsorption is achieved mostly by increasing the number of aggregates; a constant aggregation number implies that the number density of aggregates increases with increase in adsorption. In region II relatively uniform-sized aggregates (aggregation number 120 − 130) are measured on the surface.

The transition from region II to III marks the point at which most of the positive sites are filled since the zeta potential reverses its sign at this point. Adsorption in region III is therefore likely to occur through the growth of existing aggregates rather than through the formation of new ones. The growth of these aggregates can be expected to be electrostatically hindered.

4.6 Humic Acids

Humic and fulvic acids contain various types of phenolic and carboxylic functional (hydrophilic) groups as well as aromatic and aliphatic moieties which import hydrophobic properties to these substances. Fig. 4.12 gives a schematic idea on the composition of these substances. We refer to the book of Thurman (1985) and Aiken et al. (1985) for a description of the various properties of humic and fulvic acids in soils and waters and the book by Buffle (1988) for the coordinating properties of humus and humic acids.

The adsorption of humic and fulvic acids on surfaces can be interpreted along the scheme of Fig. 4.9c,d. Because of hydrophobic interaction humic and fulvic acids tend to accumulate at the solid-water interface. At the same time the adsorption is influenced by coordinative interaction, e.g., schematically,

$$\equiv Al-OH + RC\overset{O}{\underset{OH}{\diagup\diagdown}} \quad\rightleftharpoons\quad \equiv Al\,O-C\overset{O}{\diagup}-R + H_2O$$

This adsorption reaction is reflected in the pH-dependence of the adsorption. The adsorption Γ vs pH curve is very similar to that of an organic acid with a pK value in the range 3 – 5. (Fig. 4.13).

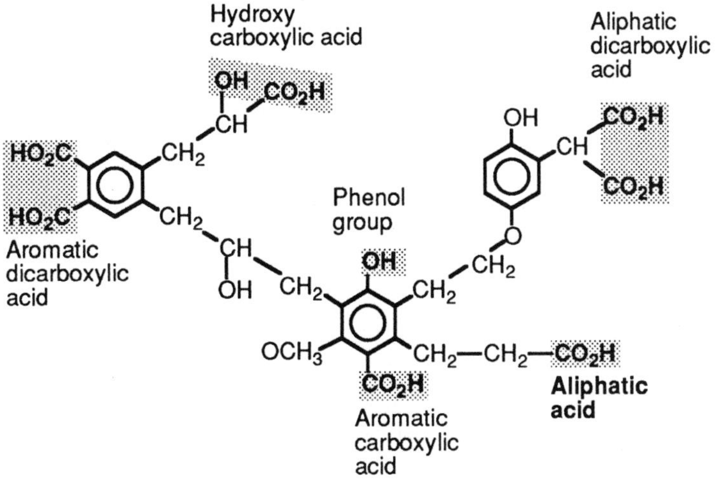

Figure 4.12

Different types of hydroxy and carboxylic groups present in natural organic substances, exemplified in a hypothetical complex polymer from Thurman (1985).

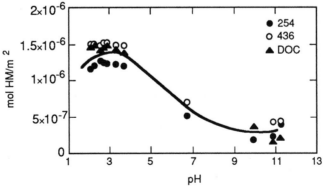

Figure 4.13

Adsorption isotherm of (Aldrich) humic acid (HM) on δ-Al$_2$O$_3$ as a function of pH. Extent of adsorption was determined both by measurements of light absorption at 254 and 436 nm, respectively and by measurements of dissolved organic carbon (DOC) of the residual HM in solution (original concentration = 25 mg per liter).

(From M. Ochs, 1991)

It is necessary to remind ourselves, that the adsorption of humic acids or fulvic acids correspond to the adsorption of a mixture of adsorbates. Adsorption equations derived for the adsorption of a single adsorbate (Langmuir, Frumkin or Gibbs Equation) cannot be used for mechanistic interpretation of the data even if these data can be fitted to such equations (Tomaić and Zutić, 1988).

Fractionation at the Interface. Since large molecules adsorb preferentially over smaller ones, a fractionation occurs at the surface. Kinetically, the small molecular weight components may adsorb first; but then eventually the larger molecules displace the smaller ones. Correspondingly adsorption to solid surfaces shifts the molecular weight distribution of the material in solution towards lower molecular weight. In case of humic substances fractionation of adsorbing humic materials may depend not only on molecular weight but also on different acidity and hydrophobicity of different humic acid fractions. The preferential adsorption of the high-molecular weight fraction onto surfaces may immobilize and increase the residence time of this fraction.

Kinetically, the adsorption of humic acids at a solid-water interface is controlled by convection or diffusion to the surface. Even at concentrations as low as 0.1 mg/ℓ near-adsorption equilibrium is attained within 30 minutes. At high surface densities, a relatively slow rearrangement of the adsorbed molecules may cause a slow attainment of an ultimate equilibrium (Ochs, Ćosović and Stumm, in preparation). The humic acids adsorbed to the particles modify the chemical properties of their surfaces, especially their affinities for metal ions (Grauer, 1989).

4.7 The Hydrophobic Effect

Adsorption of an organic solute molecule on the surface of a solid can involve removing the solute molecule from the solution, removing solvent from the solid surface, and attaching the solute to the surface of the solid. The net energy of interaction of the surface with the adsorbate may result from short-range chemical forces (covalent bonding, hydrophobic bonding, hydrogen bridges, steric or orientation effects) and long-range forces (electrostatic and van der Waals attraction forces). For some solutes, solid affinity for the solute can play a subordinate role in comparison to the affinity of the aqueous solvent.

In a simplified way the adsorption of organic matter, X, at a hydrated surface, S, may be viewed as:

$$S(H_2O)_m + X(H_2O)_n \rightleftharpoons SX(H_2O)_p + (m + n - p)H_2O \; ; \; \Delta G^o_{ads} \qquad (4.37)$$

This equation may be interpreted as the summation of the equations for the "hydration" of X,

$$X + n(H_2O) \rightleftharpoons X(H_2O)_n \qquad\qquad ; \Delta G^o_{solv}$$

and the adsorption of unhydrated X at the hydrated surface

$$S(H_2O)_m + X \rightleftharpoons SX(H_2O)_m \qquad\qquad ; \Delta G^o_1$$

and the "hydration" of the product

$$SX(H_2O)_m + (p-m)H_2O \rightleftharpoons SX(H_2O)_p \quad ; \Delta G^o_2$$

The free energy of adsorption is then given by

$$\Delta G^o_{ads} = \Delta G^o_1 + \Delta G^o_2 - \Delta G^o_{solv} \qquad\qquad\qquad (4.38)$$

where ΔG^o_{solv} is the free energy of solvation. If ΔG^o_2 is assumed to be relatively small,

$$\Delta G^o_{ads} = \Delta G^o_1 \text{ (affinity of surface) } - \Delta G^o_{solv} \text{ (affinity of solvent)} \qquad (4.39)$$

$-\Delta G^o_{solv}$ is equivalent to the free energy of displacement of X from the water. Hydrophobic substances that are sparingly soluble in water tend to be adsorbed at solid surfaces. Organic dipoles and large organic ions are preferentially accumulated at the solid-solution interface, primarily because their hydrocarbon parts have a low affinity for the aqueous phase. Simple inorganic ions (e.g., Na^+, Ca^{2+}, Cl^-), even if they are specifically attracted to the surface of the colloid, may remain in solution because they are readily hydrated. Less hydrated ions (e.g., Cs^+, organic cations, and many anions) seek positions at the interface to a larger extent than easily hydrated ions.

Such considerations are also contained in the qualitative rule that a polar adsorbent adsorbs the more polar component of nonpolar solutions preferentially, whereas a nonpolar surface prefers to adsorb a nonpolar component from a polar solution. In accord with these generalizations and a consequence of the hydrophobic effect is *Traube's rule,* according to which the tendency to adsorb organic substances from aqueous solutions increases systematically with increasing molecular weight for a homologous series of solutes. Thus, the interaction energy due to adsorption increases rather uniformly for each additional CH_2 group. Unless there is a special affinity of the surface for an organic solute (e.g., electrostatic attraction or coordinative interaction), the hydrophobic effect dominates the accumulation of organic solutes at high-energy surfaces.

The lipophilicity of a substance, that is, the tendency of a substance to become dissolved in a lipid, is often measured by the tendency of a substance to become dissolved in a nonpolar solvent, for example, by the n-octanol-water distribution coefficient. The lipophilicity of a substance is inversely proportional to its water solubility.

The Hydrophobic Effect. Hydrophobic substances, for example, hydrocarbons, are readily soluble in many nonpolar solvents but only sparingly soluble in water; thus these substances tend to reduce the contact with water and seek relatively nonpolar environments. Many organic molecules (soaps, detergents, long-chain alcohols) are of a dual nature; they contain a hydrophobic part and a hydrophilic polar or ionic group; they are amphipathic. At an oil-water interface both parts of these molecules can satisfy their compatibility with each medium. Such molecules tend to migrate to the surface or interface of an aqueous solution; they also have a tendency toward self-association *(formation of micelles)*. This is considered to be the result of *hydrophobic bonding*. This term may be misleading because the attraction of nonpolar groups to each other arises not primarily from a particular affinity of these groups for each other, but from the strong attractive forces between H_2O molecules which must be disrupted when any solute is dissolved in water. The hydrophobic effect is perhaps the single most important factor in the organization of the constituent molecules of living matter into complex structural entities such as cell membranes and organelles (Tanford, 1980).

In summary the hydrophobic effect can be viewed as a result of the attraction not of the solute to the surface or the water, but of the solvent (water) for itself, which restricts the entry of the hydrophobic solute into the aqueous phase.

The mechanism of hydrophobic adsorption is based on the fact that most *natural organic sorbents,* regardless of source, have been observed to be more or less comparable in their ability to accept nonpolar compounds. In the sense of Eq. (4.39), given above, the hydrophobic adsorption is driven by the incompatability of the nonpolar compounds with water, not by the attraction of the compounds to organic matter. Thus, as we will show below, it is possible to characterize the adsorption of a wide range of organic compounds on a wide range of organic sorbents based on a single property of the compund, its octanol-water partition coefficient, K_{OW}.

4.8 The "Sorption" of Hydrophobic Substances to Solid Materials that contain Organic Carbon

The sorption of non-polar organic hydrophobic substances to solid material that contain organic carbon must be interpreted as "absorption", i.e., a "dissolution" of the hydrophobic compound into the bulk of the organic material usually present as a component of the solid phase. The sorption of such a hydrophobic substance may be compared with the partitioning of a solute between two solvents – water and the organic phase. The partition coefficient K_P (similar to a distribution coefficient) is defined as

$$K_P = \frac{\text{mol sorbate / mass of solid}}{\text{mol of solute / volume of solution}} \left[\frac{\ell}{kg}\right] \text{ or } \left[\frac{10^{-3} \text{ m}^3}{kg}\right] \qquad (4.40)$$

The absorption increases with the content of the organic material in the solid phase and the hydrophobicity of the solute. Thus, K_P is empirically found to correlate with the organic carbon content of the solid phase, f_{OC} (weight fraction) and the hydrophobicity of the solute, K_{OW}.

The tendency of the compound to be absorbed (to dissolve) in the organic phase is related to the dissolution of the compound in n-octanol, i.e., to the chemical's octanol-water partition coefficient

$$K_{OW} = [A_{oct}] / [A_{aq}] \qquad (4.41)$$

Thus, the partition coefficient for the (ab)sorption of an organic solute onto a solid phase is given by (Karickhoff et al., 1979; Schwarzenbach and Westall, 1980)

$$K_P = bf_{OC} (K_{OW})^a \qquad (4.42)^{[1]}$$

where a and b are constants. a is often around 0.8.

Increasing water solubility corresponds to decreasing partitioning into the solid phase.

Fig. 4.14 gives a plot of data on the partition coefficient of non-polar organic compounds (aromatic chloro hydrocarbons) to various solid phases (having different contents of organic material. K_{OW} is inversely related to water solubility, S_W.

$$\log K_{OW} = a - d \log S_W.$$

The effect that hydrophobic organic substances can become sorbed to organic coatings can be used to make sorbents by coating inorganic colloids such as ferrihydrite with surfactants. Such surfactant-coated surfaces contain films of hemicelles (see 4.5) which are able to remove hydrophobic solutes, such as toluene or chlorohydrocarbons, from solution (Holsen, et al., 1991).

---

[1] This equation is only valid for $f_{OC} \geqq 0.001$.

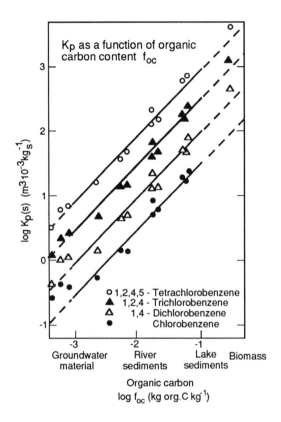

Figure 4.14

The distribution of organic substances between water and representative solid materials of different organic carbon content, f_{OC}.

(Modified from Schwarzenbach and Westall, 1980)

Organic Matter as Adsorbents

The organic material in sediments and in soils is very important in supplying sites for the interaction with solutes. Humus-like material, biogenic surfaces, biological debris and inorganic minerals covered with organic material and humus belong in this category. The molecular structure in soil and sediment organic matter have not been established completely. The functional groups are the same as these in humic acids; especially prominent are the carboxylic, the phenolic OH, alcoholic OH and amino groups. Alkalimetric titration gives some ideas on the pK values of the functional groups (Fig. 4.15). Obviously the buffering of soils by humus materials is very important.

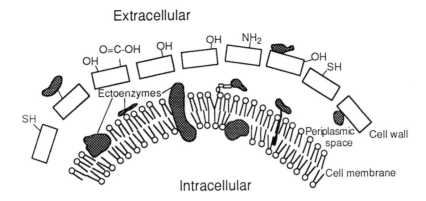

Figure 4.15a

A conceptualized cross section through a portion of the cell wall (rectangles), periplasmic space, and cell membrane (lipid bilayer with polar head groups in contact with cytoplasm and external medium, and hydrophobic hydrocarbon chains) of an aquatic microbe. Reactive functional groups (–SH, –COOH, –OH, –NH₂) present on the wall consitutents and extracellular enzymes (depicted as shaded objects) attached by various means promote and catalyze chemical reactions extracellularly.

(Modified from Price and Morel, 1990)

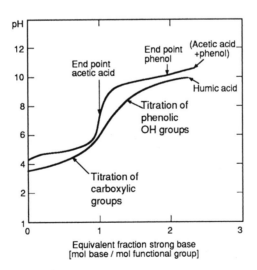

Figure 4.15b

Comparison of an alkalimetric titration curve of an equimolar (10^{-4} M) solution of acetic acid (pK = 4.8) and phenol (pK = 10) with a humic acid that contains 10^{-4} M carboxylic groups.
One may note that the titration of the humic acid – because of its polyfunctionality – is less steep than that of the acetic acid-phenol mixture.

(From Sigg and Stumm, 1991)

Humus in Soils. Humus in solid form (colloids, or coatings on mineral surfaces) can immobilize many pollutants.

1) Non-ionic organic molecules such as organo phosphates, chlorinated hydrocarbons (e.g., polychlorinated biphenyls PCB) are absorbed according to Eq. (4.42).

2) Organic compounds that contain functional groups that become positively charged when in protonated form (aliphatic and aromatic quarternary compounds, R_3N^+, commonly present in pesticide preparations, and compounds containing basic aminoacids such as in s-triazine pesticides) can become coulombically sorbed to the humus material which is negatively charged (even at low pH value). Humus is also able to enhance such attachments by hydrogen bonding.

3) Metal ions can bind to the functional groups of humus by surface complex formation, e.g.,

$$Hu - OH + Cu^{2+} \rightleftharpoons Hu - OCu^+ + H^+$$

On the other hand, one needs to be aware that *soluble* humic and fulvic acids can form *complexes with organic compounds* and render these substances more mobile (cf. Davis, 1984).

Biological Surfaces, Cell Surfaces. In most aquatic microorganisms the cell membrane or plasma membrane is surrounded by a porous cell wall composed of polysaccharides and proteins and in some cases, inorganic matrices (SiO_2 and $CaCO_3$) (Fig. 4.15a). Reactive functional groups in both the cell membrane and cell wall constituents, such as R–OH, R–COOH, RNH_2 and R–SH act as coordinating sites on the cell surface with which dissolved metals may interact. The plasma membrane, a lipid bilayer of hydrophobic interior and hydrophilic exterior with embedded proteins and carbohydrates (Singer and Nicholson, 1972) maintains the integrity of the cell. It also plays a vital role in regulating the flow of materials and energy into and out of the cell.

4.9 The Sorption of Polymers

Polymers may adsorb readily on solid surfaces. Somasundaran et al. (1964) have investigated the effect of chain length on the adsorption of alkylammonium cations and of surfactant anions on oxide surfaces. These authors consider that the adsorption is influenced by the van der Waals energy of interaction or hydrophobic bonding between CH_2 groups of adjacent adsorbed molecules. The van der Waals interactive energy per CH_2 groups is found to be approximately 1 RT (2.5 kJ mol^{-1}) (cf. Eq. 4.34). For a 12-carbon alkylamine the van der Waals cohesive energy is therefore on the order of 30 kJ mol^{-1} and exceeds in this particular case the electrostatic contribution. Each polymer molecule or polmeric ion can have many groups or segments that can potentially be adsorbed; these groups are often relatively free of mutual interaction. Usually the extent of adsorption, but not necessarily its rate, increases with molecular weight and is affected by the number and type of functional groups in the polymer molecule (see Table 4.2). For example, on a negatively charged silica surface polystyrenesulfonate is adsorbed readily while monomeric p-toluene sulfonate ($CH_3-C_6H_4-SO_3^-$) does not adsorb from 10^{-4} M solutions (Overbeek, 1976). Hydroxyl, phosphoryl, and carboxyl groups can be particularly effective in causing adsorption; the significance of such functional groups directs attention to the specificity of the chemical interactions involved in the adsorption process. Typical natural polymers are for example starch, cellulose, tannins, humic acids.

Table 4.2   Types of Polyelectrolytes

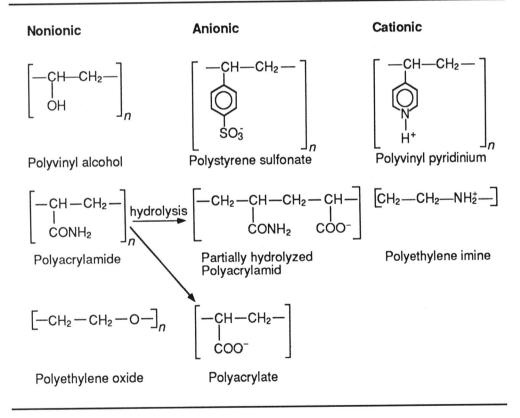

Because many segments of the polymer can be in contact with the surface, a low bonding energy per segment may suffice to render the affinity of several segments together so high that their adsorption is virtually irreversible.

Lyklema (1985) provides the following semiquantitative thermodynamic argument: Suppose that a given polymer molecule loses a conformational entropy[1] of 100 k (where k is Boltzmann's constant) upon sorption in a certain conformation, i.e., $T\Delta_{ads} S = -100$ kT. Such a molecule would not adsorb in that conformation unless the loss is overcompensated by a free energy gain from another source. This source is the interaction free energy due to contacts between segments and the adsorbent. (This interaction is of an energetic nature in the case binding takes place through Van der Waals forces, hydrogen bridges, etc., but it can have entropic terms, e.g., in the case of hydrophobic bonding. Note that these entropic terms are of a nonconfigurational nature: they are proportional to the number of segments adsorbed, independent of the conformation of the polymer as a whole.)

Lyklema (1985) also provides "for the sake of argument" a simplified numerical example: Let us assume that in the adsorbed conformation 100 segments are in contact with the surface. If the adsorption free energy per segment were −0.9 kT the contribution of segment binding to Δ_{ads} F would be −90 kT and no adsorption would ensue, but if it is −1.1 kT per segment the entropy loss would be overcompensated and strong adsorption takes place. What happens in the latter case is that the conformation in the adsorbed state adjusts itself until the overall free energy change Δ_{ads} F = 0, as required for equilibrium. This simple reasoning makes the finding of polymer adsorption theory plausible; slight changes in the adsorption free energy per segment mark the transition between "non-adsorbing" and "strongly adsorbing". That a change of only a few tenths of a kT unit per segment has such dramatic consequences is of course due to the large number of segments involved.

In modern statistical theories (see Scheutjens and Fleer, 1980, and Fleer and Lyklema, 1983) it is found that there does exist a certain critical adsorption free energy, customarily called x_s, above which the molecule adsorbs strongly and below which it does not adsorb. Typical values of x_s, (crit) range from 0.2 − 0.4 kT.

Table 4.3 represents a summary of polymer adsorption features which have been given by Lyklema (1985) to help qualitative understanding.

Polyelectrolytes. The most striking feature of polyelectrolytes is that due to the electrostatic repulsion between the segments, the formation of thick adsorbed layers is prevented. Polyelectrolytes tend to adsorb in rather flat conformations. If adsorbent and polyelectrolyte bear opposite charges, this attraction can be of an electric (coulombic) nature; if the charges have the same sign, adsorption takes place only if the non-electrostatic attraction outweighs the electrostatic repulsion (Lyklema, 1985).

---

[1] Polymer adsorption results in a reduction in the number of conformations that a coil can assume. Thus, for reasons of entropy alone adsorption would be unfavorable.

Table 4.3   Polymer Adsorption Features quoted from Lyklema (1985)

(i)	Polymer adsorption is a rather general phenomenon because only modest segment adsorption free energies are needed, provided the molecule is not too short.
(ii)	Adsorbed macromolecules are difficult to wash away with the solvent because the (free) energy required to desorb a whole molecule is so high. There is no reason to call polymer adsorption "irreversible" in the thermodynamic sense, rather the equilibrium lies extremely far to the adsorbed state.
(iii)	Admixtures containing molecules competing with the segments for sites on the surface virtually reduce x_s and hence can induce desorption. Likewise, exchange of polymers by others can be done without losing the free energy of the bound segments, hence this process is feasible. In fact, the very occurrence of such exchanges is an argument in favor of reversibility.
(iv)	An important variable is the molecular weight, M. Theoretical arguments and experimental evidence have shown that the amount of polymer adsorbed increases with M, especially if the solvent is poor, but more importantly, that the initial part of the adsorption isotherm becomes progressively steeper. In other words, with higher M the isotherm is more of the high-affinity type.
(v)	A consequence of (iv) is that from heterodisperse polymer solutions, i.e., from the great majority, the fractions with high M adsorb preferentially: the composition of the adsorbed polymer differs from that in the equilibrium solution. Polymer adsorption therefore leads to fractionation.

Electric interactions are screened by electrolytes. Hence, by adding electrolytes, the adsorption behavior is made to resemble that of uncharged macromolecules (see Fig. 4.17).

For a quantitative theory see Schee and Lyklema (1984).

Fig. 4.16 provides an illustration of the adsorption of a neutral polymer, polyvinyl alcohol, on a polar surface, and the resulting effects on the double layer properties. Adsorption of anionic polymers on negative surfaces – especially in the presence of Ca^{2+} or Mg^{2+} which may act as coordinating links between the surface and functional groups of the polymer – is not uncommon (Tipping and Cooke, 1982).

As will be discussed in Chapter 7, the adsorption of polymers may either increase or decrease colloid stability.

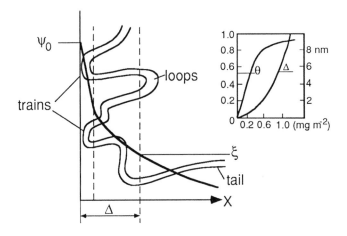

Figure 4.16

Effect of adsorbed polymer on the double-layer. Because of the presence of adsorbed train segments, the double layer is modified. The zeta-potential, ζ, is displaced because the adsorbed polymer displaces the plane of shear. The parameters for describing adsorbed polymers are the fraction of the first layer covered by segments, θ, and the effective thickness, Δ, of the polymer layer, The insert gives the distribution of segments over trains and loops for polyvinyl alcohol adsorbed on silver iodide. Results obtained from double layer and electrophoresis measurements.

(Modified from Lyklema, 1978)

Humic substances are anionic polyelectrolytes of low to moderate molecular weight (500 – 20'000); their charge is due primarily to partially deprotonated carboxylic and phenolic groups (see Fig. 4.15).

A schematic representation of the effects of pH and ionic strength on the configuration of anionic polyelectrolytes, such as humic substances, is presented in Fig. 4.17. In fresh waters at neutral and alkaline pHs, charged macromolecules assume extended shapes (large hydrodynamic radius, R_h) as a result of intramolecular electrostatic repulsive interactions. When adsorbed at interfaces under these conditions, they assume flat configurations (small hydrodynamic thickness, δ_h). At high ionic strength or at low pH, the polyelectrolytes have a coiled configuration in solution (small R_h) and extend further from the solid surface when adsorbed (large δ_h).

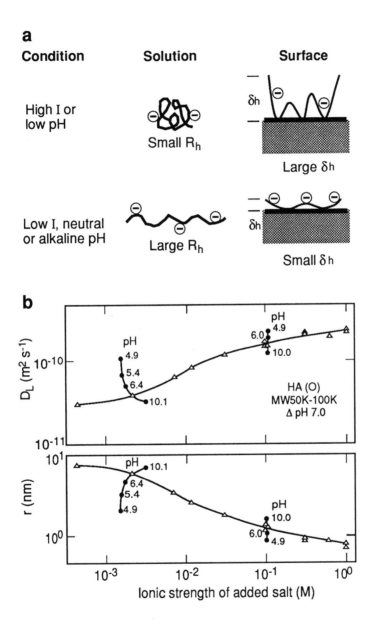

Figure 4.17

a) Schematic description of the effects of ionic strength (I) and pH on the conformations of a humic molecule in solution and at a surface. R_h denotes the hydrodynamic radius of the molecule in solution and δ_h denotes the hydrodynamic thickness of the adsorbed anionic polyelectrolyte. (Adapted from Yokoyama et al., 1989 and O'Melia (personal communication), 1991)

b) The influence of ionic strength and pH on diffusion coefficient, D_L, and on Stokes-Einstein radius of a humic acid fraction of 50'000 – 100'000 Dalton. (From Cornel, Summers and Roberts,1986)

4.10 Reversibility of Adsorption

Although adsorption and desorption together establish an adsorption equilibrium, desorption has received relatively little experimental attention.

If adsorption occurs as a bimolecular reaction, as suggested by the Langmuir equation

$$S + A \underset{k_b}{\overset{k_f}{\rightleftharpoons}} SA \tag{4.43}$$

one can postulate

$$\frac{d\,[SA]}{dt} = k_f\,[S]\,[A] - k_b\,[SA] \tag{4.44}$$

and at equilibrium, $d\,[SA]/dt = 0$,

$$\frac{[SA]}{[S]\,[A]} = \frac{k_f}{k_b} = K = K_{ads} \tag{4.45}$$

If one of the kinetic constants has been determined and K_{ads} is known, the other rate constant can be calculated.

Microscopic Reversibility

The principle we have applied here is called microscopic reversibility or principle of detailed balancing. It shows that there is a link between kinetic rate constants and thermodynamic equilibrium constants. Obviously, equilibrium is not characterized by the cessation of processes; at equilibrium the rates of forward and reverse *microscopic* processes are equal for every *elementary* reaction step. The microscopic reversibility (which is routinely used in homogeneous solution kinetics) applies also to heterogeneous reactions (adsorption, desorption; dissolution, precipitation).

In the application of the principle of microscopic reversibility we have to be careful. *We cannot apply this concept to overall reactions.* Even Eqs. (4.43) – (4.45) cannot be applied unless we know that other reaction steps (e.g., diffusional transport) are not rate controlling. In a given chemical system there are many elementary reactions going on simultaneously. Rate constants are path-dependent (which is not the case for equilibrium constants)and may be changed by catalysts. For equilibrium to be reached, *all elementary* processes must have equal forward and reverse rates and *all* species, not just reactive intermediates, must be at steady state (Lasaga, 1981). The kinetic behavior of systems near equilibrium has been shown to be linear.

Exemplification: Surface Complex Formation

Hayes and Leckie (1986) postulate on the basis of their pressure jump relaxation experiments on the adsorption-desorption of Pb^{2+} at the goethite-water interface the following mechanism:

$$SOH + Pb^{2+} \underset{k_{-1}}{\overset{k_1}{\rightleftharpoons}} SOPb^+ + H^+ \tag{4.46}$$

where k_1 and k_{-1} represent the forward and reverse rate constants. Defining intrinsic rate constants as rate constants that would be observed in the absence of an electric field, the equilibrium is represented by

$$K_1(int) = \frac{k_1(int)}{k_{-1}(int)} = \frac{k_1}{k_{-1}} \exp\left(\frac{F\psi_0}{RT}\right) \tag{4.47}$$

The principle of the pressure-jump method is based on the pressure dependence of the equilibrium constant, i.e.,

$$\left(\frac{\delta \ln K}{\delta P}\right)_T = -\frac{\Delta V}{RT} \tag{4.48}$$

where ΔV is the standard molar volume change of the reaction, and P the pressure.

A pressure perturbation results in the shifting of the equilibrium; the return of the system to the original equilibrium state (i.e., the relaxation) is related to the rates of all elementary reaction steps. The relaxation time constant associated with the relaxation can be used to evaluate the mechanism of the reaction. During the shift in equilibrium (due to pressure-jump and relaxation) the composition of the solution changes and this change can be monitored, for example by conductivity. A description of the pressure-jump apparatus with conductivity detection and the method of data evaluation is given by Hayes and Leckie (1986).

A representative result of their data on Pb adsorption on goethite is given in Table 4.4.

Table 4.4   Intrinsic Rate Constants for the Reaction

$$=FeOH + Pb^{2+} \overset{k_1}{\underset{k_{-1}}{\rightleftharpoons}} =FeOPb^+ + H^+$$

(From Hayes and Leckie, 1986)

ΔP	$k_1(int)$	$k_{-1}(int)$	$K_1(int)$
(atm)	$M^{-1}s^{-1}$	$M^{-1}s^{-1}$	
140	1.7×10^5	4.2×10^2	4.0×10^2
100	2.4×10^5	6.1×10^2	4.0×10^2
70	3.6×10^5	8.9×10^2	4.0×10^2

Evidence for Reversibility for Sorption on Illite

Fig. 4.18 shows the result of Cd^{2+} adsorption on illite in presence of Ca^{2+} (Comans, 1987). The data are fitted by Freundlich isotherms after an equilibration time of 54 days. It was shown in the experiments leading to these isotherms that adsorption approaches equilibrium faster than desorption. Comans has also used ^{109}Cd to assess the isotope exchange; he showed that at equilibrium (7 – 8 weeks equilibration time) the isotopic exchangeabilities are approximately 100 %; i.e., all adsorbed Cd^{2+} is apparently in kinetic equilibrium with the solution. The available data do not allow a definite conclusion on the specific sorption mechanism.

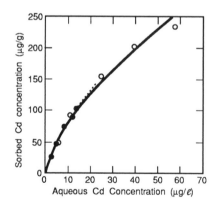

Figure 4.18

Adsorption-desorption equilibrium for Cd(II) on illite after 54 days of equilibration. The solution contains HCO_3^-, 2×10^{-3} M Ca^{2+} and has a pH = 7.8. Freundlich isotherms based on separate adsorption (●) and desorption (○) data are given from Comans (1987).

In a similar study (Comans et al., 1990), the reversibility of Cs^+ sorption on illite was studied by examining the hysteresis between adsorption and desorption isotherms and the isotopic exchangeability of sorbed Cs^+. Apparent reversibility was found to be influenced by slow sorption kinetics and by the nature of the competing cation. Cs^+ migrates slowly to energetically favorable interlayer sites from which it is not easily released.

4.11 Ion Exchange

Ion Exchange Capacity. There are different meanings of "ion exchange". In a *most general sense*, any replacement of an ion in a solid phase in contact with a solution by another ion can be called ion exchange. This includes reactions such as

$$CaCO_3(s) + Sr^{2+} \rightleftharpoons SrCO_3(s) + Ca^{2+}$$
$$NaAlSi_3O_8(s) + K^+(aq) \rightleftharpoons KAlSi_3O_8(s) + Na^+$$
$$Fe(OH)_3(s) + HPO_4^{2-} + 2\,H^+ \rightleftharpoons FePO_4(s) + 3\,H_2O$$

In a *more restrictive sense,* the term "ion exchange" is used to characterize the replacement of one *adsorbed, readily exchangeable* ion by another. This circumscription, used in soil science (Sposito, 1989), implies a surface phenomenon involving charged species in *outer-sphere complexes* or in the *diffuse ion swarm*. It is not possible to adhere rigorously to this conceptualization because the distinction between inner-sphere and outer-sphere complexation is characterized by a continuous transition, (e.g., H^+ binding to humus).

A further difficulty is the distinction between a concept and an operation, for example in the definition of ion exchange capacity. Operationally, "the ion exchange capacity of a soil (or of soil-minerals in waters or sediments) is the number of moles of adsorbed ion charge that can be desorbed from unit mass of soil, under given conditions of temperature, pressure, soil solution composition, and soil-solution mass ratio" (Sposito, 1989). The measurement of an ion exchange capacity usually involves the replacement of (native) readily exchangeable ions by a "standard" cation or anion.

The cation exchange capacity of a negative double layer may be defined as the excess of counter ions that can be exchanged for other cations. This ion exchange capacity corresponds to the area marked σ_+ in Fig. 3.2d. It can be shown that σ_+/σ_- (where σ_- is the charge due to the deficiency of anions) remains independent of electrolyte concentration only for constant potential surfaces. However, for constant charge surfaces, such as double layers on clays, σ_-/σ_+ increases with increasing electrolyte concentration, Hence, the cation exchange capacity (as defined by σ_+) increases with dilution and becomes equal to the total surface charge at great dilutions.

Since cations are adsorbed electrostatically not only due to the permanent structural charge, σ_o, (caused by isomorphic substitution) but also due to the proton charge, σ_H, (the charge established because of binding or dissociating protons – see Chapter 3.2) the ion exchange capacity is pH-dependent (it increases with pH). Furthermore, the experimentally determined capacity may include inner-spherically bound cations.

Standard cations used for measuring cation exchange capacity are Na^+, NH_4^+, and Ba^{2+}. NH_4^+ is often used but it may form inner-sphere complexes with 2:1 layer clays and may substitute for cations in easily weathered primary soil minerals. In other words, one has to adhere to detailed operational laboratory procedures; these need to be known to interpret the data; and it is difficult to come up with an operationally determined "ion exchange capacity" that can readily be conceptualized unequivocally.

Simple Models. The surface chemical properties of clay minerals may often be interpreted in terms of the surface chemistry of the structural components, that is, sheets of tetrahedral silica, octahedral aluminum oxide (gibbsite) or magnesium hydroxide (brucite). In the discrete site model, the cation exchange framework, held together by lattice or interlayer attraction forces, exposes fixed charges as anionic sites.

In clays such as montmorillonites, because of the difference in osmotic pressure between solution and interlayer space, water penetrates into the interlayer space. Depending upon the hydration tendency of the counterions and the interlayer forces, different interlayer spacings may be observed. A composite balance among electrostatic, covalent, van der Waals, and osmotic forces influences the swelling pressure in the ion-exchange phase and in turn also the equilibrium position of the ion-exchange equilibrium. One often distinguishes between inner-crystalline swelling (which is nearly independent of ionic strength) and osmotic swelling (which is strongly dependent on ionic strength (Weiss, 1954). The extent of swelling increases with the extent of hydration of the counterion. Furthermore, the swelling is much less for a bivalent cation than for a monovalent counterion. Because only half as many Me^{2+} are needed as Me^+ to neutralize the charge of the ion exchanger, the osmotic pressure difference between the solution and the ion-exchange framework becomes smaller. Thus, if strongly hydrated Na^+ replaces less hydrated Ca^{2+} and Mg^{2+} in soils, the resulting swelling adversely affects the permeability of soils. Similarly, in waters of high relative $[Na^+]$, the bottom sediments are less permeable to water. Surface complex formation between the counterions and the anionic sites (oxo groups) also reduces the swelling pressure.

From a geometric point of view, clays can be packed rather closely. Muds containing clays, however, have a higher porosity than sand. The higher porosity of the clays is caused in part by the high water content (swelling), which in turn is related to the ion-exchange properties.

Ion-Exchange Equilibria. The double-layer theory predicts qualitatively correctly that the affinity of the exchanger (the ion exchanging solid – clay, humus, ion exchange resin, zeolite) for bivalent ions is larger than that for monovalent ions and that this selectivity for ions of higher valency decreases with increasing ionic strength of the solution. However, according to the Gouy theory, there should be no ionic selectivity of the exchanger between different equally charged ions.

Relative affinity may be defined quantitatively by *formally* applying the mass law to exchange reactions:

$$\{Na^+ R^-\} + K^+ \quad = \{K^+ R^-\} + Na^+ \tag{4.49}$$

$$2\{Na^+ R^-\} + Ca^{2+} = \{Ca^{2+} R_2^-\} + 2 Na^+ \tag{4.50}^{[1]}$$

where R^- symbolizes the negatively charged network of the cation exchanger. A selectivity coefficient, Q, can then be defined by

$$Q_{(NaR \to KR)} \quad = \frac{X_{KR} [Na^+]}{X_{NaR} [K^+]} \tag{4.51}$$

$$Q_{(NaR \to CaR)} \quad = \frac{X_{CaR} [Na^+]^2}{X_{NaR}^2 [Ca^{2+}]} \tag{4.52}$$

where X represents the equivalent fraction of the counterion on the exchanger (e.g., $X_{CaR} = 2 [CaR_2] / (2 [CaR_2] + [NaR])$). The selectivity coefficients may be treated as mass-law constants for describing, at least in a semiquantitative way, equilibria for the interchange of ions, but these coefficients are neither constants nor are they thermodynamically well defined. Because the activities of the ions within the lattice structure are not known and vary depending on the composition of the ion-exchanger phase, the coefficients tend to deviate from constancy. Nevertheless, it is expedient to use Eq. (4.52) for illustrating the concentration dependence of the selectivity for more highly charged ions. A theory on the termodynamics of ion exchange was given by Sposito (1984). The calculations for the distribution of ions between the ion-exchanging material and the solution involves the same procedures as in other calculations or equilibrium speciation. In addition to the mass law expression one needs the corresponding mass balance equations for [TOT–R], [TOT–Na] and [TOT–Ca].

In Fig. 4.19 the mol fraction of Ca^{2+} on the exchanger is plotted as a function of the mol fraction of Ca^{2+} in the solution. For a hypothetical exchange with no selectivity, the exchange isotherm is represented by the dashed line. In such a case the ratio

---

[1] This reaction could also be formulated as $Na_2R + Ca^{2+} = CaR + 2 Na^+$.

of the counterions is the same for the exchanger phase as in the solution. The selectivity of the exchanger for Ca^{2+} increases markedly with increased dilution of the solution; in solutions of high concentration the exchanger loses its selectivity. The representation in this figure makes it understandable why a given exchanger may contain predominantly Ca^{2+} in equilibrium with a fresh water; in seawater, however, the counterions on the exchanger are predominantly Na^+. Fig. 4.19 illus-

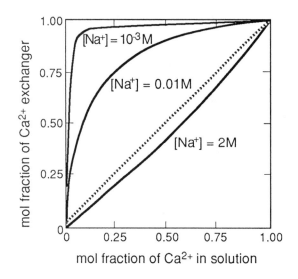

Figure 4.19

Typical exchange isotherms for the reactions $Ca^{2+} + 2\,\{Na^+\,R^-\} \rightleftharpoons \{Ca^{2+}\,R_2^{2-}\} + 2\,Na^+$. In dilute solutions the exchanger shows a strong preference for Ca^{2+} over Na^+. This selectivity decreases with increasing ion concentration. The 45° line represents the isotherm with no selectivity.

Table 4.5   Ion Exchange of Clays with Solutions of $CaCl_2$ and KCl of Equal Equivalent Concentration[a]

		Ca^{2+} / K^+ Ratios on clay			
	Exchange capacity	Concentration of solution $2\,[Ca^{2+}] + [K^+]$ (meq liter^{-1})			
Clay	(meq g^{-1})	100	10	1	0.1
Kaolinite	0.023	–	1.8	5.0	11.1
Illite	0.162	1.1	3.4	8.1	12.3
Montmorillonite	0.810	1.5	–	22.1	38.8

[a] From L. Wiklander, *Chemistry of the Soil*, F.E. Bear, Ed., 2nd ed., Van Nostrand Reinhold, New York, 1964.

trates why, in the technological application of synthetic ion exchangers such as water softeners, Ca^{2+} can be selectively removed from dilute water solutions, whereas an exhausted exchanger in the Ca^{2+} form can be reconverted into a {NaR} exchanger with a concentrated brine solution or with undiluted seawater.

Table 4.5 gives the experimentally determined distribution of Ca^{2+} and K^+ for three clay minerals at various equivalent concentrations of Ca^{2+} and K^+. The results demonstrate the concentration dependence of the selectivity; they also show that marked differences exist among various clays.

With the help of selectivity coefficients, such as in Eqs. (4.51) and (4.52) a general order of affinity can be given. For most clays the Hofmeister series (the same series as was given in Eq. (2.17) for the affinity of ions to oxide surfaces)

$$Cs^+ > K^+ > Na^+ > Li^+$$
$$Ba^{2+} > Sr^{2+} > Ca^{2+} > Mg^{2+}$$

(4.53)

is observed; i.e., the affinity increases with the (non-hydrated) radius of the ions. In other words, the ion with the larger hydrated radius tends to be displaced by the ion of smaller hydrated radius. This is an indication that inner-sphere interaction occurs in addition to Coulombic interaction (Coulombic interaction alone would give preference to the smaller unhydrated ion). For some zeolites and some glasses, a reversed selectivity (rather than that of 4.53) may be observed.

The dependence of selectivity upon electrolyte concentration has important implications for analytical procedures for determining ion-exchange capacity and the composition of interstitial waters. The rinsing of marine samples with distilled water shifts the exchange equilibria away from true seawater conditions; the influence of rinsing increases the bivalent/monovalent ratio, especially the Mg^{2+}/Na^+ ratio. Sayles and Mangelsdorf (1979) have shown that the net reaction of fluvial clays and seawater is primarily an exchange of seawater Na^+ for bound Ca^{2+}. This process is of importance in the geochemical budget of Na^+.

Sodium Adsorption Ratio. Because of the swelling effects of Na^+, the relative amount of sodium (sodicity) in the water quality especially in the irrigation water quality is an important measurement in soil science. Decreased permeability can interfer with the drainage required for normal salinity control and with the normal water supply and aeration required for plant growth. The relative sodium status of irrigation waters and soil solutions is often expressed by the *sodium adsorption ratio (SAR)*

$$SAR = \frac{[Na^+]}{([Ca^{2+}]/2 + [Mg^{2+}]/2)^{1/2}}$$

Combining [Ca^{2+}] and [Mg^{2+}] in this expression, derived from ion exchange equilibrium considerations, is not strictly valid but cause litte deviation from more exact formulations and is justified because these two bivalent cations behave similarily during cation exchange.

4.12 Transport of Adsorbable Constituents in Ground Water and Soil Systems

Soil and groundwater carriers can be looked at as giant chromatography columns. The concepts on transport of chemicals accompanied by concommittant adsorption and desorption can be borrowed from chromatography theory. Nevertheless, there are a few differences of importance:

1) Particles of the soil matrix are usually very polydisperse, and heterogeneously packed;
2) one has to distinguish between transport in groundwater aquifers which are saturated with water and in the unsaturated subsurface zone where pores contain water and air (this might lead to very heterogeneous flow structure);
3) cracks and root zones may lead to preferential flow paths resulting in channeling (Borkovec et al., 1991).

Changes in concentration can occur because of
- chemical reactions within the aqueous phase [1]
- transfer of solute to (or from) solid surfaces in the porous medium or to or from the gas phase in the unsaturated zone.

We will concentrate on adsorption-desorption. The one-dimensional form for homogeneous saturated media of the advection-dispersion equation can then be written as

$$-u\frac{\delta c_i}{\delta x} \quad + \quad D\frac{\delta^2 c_i}{\delta x^2} \quad - \quad \frac{\rho}{\theta}\frac{\delta S_i}{\delta t} \quad = \quad \frac{\delta c_i}{\delta t} \qquad\qquad (4.54)$$

   Advection   Dispersion   Sorption

where   u   = linear velocity   [cm s^{-1}]
        D   = dispersion coefficient   [cm^2 s^{-1}]
        c_i = concentration of species i   [mol ℓ^{-1}]
        x   = distance (in direction of flow)   [cm]
        S_i = species i sorbed   [mol kg^{-1}]
        ρ = bulk density   [kg ℓ^{-1}]   [2]
        θ = porosity   [–] (volume of voids/volume total)

---

[1] Acid-base, complexation, solution precipitation, redox reactions, biodegradation, radioactive decay, hydrolysis.

The third term on the left hand side of Eq. (4.54) represents the change in concentration in the water caused by adsorption or desorption; this term can be interpreted in terms of

$$\frac{\rho}{\theta}\frac{\delta S_i}{\delta t} = \frac{\rho}{\theta}\frac{\delta S_i}{\delta c_i}\frac{\delta c_i}{\delta t} \tag{4.55}$$

where $(\delta S_i/\delta c_i)$ can be interpreted in terms of a linear sorption constant

$$\frac{\delta S_i}{\delta c_i} = K_p \; [\ell\,\text{kg}^{-1}] \tag{4.56[3]}$$

If we set the dispersion (second term in Eq. 4.54) equal to zero

$$D\frac{\delta^2 c_i}{\delta x^2} = 0 \tag{4.57}$$

we can rewrite the transport-sorption Eq. (4.54) [using (4.55) and (4.56)] as the *"retardation equation"*.

$$-u\frac{\delta c_i}{\delta x} = \frac{\delta c_i}{\delta t}\left(1 + \frac{\rho}{\theta}K_p\right) \tag{4.58}$$

where the expression in paranthesis is the retardation factor, t_R

$$t_R = 1 + \frac{\rho}{\theta}K_p = u_R^{-1} = \frac{\bar{u}}{\bar{u}_i} \tag{4.59}$$

where $\bar{u}$ and $\bar{u}_i$ are, respectively, the average linear velocity of the ground water and of the retarded constituent. Both velocities are measured where $c/c_0 = 0.5$ in the concentration profile (see Fig. 4.20a).

Typical values of ρ are $1.6 - 2.1$, kg ℓ^{-1} (corresponding to a density of the solid ma-

---

[2] The bulk density, ρ, refers to the density of the entire assemblage of solids plus voids, i.e., mass of rock per volume of rocks plus voids. (If we refer to the density of the solid material itself, i.e., of the "rock": $\rho' = \frac{\text{mass of rock}}{\text{volume of rock}}$, we can replace $\frac{\rho}{\theta}$ in Eq. (4.54) by $\rho'\left(\frac{1-\theta}{\theta}\right)$.)

[3] Eq. (4.56) is typically observed for absorption reactions, for example for the (ab)sorption of non-polar organic substances in a solid matrix containing organic (humus-like) material. Eq. (4.56) corresponds also to the initial linear portion of a Langmuir isotherm or to a Freundlich equation $S = K_p\,c^b$ where $b \cong 1$.

terial itself of 2.65) and thus $\rho/\theta \approx 4 - 10$ kg ℓ^{-1}. Thus, representative values of $\theta = 0.2 - 0.4$ we can rewrite the retardation factor (Eq. 4.59)

$$\frac{\bar{u}}{\bar{u}_i} = (1 + 4\,K_p) \text{ to } (1 + 10\,K_p)$$ (4.60)

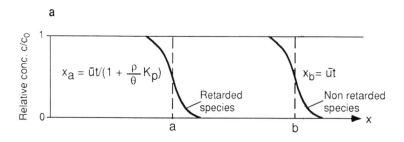

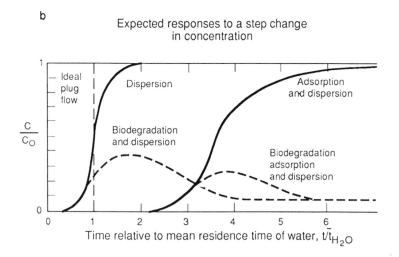

Figure 4.20

a) Advances of adsorbed and non-adsorbed solutes through a column of porous materials. Partitioning of species between solid material and water is described by K_p (Eq. 4.56). The relative velocity is given by

$$\frac{\bar{u}_i}{\bar{u}} = 1/[1 + (\rho/\theta)K_p].$$

Solute inputs are at concentration c_0 at time $t > 0$ (modified from Freeze and Cherry, 1979).

b) Illustration of the effects of dispersion, sorption, and biodecomposition on the time change in concentration of an organic compound at an aquifer observation well following the initiation of water injection into the aquifer at some distance away from the observation well. c represents the observed concentration and c_0 the concentrations in the injection water (from McCarty and Rittmann, 1980).

Fig. 4.20b shows how, in addition to adsorption, biodegradation can affect the movement of organic compounds (note that the abscissa in Fig. 20b is different from that in Fig. 20a). While a conservative contaminant would behave just as the water in which it is contained, an adsorbable solute (as in Fig. 20a) is retarded and arrives at the observation well later than the conservative contaminant. Biodegradation would act to destroy the organic contaminant, so it's concentration at the observation well would never reach C_0; its concentration would initially be governed by dispersion alone, until sufficient time had passed to establish an acclimated bacterial population with sufficient mass to change the organic concentration (McCarty et al., 1981).

K_p Values for Non-polar Organic Substances

As we have seen in (Eq. 4.40) and in Fig. 4.14, the distribution (or partition) coefficient for the (ab)sorption of a non-polar organic solute onto a solid phase can be estimated from the octanol-water coefficient K_{OW} (see Eq. 4.42)

$$K_p = b \, f_{OC} \, (K_{OW})^a \tag{4.61}$$

where f_{OC} is organic carbon (weight fraction) constant of the solid material and a and b are constants.

Table 4.6 gives a few representative values for K_{OW} and K_p for non-polar organic substances on typical soil material and Table 4.7 gives estimates on typical retardation factors estimated for an aquifer. The data show that many non-polar organic substances, with the possible exception of very lipophilic substances such as hexachlorobenzene, are not markedly retarded in aquifers that contain little organic material ($f_{OC} = 0.001 - 0.005$). On the other hand, such substances are effectively retained in soils rich in organic carbon.

Fig. 4.21 gives mean values for some organic substances in the Glatt infiltration system. The figure illustrates that tetrachlorethylene is not eliminated during the infiltration process. In the case of 1,4-dichlorbenzene and 1,3-dimethylbenzene (m-xylene) elimination occurs but this elimination is caused by biodegradation.

The conclusions of retardation are applicable to a wide range of sorption and transport problems including artificial groundwater recharge, and leaching of pollutants from landfills.

Surface Complex Formation, Ion Exchange and Transport in Ground-Water and Soil Systems

The retardation equation can also be applied to inorganic soluble substances (ions, radionuclides, metals). But here we have to consider, in addition to the sorption or ion exchange process, that the speciation of metal ions or ligands in a multi-

Table 4.6   Octanol-water Coefficients and Partition Coefficients of Organic
Substances on Natural Soil Materials
(Modified from Schwarzenbach et al., 1983)

Substance	log K_{OW}	$K_p \pm s$ [a]
Toluene	2.69	0.37 ± 0.12
1,4-Dimethylbenzene	3.15	0.50 ± 0.10
1,3,5-Trimethylbenzene	3.60	1.00 ± 0.16
1,2,3-Trimethylbenzene	3.60	0.95 ± 0.11
1,2,4,5-Tetramethylbenzene	4.05	1.96 ± 0.45
n-Butylbenzene	4.13	3.69 ± 0.98
Tetrachlorethylene	2.60	0.56 ± 0.09
Chlorbenzene	2.71	0.39 ± 0.12
1,4-Dichlorbenzene	3.38	1.10 ± 0.16
1,2,4-Trichlorbenzene	4.05	3.52 ± 0.39
1,2,3-Trichlorbenzene	4.05	3.97 ± 0.64
1,2,4,5-Tetrachlorbenzene	4.72	12.74 ± 2.52
1,2,3,4-Tetrachlorbenzene	4.72	10.48 ± 1.66

[a] Units 10^{-3} m^3 kg^{-1}; values experimentally determined in groundwater material (sieve fraction 63 – 123 μm) from the Glatt aquifer.

Table 4.7   Estimated mean Retardation Factors for Organic Substances in the
River-Groundwater Infiltration System of the Glatt River
(Modified from Schwarzenbach et al., 1983)

	Calculated retardation factor t_R		
	Riverbed	Aquifer	
		close to river	some distance from river
Substance	(~ 0.1 m)[a]	(< 5 m)[a]	(> 5 m)[a]
Chloroform	3	2	1.2
1,1,1-Trichlorethane	6	2	1.2
Trichlorethylene	7	3	1.3
Tetrachlorethylene	16	5	1.8
1,4-Dichlorbenzene	36	10	2.7
Hexachlorbenzene	2950	740	150
Methylbenzene	12	4	1.6
1,3-Dimethylbenzene	25	7	2.2
Naphtalene	31	9	2.5
2-Chlorphenol	2	1	1.1
2,3-Dichlorphenol	10	3	1.4
2,4,6-Trichlorphenol	5	2	1.2

[a] Distance from river

Information used for calculation: ρ' $= 2.5$ g cm^{-3},
θ $= 0.2$,
f_{OC} $= 0.001 - 0.005$,
pH $= 7.5$.

Only 40 % of the material had $\varnothing < 125$ μm and was assumed to cause adsorption.

(From Schwarzenbach et al., 1983)

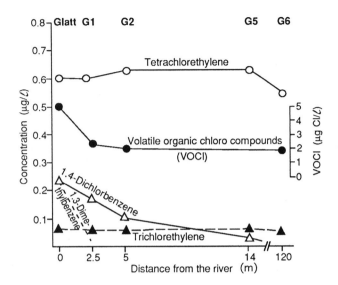

Figure 4.21

Mean values for concentrations of volatile organic substances in the Glatt infiltration system. (From Schwarzenbach et al., 1983)

component system influences the specific sorption process and varies during the pollutant transport in the groundwater; chemistry then becomes an important part of the transport.

How does the breakthrough of a solute which interacts by surface complex formation (or an equivalent Langmuir isotherm) with the surfaces look like? The response of the column is no longer linear and will depend on the concentration of the solute. In line with the surface complex formation equilibrium (Γ_M vs [M] shows saturation), the distribution coefficient decreases with increasing residual concentration and thus leading to a decreasing retardation factor. This leads to a self-sharpening front, a travelling non-linear wave); i.e., in a c/c_o vs distance plot (cf. Fig. 4.20a) a sharp abrupt change of c/c_o occurs (instead of the smooth symmetrical curve for linear adsorption) is observed. Furthermore, the higher the concentration, the earlier the breakthrough (Borkovec et al., 1991).

Mass transport models for multicomponent systems have been developed where the equilibrium interaction chemistry is solved independently of the mass transport equations which leads to a set of algebraic equations for the chemistry coupled to a set of differential equations for the mass transport. (Cederberg et al., 1985).

Radionuclides. To mitigate radioactive contamination, it is important to understand the processes and mechanisms of interactions between radionuclides and the solid materials of aquifers. Cationic species of radionuclides may be sorbed on the aquifer material by processes such as ion exchange, or surface complex formation, thus, retarding their transport by groundwater.

Sorption depends on Sorption Sites. The sorption of alkaline and earth-alkaline cations on expandable three layer clays – smectites (montmorillonites) – can usually be interpreted as stoichiometric exchange of *interlayer* ions. Heavy metals however are sorbed by surface complex formation to the OH-functional groups of the outer surface (the so-called broken bonds). The non-swellable three-layer silicates, micas such as illite, can usually not exchange their interlayer ions; but the outside of these minerals and the weathered crystal edges ("frayed edges") participate in ion exchange reactions.

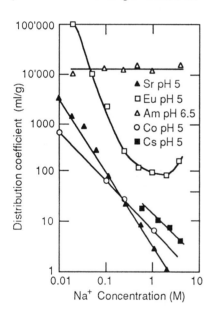

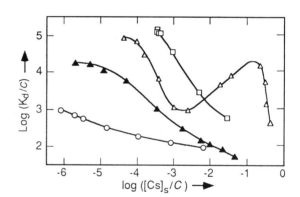

Figure 4.22

Ion exchange of various radionuclides on Na^+-montmorillonites (from Shiao et al., 1979).
(The relationship plotted can be derived from Eqs. (4.51) and (4.52); e.g., for the exchange of Sr^{2+} on Na^+-montmorillonites

$$K_p = \frac{X_{SrR}}{[Sr^{2+}]} = Q\frac{X_{NaR}^2}{[Na^+]^2}, \text{ or}$$

$$\log K_p = \log Q + 2\log(X_{NaR} - [Na^+])$$

Since X_{NaR} remains relatively constant, $\log K_p$ vs $\log[Na^+]$ should plot with a slope of -2.

Figure 4.23

Sorption of cesium in synthetic groundwater on clay minerals:

○ Montmorillonite   □ Illite

△ < 40 μm Chlorite   ▲ < 2-μm Chlorite

The data are "normalized" with regard to the ion exchange capacity C of the sorbents. The sorption curves of the illite and of the < 40-μm chlorite are strongly non-linear, whereas that of the montmorillonite approaches linearity.

(From Grütter, von Gunten, Kohler and Rössler, 1990)

For the understanding of the binding of heavy metals on clays one needs to consider – in addition to ion exchange – the surface complex formation on end-standing functional OH-groups. Furthermore, the speciation of the sorbate ion (free hydroxo complex, carbonato- or organic complex) and its pH-dependence has to be known.

The surface characteristics of *kaolinite* was discussed in Chapter 3.4 and in Fig. 3.9. While the siloxane layer may – to a limited extent – participate in ion exchange reactions. The functional OH-groups at the gibbsite and edge surfaces are able to surface complex heavy metal ions. (Schindler et al., 1987).

As the figure shows the exchange of Sr^{2+} on Na^+-montmorillonite fits the ion exchange theory very well. But the adsorption of heavy metals cannot be accounted for by this theory. Co(II) behaves as if it were monovalent; K_D for americium is independent of $[Na^+]$ (americium occurs at pH = 6.5 as a hydroxo complex).

The Adsorption of Cs^+ on Clays – an ion with a simple solution chemistry (no hydrolysis, no complex formation) – can be remarkably complex. Grütter et al. (1990) have studied adsorption and desorption of Cs^+ on glaciofluvial deposits and have shown that the isotherms for sorption and exchange on these materials are non-linear. Part of this non-linearity can be accounted for by the collaps of the c-spacing of certain clays (vermiculite, chlorite). As illustrated in Fig. 4.23 the Cs^+ sorption on illite and chlorite is characterized by non-linearity.

Appendix

A.4.1 Contact Angle, Adhesion and Cohesion

In Chapter 4.2 we introduced the interfacial (surface) tension (equivalent to surface or interfacial energy) as the minimum work required to create a differential increment in surface area. The interfacial energy, equally applicable to solids and liquids, was referred to as the interfacial Gibbs free energy (at constant temperature, pressure and composition) (n refers to the composition other than the surfactant under consideration).

$$\gamma = \left(\frac{\delta G}{\delta \bar{A}}\right)_{T,p,n} \tag{A.4.1}$$

where γ is the interfacial tension usually expressed in J m^{-2} or N cm^{-1} (N = Newton).

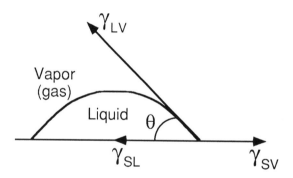

Figure A.4.1

Components of interfacial tension (energy) for the equilibrium of a liquid drop on a smooth surface in contact with air (or the vapor) phase. The liquid (in most instances) will not wet the surface but remains as a drop having a definite *angle of contact* between the liquid and solid phase.

The Contact Angle. Three phases are in contact when a drop of liquid (e.g., water) is placed on a perfectly smooth solid surface and all three phases are allowed to come to equilibrium.

The change in surface free energy, ΔG^s, accompanying a small displacement of the liquid such that the change in area of solid covered, ΔA, is

$$\Delta G^s = \Delta A \left(\gamma_{SL} - \gamma_{SV}\right) + \Delta A\, \gamma_{LV} \cos\left(\theta - \Delta\theta\right) \tag{A.4.2}$$

at equilibrium

$$\lim_{\Delta A \to 0} (\Delta G^s / \Delta A) = 0, \text{ and}$$

$$\gamma_{LV} \cos \theta = \gamma_{SV} - \gamma_{SL} \qquad\qquad (A.4.3)^{1)}$$

This equation, called the *Young Equation,* is in accord with the concept that the various surface forces can be represented by surface tensions acting in the direction of the surfaces. Eq. (A.4.3) results from equating the horizontal components of these tensions.

For practical purposes, if the contact angle is greater than 90° the liquid is said not to wet the solid (if the liquid is water one speaks of a hydrophobic surface); in such a case drops of liquids tend to move about easily and not to enter capillary pores. If $\theta = 0$, (ideal perfect wettability) Eq. (A.4.3) no longer holds and a *spreading coefficient,* $S_{LS(V)}$, reflects the imbalance of surface free energies.

$$S_{LS(V)} = \gamma_{SV} - (\gamma_{LV} + \gamma_{SL}) \qquad\qquad (A.4.4)$$

Young's equation is a plausible, widely used result, but experimental verification is often rendered difficult; e.g., the two terms which involve the interface between the solid and the two other phases cannot be measured independently[2]. Furthermore, many complications can arise with contact angle measurements; γ_S values of ionic solids based on contact angle measurements are different from those estimated from solubility (Table 6.1) (cf. Table A.4.1).

The measurement of contact angles for a sessile drop or bubble resting on or against a plane solid surface can be measured by direct microscopic examination.

---

[1] γ_{SV} (the surface tension of the solid in equilibrium with the vapor of the liquid) is often replaced by γ_S, the surface energy of a solid (hypothetically in equilibrium with vacuum or its own vapor). The difference between γ_{SV} and γ_S is the equilibrium film pressure π ($\pi = \gamma_{SV} - \gamma_S$) which depends on the adsorption (which lowers surface tension – cf. Eq. (4.3) – of the vapor of the liquid on the solid-gas interface).

[2] As shown by Fowkes (1968) the interfacial energy between two phases (whose surface tensions – with respect to vacuum – are γ_1 and γ_2) is subject to the resultant force field made up of components arising from attractive forces in the bulk of each phase and the forces, usually the London dispersion forces (cf. Eq. 4.2) operating accross the interface itself. Then the interfacial tension (energy) between two phases γ_{12} is given by

$$\gamma_{12} = \gamma_1 + \gamma_2 - 2 \left(\gamma_{1(L)} \gamma_{2(L)} \right)^{1/2} \qquad\qquad (A.4.5)$$

The geometric mean of the dispersion force components $\gamma_{1(L)}$ and $\gamma_{2(L)}$ may be interpreted as a measure of the interfacial attraction resulting from dispersion forces between adjacent dissimilar phases (see Table 4.A.1).

Another method is to adjust the angle of a plate immersed in the liquid so that the liquid surface remains perfectly flat right up to the solid surface.

Adhesion and Cohesion

Processes related to γ and θ are adhesion, cohesion and spreading. We consider two phases A and B without specifying their physical state; their common interface is AB. We can distinguish the following processes as they affect a unit area using a connotation given by Hiemenz (1986).

1) Cohesion:
    no surface $\longrightarrow$ 2 A (or B) surfaces   (A.4.6)

2) Adhesion:
    1 AB surface $\longrightarrow$ 1 A + 1 B surface   (A.4.7)

Fig. A.4.2 gives a schematic illustration.

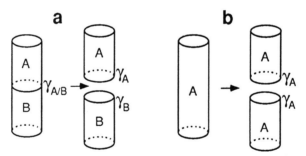

Figure A.4.2

Work of adhesion (a) and work of cohesion (b)

The work of *adhesion*, W_{AB}, is the work to separate 1 m^2 of AB interface into two separate A and B interfaces, and is given by

$$\Delta G = W_{AB} = \gamma_{final} - \gamma_{initial} = \gamma_A + \gamma_B - \gamma_{AB} \qquad (A.4.8)$$

(If AB is a solid-liquid interface, A and B are the A and B interfaces with vapor and, in an exact sense, γ_A and γ_B, are γ_{AV} and γ_{BV}, respectively.)

The work of *cohesion*, e.g., of a pure liquid, consists of producing two new interfaces, each of 1 m^2; it measures the attraction between the molecules of this phase

$$\Delta G = W_{AA} = 2\gamma_A \qquad (A.4.9)$$

Table A.4.1 Attractive Forces at Interfaces-surface Energy, γ, and London-van
der Waals Dispersion Force Component of Surface Energy, $\gamma_{(L)}$ [a)]

	γ $mJ\ m^{-2}$	$\gamma_{(L)}$ $mJ\ m^{-2}$
Liquids [b)]		
Water	72.8	21.8
Mercury	484.0	200.0
n-Hexane	18.4	18.4
n-Decane	23.9	23.9
Carbon tetrachloride	26.9	–
Benzene	28.9	–
Nitrobenzene	43.9	–
Glycerol	63.4	37.0
Solids [c)]		
Paraffin wax	–	25.5
Polyethylene	–	35.0
Polystyrene	–	44.0
Silver	–	74.0
Lead	–	99.0
Anatase (TiO_2)	–	91.0
Rutile (TiO_2)	–	143.0
Ferric oxide	–	107.0
Silica	–	123.0 (78)
Graphite	–	110.0

[a)] Based on information provided by Fowkes (1968). (20° C) a dash indicates that no value is available

[b)] $\gamma_{(L)}$ values for water and mercury have been determined by measuring the interfacial tension of these liquids with a number of liquid saturated hydrocarbons. The intermolecular attraction in the liquid hydrocarbons is entirely due to London-van der Waals dispersion forces for all practical purposes. $\gamma_{(L)}$ was derived from contact angle measurements.

[c)] $\gamma_{(L)}$ of solids were derived from contact angle measurements or from measurements of equilibrium film pressures of adsorbed vapor on the solid surface.

Wetting, Water Repellency, and Detergency

Obviously, the spreading of water on a hydrophobic solid is helped by adding a surfactant. γ_{LV} and γ_{SL} are reduced, thus, on both accounts cos θ (Eq. A.4.3) increases and θ decreases.

Water repellency is achieved by making the surface hydrophobic – γ_{SV} has to be reduced as much as possible if $\gamma_{SV} - \gamma_{SL}$ is negative, θ > 90° C – water repelling materials include waxes, petroleum residues and silicones; for example, if $\equiv SiOH$ groups on glass or SiO_2 surfaces have reacted with silanes to form $\equiv SiO–Si–Alkyl$

groups, these surfaces have a hydrocarbon type of surface. If θ is greater than 90°, the water will tend not to penetrate into the hollows or pores in the solid; under these conditions gas bubbles will attach to the surfaces. The contact angle plays an important role in flotation (see Chapter 7.7) and in detergency. In the latter the action of the detergent is to lower the adhesion between a dirt particle and a solid surface (Fig. A.4.3). While we use here the arguments used in problems of detergency, the same considerations apply to illustrate that surfactants can reduce the adhesion of particles to solid surfaces or to other particles (see Fig. A.3).

Figure A.4.3

The adhesion of solid particles to solid surfaces (or to other particles) can be reduced by surfactants (or detergents). The adhesion is given (cf. Eq. A.4.8) by $W_{SP} = \gamma_{PW} + \gamma_{SW} - \gamma_{SP}$. Thus, γ_{PW} and γ_{SW} have to be lowered, to reduce the adhesion.

Table A.4.2   Some Contact Angle Values for Solid-Liquid-Air Interfaces [1] (25° C)

Solid	Liquid	Contact Angle
Glass	water	~ 0
TiO_2	water	~ 0
graphite	water	86
paraffine	water	106
PTFF [2]	water	108
PE [3]	water	94

[1]   Data given are quoted from Adamson (1990) where references are given
[2]   polytetrafluorethylene
[3]   polyethylene

The Oil-Water Interface

Although the oil-water interface does not belong into the realm of this book, we simply like to illustrate briefly, that the elementary concepts discussed here are directly applicable to this interface and can be used to understand some of the phenomena associated with oil spills.

In Fig. A.4.4, adhesion and cohesion are illustrated for the oil-water interface.

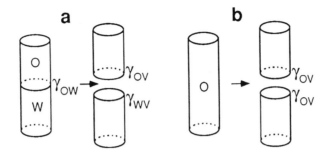

Figure A.4.4

Work of adhesion (a) and of cohesion (b) for the oil-water interface.

Adhesion: $W_{OW} = \gamma_{OV} + \gamma_{WV} - \gamma_{OW}$ (i)

Cohesion for oil and water: $W_{OO} = 2\gamma_{OV};$ $W_{WW} = 2\gamma_{WW}$ (ii)

Spreading: $S_{O/W} = \gamma_{WV} - \gamma_{OV} - \gamma_{OW}$ (iii)

The work of adhesion is influenced by the orientation of the molecules at the interface. For example, with the help of Table A.4.1 and Eq. (A.4.8), the work of adhesion of n-decane-water (corresponding to a paraffinic oil-water system) and of glycerol-water can be computed to be 40×10^{-3} J m^{-2} and 56×10^{-3} J m^{-2}, respectively. It requires more work to separate the polar glycerol molecules (oriented with the OH groups toward the water) from the water phase than the nonpolar hydrocarbon molecules. For paraffinic oils W_{OO} is about 44 mJ m^{-2}, for water W_{WW} is 144 mJ m^{-2}, and for glycerol W_{OO} is 127 mJ m^{-2}.

Spreading (Eq. (iii) of Fig. A.4.4) occurs when the oil adheres to the water more strongly than it coheres to itself; this is generally the case when a liquid of low surface tension is placed on one of high surface tension. This mineral oil spreads on water, but water cannot spread on this oil. The initial spreading coefficient does not consider that the two liquids will, after contact, become mutually saturated. The addition of surfactants which lower γ_{OW} and γ_{SW} (cf. Fig. A.4.3) cause the dispersion of the oil into droplets.

A.4.2 Electrocapillarity

Researches carried out in electrochemistry on solid electrodes and especially on the mercury-water interface have made a significant contribution to an understanding of interfacial phenomena. Although the electrode-water interfaces are typically not encountered in natural systems, they are often more readily amenable to pre-cise experimental exploration and are of reference value to the appreciation of the properties of constant potential surfaces, e.g., silver halogenides in presence of potential determining Ag^+ on halogenide ions, or hydrous oxide in equilibrium with potential determining H^+ ions.

The polarized Electrode-Electrolyte and the reversible Solid-Electrolyte Interface

A polarizable Interface is represented by a (polarizable) electrode where a poten-tial difference across the double layer is applied *externally*, i.e., by applying be-tween the electrode and a reference electrode using a potentiostat. At a *reversible interface* the change in electrostatic potential across the double layer results from a chemical interaction of solutes (potential determining species) with the solid. The characteristics of the two types of double layers are very similar and they differ primarily in the manner in which the potential difference across the interface is established.

Electro Capillarity and the dropping Mercury Electrode. The term electro capillarity derives from the early application of measurements of interfacial tension at the Hg-electrolyte interface. The interfacial tension, γ, can be measured readily with a dropping mercury electrode. E.g., the life time of a drop, t_{max}, is directly proportional to the interfacial tension γ. Thus, γ is measured as a function of ψ in presence and absence of a solute that is adsorbed at the Hg-water interface; this kind of data is amenable to thermodynamic interpretation of the surface chemical properties of the electrode-water interface.

A schematic example is given in Fig. A.4.5. The slope of the electrocapillary curve depends on the nature of the solution or the equilibrium structure of the double layer and on the specific sorbability of dissolved substances. In line with the Gibbs equation (Eq. 4.3), sorbable species depress the interfacial tension.

$$\Gamma_i = -\frac{1}{RT}\left(\frac{\delta\gamma}{\delta \ln a_i}\right) \tag{A.4.10}$$

where a_i is the activity (concentration) of the sorbable substance in the solution. The equation is valid for constant temperature and pressure and composition other than species i. This equation permits the estimation of the extent of adsorption as a function of potential. (The depression of interfacial tension by adsorption is deter-mined by substrating the values of its solution from that of the background solution (Fig. A.4.5b).) The adsorption of active anions is promoted at potentials more posi-tive than the electrocapillary maximum; on the other hand, at potentials more ne-gative than the electrocapillary zero, surface active cationic species are preferen-tially adsorbed.

At positive and negative polarization, i.e., at potentials, positive and negative of the electrocapillary maximum, the two curves coincide; thus, indicating total desorption of the surface active species. (Camphor used as an adsorbate in Fig. A.4.5 may be characterized as a somewhat polar molecule behaving in a slightly cationic way.)

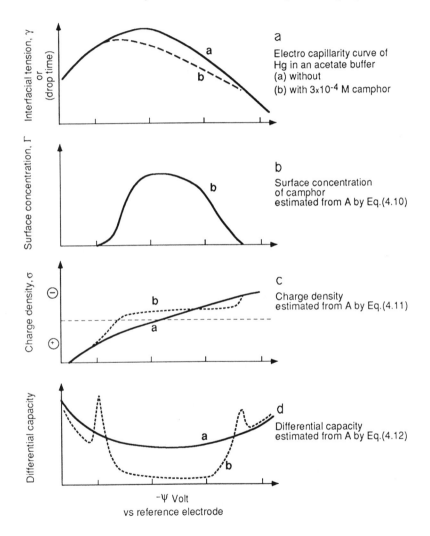

a
Electro capillarity curve of Hg in an acetate buffer (a) without (b) with 3×10^{-4} M camphor

b
Surface concentration of camphor estimated from A by Eq.(4.10)

c
Charge density estimated from A by Eq.(4.11)

d
Differential capacity estimated from A by Eq.(4.12)

Figure A.4.5

Electrocapillary phenomena on Hg-electrode in presence and absence of an adsorbate (camphor). From a measurement of interfacial tension (a) (e.g., from droptime of a Hg-electrode) or of differential capacity (d) (e.g., by an a.c-method) as a function of the electrode potential (established by applying a fixed potential across the Hg-electrolyte interface) one can calculate the extent of adsorption (b) (from (a) by the Gibbs Equation) and of the structure of the interface as a function of the surface potential. Figs. a, c and d are interconnected through the Lippmann Equations.

The Lippmann Equation. It can be shown thermodynamically that the slope of the electrocapillary curve is equal to the charge density, σ, in the electric double layer (First Lippmann Equation).

$$\sigma = -\left(\frac{\delta\gamma}{\delta\psi}\right)_{T,p, \text{composition}} \tag{A.4.11}$$

At the electrocapillary maximum, the charge density, σ, is zero (point of zero charge) (Fig. A.4.5c). By definition, the differential capacity of the double layer, C_d, is equal (Second Lippmann Equation).

$$C_d = -\left(\frac{\delta\sigma}{\delta\psi}\right)_{T,p, \text{composition}} = +\left(\frac{\delta^2\gamma}{\delta\psi^2}\right)_{T,p, \text{composition}} \tag{A.4.12}$$

The second differential in the electro capillary curve gives the differential capacity of the electric double layer. The differential capacity curve (Fig. A.4.5d), obtained in the presence of adsorbed species, shows peaks at high polarizations at potentials where sudden changes occur in the charge density curves (see Figs. A.4.5 c and d). These changes indicate the potential range where desorption of the adsorbate species take place. When this occurs, material of high dielectric constant is displaced by material of high dielectric constant in a manner analogous to the salting-out effect. The differential capacity can be measured directly from the build-up and decay of electrode potentials. (C_d is most frequently determined by an alternating current method.)

Experimentally either the electrocapillary curve or the differential capacity (as a function of ψ) is determined. From either set of data, the interfacial properties (adsorption and/or charge) as a function of ψ and a quantitative description of the structure of the interface can be obtained.

A comparison with the *reversible* interface can be made. The reversible solid electrolyte interface can be used in a similar way to explore the distribution of charge components at solid-water interfaces. As we have seen, the surface charge density, σ, (Eqs. (3.1) and (iii) in Example 2.1) can be readily determined experimentally (e.g., from an alkalimetric titration curve). The Lippmann equations can be used as with the polarized electrodes to obtain the differential capacity from

$$\frac{F}{RT}\left(\frac{\delta\sigma}{\delta pH}\right)_{T,p, \text{composition}} = -C_d \tag{A.4.13}$$

and to infer the change in interfacial tension as a function of pH by integration (graphically) the σ vs pH curve

$$\gamma = \frac{RT}{F} \int \sigma dpH + K \qquad\qquad (A.4.14)$$

where K is the integration constant. The information provided by experimental values of σ as a function of pH and of composition is thermodynamically inter-connected (Gibbs adsorption equation) with the composition and pH-dependent sorption of (ionic) solute components. Fig. A.4.6 gives a schematic comparison between the polarized electrode-electrolyte and the reversible oxide electrode-electrolyte interface.

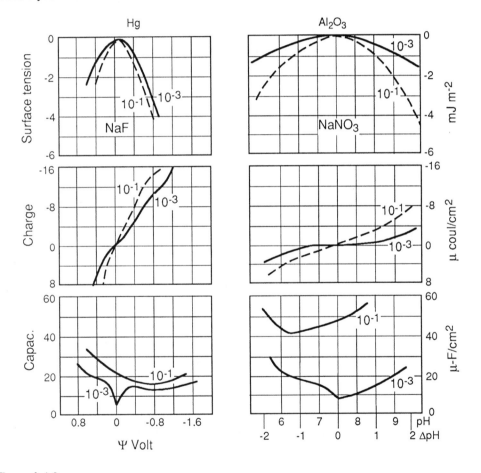

Figure A.4.6

Comparison between the polarized electrode-electrolyte interface and the reversible (Al_2O_3) oxide-electrolyte interface. Surface tension (interfacial) tension, charge density and differential capacity, respectively, are plotted as a function of the rational potential ψ (at pzc ψ is set = 0) in the case of Hg and as a function of ΔpH (pH-pH_{pzc}) in the case of Al_2O_3 (pH = pH_{pzc} when σ = 0).

(Modified from Stumm, Huang and Jenkins, 1970)

Eqs. (A.4.13) and (A.4.15) imply that the pH axis is equivalent to the potential axis, ideally that the Nernst Equation (Eq. 3.24)

$$\psi_0 = \frac{2.3\,RT}{F}\,\Delta pH \qquad\qquad (A.4.15)$$

is fulfilled. As we have seen in Figs. 3.4 and 3.17, the applicability of the Nernst Equation for some oxides is justified.

The right hand side of Fig. A.4.6 is contained in Fig. 3.3. Capacity measurements can readily be made at solid electrodes to study adsorption behavior. For a review see Parsons (1987). As Fig. A.4.7 illustrates, capacity potential curves of three low-index phases of silver, in contact with a dilute aqueous solution of NaF, show different minimum capacities (corresponding to the condition $\sigma = 0$) and therefore remarkably different potentials of pzc. The closest packed surface (111) has the highest pzc and the least close-packed (110) has the lowest pcz; these values differ by 300 mV. Such complications observed with single crystal electrodes, seem likely to have their parallel at other solid surfaces. For example, it is to be expected that a crystalline oxide will have different pzc values at its various types of exposed faces.

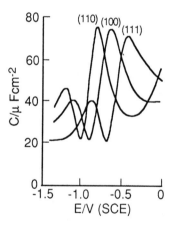

Figure A.4.7

Capacity-potential curves for three low-index planes of silver in contact with aqueous 0.01 M NaF at 25 °C.

(Redrawn from Valette and Hamelin, 1973)

Problems

1) To aliquots of an iron(III) oxide suspension (10 mg/ℓ) increasing quantities of phthalic acid H_2P_T are added at pH = 6.0 and at a constant ionic strenght of 10^{-3} M (25° C).

Added H_2P_T [M]	Residual H_2P_T [M]
1×10^{-7}	0.43×10^{-7}
3×10^{-7}	1.4×10^{-7}
1×10^{-6}	4.8×10^{-7}
3×10^{-6}	18.5×10^{-7}
1×10^{-5}	78×10^{-7}
3×10^{-5}	290×10^{-7}
1×10^{-4}	890×10^{-7}

a) Fit the data in terms of a Freundlich or a Langmuir isotherm. Estimate the maximum capacity for adsorption of H_2P_T by iron(III) oxide.

b) Explain why the data give a perfect (or not so perfect) fit to the isotherm.

c) Separate experiments on the iron oxide have shown that it is characterized by a specific surface area of 40 m^2 g^{-1}. How many surface sites ($\equiv$Fe-OH) per nm^2 are available for the interaction with phthalic acid?

d) Phthalic acid is a two-protic acid with pK_{a1} and pK_{a2} values of 2.9 and 5.3, respectively. Can we interpret the adsorption isotherm in terms of the adsorbate H_2P_T; or should we formulate in terms of $[P^{2-}]$?

e) If this adsorption is interpreted in terms of a ligand exchange, what would be the representative equilibrium constant? $\frac{\equiv P^{2-}}{\Gamma}$

2) What is the difference between ion exchange, e.g., the binding of Ca^{2+} in exchange of Na^+ and surface complex formation, e.g., the binding of Ca^{2+} on a MnO_2 surface? What experiments could be carried out to distinguish between these two processes?

3) How could one distinguish between phosphate binding (ligand exchange) to a hydrous oxide surface and the precipitation of iron phosphate?

4) Adsorption isotherms are often plotted for mixtures of adsorbates (humic acid, fatty acids etc.) using collective parameters such as organic carbon. Enumerate the various reasons that make the rational interpretation of such adsorption isotherms difficult.

5) Surface activity has been defined as a tendency of a solute to concentrate at an interface. What kind of configurations are necessary for surface activity? What substances desorb from the interface?

6) Explain the relationship between surface tension, contact angle and capillary rise.

7) Compare the data given in Table A.4.2 and interprete them in terms of interfacial energy, adhesion, and cohesion.

8) How could one distinguish experimentally in the interaction of a hydrous oxide surface with a fatty acid, whether the interaction is due to hydrophobic bonding or to coordinative interaction (ligand exchange of the carboxyl group with the surface functional groups of the hydrous oxide)?

9) a) Can one distinguish between adsorption and precipitation from an analysis of the change in solution composition? Specifically test the following criteria:

   i) IAP < K_{S0}, therefore adsorption;
   ii) IAP > K_{S0}, therefore precipitation;
   iii) a good fit to a Langmuir (or Freundlich) isotherm;

   and show that all three ciriteria do not suffice to distinguish between precipitation and adsorption.

   b) What techniques are able to "prove" precipitation (or coprecipitation)?

10) Estimate the half time for the adsorption of Ni^{2+} at an Al_2O_3 (10^{-4} surface sites per liter).

11) Explain why anionic polyelectrolytes may become adsorbed on negatively charged surfaces.

Reading Suggestions

Bolt, G. H., and W. H. Van Riemsdijk (1987), "Surface Chemical Processes in Soil", in W. Stumm, Ed., *Aquatic Surface Chemistry*, Wiley-Interscience, New York, pp. 127-164.

Cederberg, G. A., R. L. Street, and J. O. Leckie (1985), "A Groundwater Mass Transport and Equilibrium Chemistry Model for Multicomponent Systems", *Water Resources Research* 21, 1095-1104.

Ćosović, B. (1990), "Adsorption Kinetics of the Complex Mixture of Organic Solutes at Model and Natural Phase Boundaries", in W. Stumm, Ed., *Aquatic Chemical Kinetics*, Wiley-Interscience, New York, pp. 291-310.

Freeze, R. A., and J. A. Cherry (1979), *Groundwater*, Prentice Hall, NJ, 604 pp.

Hering, J. G., and F. M. M. Morel (1990), "Kinetics of Trace Metal Complexation: Ligand-Exchange Reactions", *Environm. Sci. Technol. 1* 24, 242-252.

Hiemenz, P. C. (1986), *Principle of Colloid and Surface Chemistry, 2nd Ed.*, M. Dekker, New York.

Lyklema, J. (1985), "How Polymers Adsorb and Affect Colloid Stability, Flocculation, Sedimentation, and Consolidation"; Proceedings of the Engineering Foundation Conference, Sea Island, Georgia, in B. M. Moudgil and P. Somasundaran, Eds., pp. 3-21.

Maity, N., and G. F. Payne (1991), "Adsorption from Aqueous Solutions Based on a Combination of Hydrogen Bonding and Hydrophobic Interaction", Langmuir 7, pp. 1247-1254.

McCarty, P. L, M. Reinhard, and B. E. Rittmann (1981), "Trace Organics in Groundwater", Environ. Sci. & Technol. **15/1**, 40-51.

Parks, G. A. (1990), "Surface Energy and Adsorption at Mineral/Water Interfaces: An Introduction", in M. F. Hochella Jr. and A. F. White, Eds., *Mineral-Water Interface Geochemistry*, Mineralogical Society of America, pp. 133-175.

Schwarzenbach, R., D. Imboden, and Ph. M. Gschwend (1992), *Environmental Organic Chemistry*, Wiley Interscience, New York.

Sposito, G. (1984), *The Surface Chemistry of Soils*, Oxford University Press, New York.

Sposito, G. (1990), "Molecular Models of Ion Adsorption on Mineral Surfaces", in M. F. Hochella Jr. and A. F. White, Eds., *Mineral-Water Interface Geochemistry*, Mineralogical Society of America, Washington, D.C., pp. 261-279.

Thurman, E. M. (1985), *Organic Geochemistry of Natural Waters*, Nijhoff, Boston.

Ulrich, H-J., W. Stumm, and B. Ćosović (1988), "Adsorption of Aliphatic Fatty Acids on Aquatic Interfaces. Comparison between 2 Model Surfaces: The Mercury Electrode and δ-Al_2O_3 Colloids", *Environ. Sci. & Techn.* 22, 37-41.

Zutić, V., and J. Tomaić (1988), "On the Formation of Organic Coatings on Marine Particles: Interactions of Organic Matter at Hydrous Alumina/Seawater Interfaces", *Mar. Chem.* 23, 51-67.

Chapter 5

The Kinetics of Surface Controlled Dissolution of Oxide Minerals; an Introduction to Weathering

5.1 Introduction
Chemical Weathering [1]

Chemical weathering is one of the major processes controlling the global hydrogeochemical cycle of elements. In this cycle, water operates both as a reactant and as a transporting agent of dissolved and particulate components from land to sea. The atmosphere provides a reservoir for carbon dioxide and for oxidants required in the weathering reactions. The biota assists the weathering processes by providing organic ligands and acids and by supplying locally, upon decomposition, increased CO_2 concentrations.

During chemical weathering, rocks and primary minerals become transformed to solutes and soils and eventually to sediments and sedimentary rocks.

When a mineral dissolves, several successive elementary steps may be involved:
1) mass transport of dissolved reactants from bulk solution to the mineral surface,
2) adsorption of solutes,
3) interlattice transfer of reacting species,
4) chemical reactions,
5) detachment of reactants from the surface, and
6) mass transport into the bulk of the solution.

Under natural conditions the rates of dissolution of most minerals are too slow to depend on mass transfer of the reactants or products in the aqueous phase. One can thus restrict the discussion to the case of weathering reactions where the rate-controlling mechanism is the mass transfer of reactants and products in the solid phase or to reactions controlled by a surface process and the related detachment process of reactants.

Calcareous minerals and evaporite minerals (halides, gypsum) are very soluble and dissolve rapidly and, in general, congruently (i.e., yielding upon dissolution the same stoichiometric proportions in the solution as the proportions in the dissolving mineral and without forming new solid phases). Their contribution to the total dissolved load in rivers can be estimated by considering the mean composition of river

---

[1] In writing this section, I depended largely on a review by Stumm and Wollast (1990).

water and the relative importance of various rocks to weathering (Garrels and Mackenzie, 1971). Recent estimates (Holland, 1978; Meybeck, 1979; Wollast and Mackenzie, 1983) indicate that evaporites and carbonates contribute approximately 17 % and 38 %, respectively, of the total dissolved load in the world's rivers. The remaining 45 % is due to the weathering of silicates, underlining the significant role of these minerals in the overall chemical denudation of the Earth's surface.

There are no unequivocal weathering reactions for the silicate minerals. Depending on the nature of parent rocks and hydraulic regimes, various secondary minerals like gibbsite, kaolinite, smectites, and illites are formed as reaction products. Some important dissolution processes of silicates are given, for example, by the following reactions:

$$CaAl_2Si_2O_8 + 2\,CO_2 + 3\,H_2O$$
anorthite

$$= Al_2Si_2O_5(OH)_4 + Ca^{2+} + 2\,HCO_3^- \qquad (5.1a)$$
kaolinite

$$2\,NaAlSi_3O_8 + 2\,CO_2 + 6\,H_2O$$
albite

$$= Al_2Si_4O_{10}(OH)_2 + 2\,Na^+ + 2\,HCO_3^- + 2\,H_4SiO_4 \qquad (5.1b)$$
montmorillonite

$$KMgFe_2AlSi_3O_{10}(OH)_2 + \tfrac{1}{2}\,O_2 + 3\,CO_2 + 11\,H_2O$$
biotite

$$= Al(OH)_3 + 2\,Fe(OH)_3 + K^+ + Mg^{2+} + 3\,HCO_3^- + 3\,H_4SiO_4 \qquad (5.1c)$$
gibbsite   goethite

In all cases, water and carbonic acid, which is the source of protons, are the main reactants. The net result of the reaction is the release of cations (Ca^{2+}, Mg^{2+}, K^+, Na^+) and the production of alkalinity via HCO_3^-. When ferrous iron is present in the lattice, as in biotite, oxygen consumption may become an important factor affecting the weathering mechanism and the rate of dissolution.

These reactions, however, are complex and generally proceed through a series of reaction steps. The rate of weathering of silicates may vary considerably depending on the arrangement of the silicon tetrahedra in the mineral and on the nature of the cations.

The Role of Weathering in Geochemical Processes in Oceanic and Global Systems. As indicated in the stoichiometric equations given above, the rate of chemical weathering is important in determining the rate of CO_2 consumption. Furthermore, the global weathering rate is most likely influenced by temperature and is proportional to total continental land area and the extent of its coverage with vegetation; the latter dependence results from CO_2 production in soils, which is a consequence of plant respiration and the decay of organic matter as well as the release of complex-forming substances (ligands) (dicarboxylic acids, hydroxycarboxylic acids, and phenols), i.e., anions that form soluble complexes with cations that originate from the lattice or form surface complexes with the surface of oxide minerals. Regionally, the extent of weathering is influenced by acid rain (Schnoor, 1990; Schnoor and Stumm, 1985).

It has been shown by Berner et al. (1983) and by Berner and Lasaga (1989) that silicate weathering is more important than carbonate mineral weathering as a long-term control on atmospheric CO_2. The HCO_3^- and Ca^{2+} ions produced by weathering of $CaCO_3$,

$$CaCO_3 + CO_2 + H_2O = Ca^{2+} + 2\,HCO_3^- \tag{5.2}$$

precipitate in the ocean (through incorporation by marine organisms) as $CaCO_3$. The CO_2 consumed in carbonate weathering is released again upon formation of $CaCO_3$ in the ocean (reversal of the reaction given above). Thus, globally, carbonate weathering results in no net loss of atmospheric CO_2. The weathering of calcium silicates, e.g., Eq. (5.1a) or, in a simplified way,

$$CaSiO_3 + 2\,CO_2 + 3\,H_2O = Ca^{2+} + 2\,HCO_3^- + H_4SiO_4 \tag{5.3}$$

also produces Ca^{2+} and HCO_3^-, which form $CaCO_3$ in the sea, but only half of the CO_2 consumed in the weathering is released and returned to the atmosphere upon $CaCO_3$ formation, Thus, silicate weathering results in a net loss of atmospheric CO_2. Of course, ultimately, the cycle is completed by metamorphic and magmatic breakdown, deep in the Earth, of $CaCO_3$ with the help of SiO_2, a reaction that may be represented in a simplified way as $CaCO_3 + SiO_2 = CaSiO_3 + CO_2$. Knowledge of the rate of dissolution of minerals is also necessary for the quantitative evaluation of geochemical processes in the oceanic system.

Diffusion-Controlled Versus Surface-Controlled Mechanisms

Among the theories proposed, essentially two main mechanisms can be distinguished; these are that the rate-determining step is a transport step (e.g., a transport of a reactant or a weathering product through a layer of the surface of the mineral) or that the dissolution reaction is controlled by a surface reaction. The rate equation corresponding to a transport-controlled reaction is known as the parabolic rate law when

$$r = \frac{dC}{dt} = k_p t^{-\frac{1}{2}} \ [M\ s^{-1}] \tag{5.4a}$$

where k_p is the reaction rate constant $[M\ s^{-\frac{1}{2}}]$. By integration the concentration in solution, C [M], increases with the square root of time:

$$C = C_0 + 2\,k_p t^{\frac{1}{2}} \tag{5.4b}$$

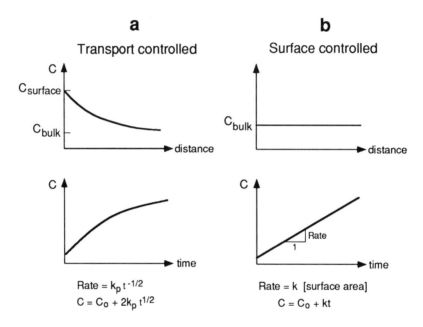

a
Transport controlled

b
Surface controlled

Rate $= k_p\,t^{-1/2}$
$C = C_0 + 2k_p\,t^{1/2}$

Rate $= k$ [surface area]
$C = C_0 + kt$

Figure 5.1

Transport vs surface controlled dissolution

Schematic representation of concentration in solution, C, as a function of distance from the surface of the dissolving mineral. In the lower part of the figure, the change in concentration (e.g., in a batch dissolution experiment) is given as a function of time.

a) Transport controlled dissolution. The concentration immediately adjacent to the mineral reflects the solubility equilibrium. Dissolution is then limited by the rate at which dissolved dissolution products are transported (diffusion, advection) to the bulk of the solution. Faster dissolution results from increased flow velocities or increased stirring. The supply of a reactant to the surface may also control the dissolution rate.

b) Pure *surface controlled dissolution* results when detachment from the mineral surface via surface reactions is so slow that concentrations adjacent to the surface build up to values essentially the same as in the surrounding bulk solution. Dissolution is not affected by increased flow velocities or stirring.

A situation, intermediate between a) and b) – a mixed transport-surface reaction controlled kinetics – may develop.

Alternatively, if the reactions at the surface are slow in comparison with diffusion or other reaction steps, the dissolution processes are controlled by the processes at the surface. In this case the concentrations of solutes adjacent to the surface will be the same as in the bulk solution. The dissolution kinetics follows a zero-order rate law if the steady state conditions at the surface prevail:

$$r = \frac{dC}{dt} = kA \ [M \ s^{-1}] \qquad (5.5)$$

where the dissolution rate r $[M \ s^{-1}]$ is proportional to the surface area of the mineral, A $[m^2]$; k is the reaction rate constant $[M \ m^{-2} \ s^{-1}]$. Fig. 5.1 compares the two control mechanisms. Since most important dissolution reactions are surface controlled we will concentrate on the kinetics of surface controlled reactions.

Fig. 5.2a shows examples of the results obtained on the dissolution of δ-Al_2O_3. In batch experiments where pH is kept constant with an automatic titrator, the concentration of Al(III)(aq) (resulting from the dissolution) is plotted as a function of time. The linear dissolution kinetics observed for every pH is compatible with a process whose rate is controlled by a surface reaction. The rate of dissolution is obtained from the slope of the plots.

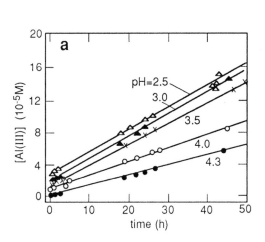 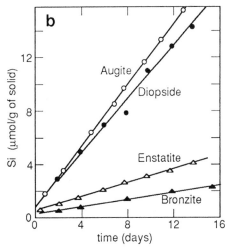

Figure 5.2

a) Linear dissolution kinetics observed for the dissolution of δ-Al_2O_3, representative of processes whose rates are controlled by a surface reaction and not by a transport step (data from Furrer and Stumm, 1986).

b) Linear dissolution kinetics of frame silicates. Minerals used were pyroxenes and olivines; their essential structural feature is the linkage of SiO_4 tetrahedra, laterally linked by bivalent cations (Mg^{2+}, Fe^{2+}, Ca^{2+}). Plotted are amounts of silica released versus time for the dissolution of etched enstatite, bronzite (p_{O_2} = 0), diopside, and augite at pH 6 (T = 20° C for bronzite; T = 50° C for the other minerals) (from Schott and Berner, 1985).

Fig. 5.2b gives results by Schott and Berner (1985) on the dissolution rate of iron-free pyroxenes and olivines, as measured by the silica release.

5.2 A General Rate Law for Surface Controlled Dissolution

We would like to provide the reader first with a qualitative understanding of the subject of dissolution kinetics.

The reactivity of the surface (Fig. 5.3), i.e., its tendency to dissolve, depends on the type of surface species present; e.g., an inner-sphere complex with a ligand such as that shown for oxalate

$$
\text{Me}\begin{matrix}\diagdown\diagup\,OH_2\\ \diagup\diagdown\,OH\end{matrix} + C_2O_4^{2-} + H^+ \rightleftharpoons \text{Me}\begin{matrix}\diagdown\,O\diagdown C \diagup O^-\\ \vert \\ \diagup\diagdown\,O\diagup C\diagdown O\end{matrix} + 2\,H_2O
$$

or other dicarboxylates, dihydroxides or hydroxy-carboxylic acids,

$$
R{<}\begin{matrix}COOH\\COOH\end{matrix}, \quad R{<}\begin{matrix}OH\\OH\end{matrix} \quad or \quad R{<}\begin{matrix}COOH\\OH\end{matrix}
$$

facilitates the detachment of a central metal ion and enhances the dissolution. This is readily understandable, because the ligands shift electron density towards to the central metal ion at the surface and bring negative charge into the coordination sphere of the Lewis acid center and enhance simultaneously the surface protonation and can labilize the critical Me-oxygen lattice bonds, thus, enabling the detachment of the central metal ion into solution.

Similarly, *surface protonation* tends to increase the dissolution rate, because it leads to highly polarized interatomic bonds in the immediate proximity of the surface central ions and thus facilitates the detachment of a cationic surface group into the solution. On the other hand, a surface coordinated metal ion, e.g., Cu^{2+} or Al^{3+}, may block a surface group and thus retard dissolution. An outer-sphere surface complex has little effect on the dissolution rate. Changes in the oxidation state of surface central ions have a pronounced effect on the dissolution rate (see Chapter 9).

The ideas developed here are largely based on the concept of the coordination at the (hydr)oxide interface; the ideas apply equally well to silicates. Somewhat modified concepts for the surface chemistry of carbonate, phosphate, sulfide and disulfide minerals have to be developed.

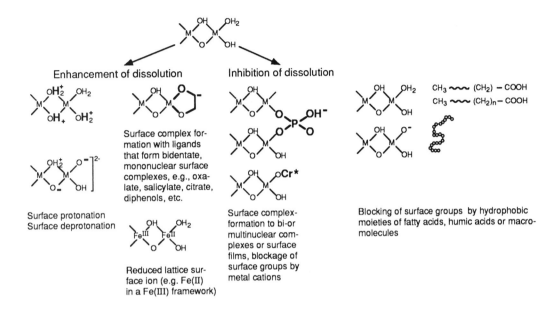

Figure 5.3

Effect of protonation, complex formation with ligands and metal ions and reduction on dissolution rate. The structures given here are schematic short hand notations to illustrate the principal features (they do not reveal the structural properties nor the coordination numbers of the oxides under consideration; charges given are relative).

In the dissolution reaction of an oxide mineral, the coordinative environment of the metal changes; for example, in dissolving an aluminum oxide layer, the Al^{3+} in the crystalline lattice exchanges its O^{2-} ligand for H_2O or another ligand L. In line with Fig. 5.3 the most important reactants participating in the dissolution of a solid mineral are H_2O, H^+, OH^-, ligands (surface complex building), and reductants and oxidants (in the case of reducible or oxidizable minerals).

Thus, the reaction occurs schematically in two sequences:

$$\text{surface sites + reactants (}H^+\text{, }OH^-\text{, or ligands)} \xrightarrow{\text{fast}} \text{surface species} \quad (5.6)$$

$$\text{surface species} \xrightarrow[\text{detachment of Me}]{\text{slow}} \text{Me(aq)} \quad (5.7)$$

where Me stands for metal. Although each sequence may consist of a series of smaller reaction steps, the rate law of surface-controlled dissolution is based on the idea

1) that the attachment of reactants to the surfaces sites is fast and
2) that the subsequent detachment of the metal species from the surface of the crystalline lattice into the solution is slow and thus rate limiting.

In the first sequence the dissolution reaction is initiated by the surface coordination with H^+, OH^-, and ligands which polarize, weaken, and tend to break the metal-oxygen bonds in the lattice of the surface. Since reaction (5.7) is rate limiting and using a steady state approach the rate law on the dissolution reaction will show a dependence on the concentration (activity) of the particular surface species, C_j [mol m^{-2}]:

$$\text{dissolution rate} \propto \langle \text{surface species} \rangle \qquad (5.8a)$$

We reach the same conclusion (Eq. 5.8a) if we treat the reaction sequence according to the activated complex theory (ACT), often also called the transition state theory. The particular surface species that has formed from the interaction of H^+, OH^-, or ligands with surface sites is the *precursor* of the activated complex (Fig. 5.4):

$$\text{dissolution rate} \propto \langle \text{precursor of the activated complex} \rangle \qquad (5.8b)$$

$$R = k\, C_j \qquad (5.8c)$$

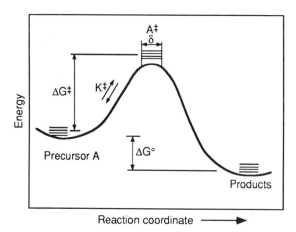

Figure 5.4

Activated complex theory for the surface-controlled dissolution of a mineral far from equilibrium. A is the precursor, i.e., a surface site that can be activated to $A^{\ddagger}$. The latter is in equilibrium with the precursor. The activation energy for the conversion of the precursor into the product is given by $\Delta G^{\ddagger}$.

where R is the dissolution rate, e.g., in mol m^{-2} h^{-1}, C_j = the surface concentration of the precursor [mol m^{-2}].

The surface concentration of the particular surface species, C_j, corresponds to the concentration of the precursor of the activated complex. Note that we use braces ⟨ ⟩ and brackets [ ] to indicate surface concentrations [mol m^{-2}] and solute concentrations [M], respectively.

5.3 Ligand Promoted Dissolution

We will first describe a relatively simple scenario for the enhancement of the dissolution of Al_2O_3 by a (complex-forming) ligand. As we have seen ligands tend to become adsorbed specifically and to form surface complexes with the Al(III) Lewis acid centers of the hydrous oxide surface. They also usually form complexes with Al(III) in solution. Complex formation in solution increases the solubility. This has no direct effect on the dissolution rate, however, since the dissolution is surface-controlled.

In nature, ligands that enhance the dissolution reaction such as oxalate, citrate, diphenols, hydroxy carboxylic acids are formed, as a byproduct of biodegradation of organic matter. Such ligands are also among the exudation products of plants and trees released through the roots.

The enhancement of the dissolution rate by a ligand in a surface-controlled reaction implies that surface complex formation facilitates the release of ions from the surface to the adjacent solution. These ligands can bring electron density or negative charge into the coordination sphere of the surface Al-species, lowering their Lewis acidity. This may labilize the critical Al-oxygen bonds, thus facilitating the detachment of the Al from the surface. It has been shown that bidentate ligands (i.e., ligands with two donor atoms) such as dicarboxylates and hydroxy-carboxylates, e.g., oxalate (see Fig. 5.5a), can form relatively strong surface chelates, i.e., ring-type surface complexes.

Trans Effect. The labilizing effect of a ligand on the bonds in the surface of the solid oxide phase of the central metal ions with oxygen or OH can also be interpreted in terms of the trans effect, i.e., the influence of the ligand on the strength of the bond that is trans to it, e.g., in our example the effect of a ligand such as a bicarboxylate on the strength of the Al-oxygen bonds

The trans influence is a ground state property; it is attributable to the fact that ligands trans to each other both participate in the orbital of the metal ion (in our example, Al); the more one ligand preempts this orbital, the weaker will the bond to the other ligand be.

In a similar way a sigma bond exerted by a functional surface OH-group to a metal ion causes a trans effect on the ligands bound to the metal ions (cf. Chapter 9.1).

Ligand Adsorption enhances Surface Protonation. In Fig. 3.5 it is shown that the binding of a ligand to an oxide surface decreases the net surface charge but increases surface protonation. This is illustrated in some detail in Example 5.1 and in Fig. 5.14 (at the end of Chapter 5.6). Since protons bound to O and OH in the surface lattice enhance the dissolution rate (Chapter 5.4) the effect of surface protonation induces by the adsorption of a ligand, at any given (constant) pH, may explain *part* of the ligand promotion effect. Since the extent of surface protonation, $\Gamma_H - \Gamma_{OH}$, is proportional to the surface concentration of the ligand, C_L^s, the rate law can be formulated in terms of C_L^s, although the precursor to the dissolution is a surface complex attached to protonated O and OH atoms.

In Fig. 5.5a a simple scheme of reaction steps is proposed. Some of the assumptions of our model are summarized in Table 5.1. The short-hand representation of a surface site is a simplification that does not take into account either detailed structural aspects of the oxide surface or the oxidation state of the metal ion and its coordination number. It implies (model assumption 2 in Table 5.1) that all functional surface groups, such as those in a cross-linked polyhydroxo-oxo acid, are treated as if they were identical.

The scheme in Fig. 5.5 indicates that the ligand, for example, oxalate, is adsorbed very fast in comparison to the dissolution reaction; thus, adsorption equilibrium may be assumed. The surface chelate formed is able to weaken the original Al-oxygen bonds on the surface of the crystal lattice. The detachment of the oxalato-aluminum species is the slow and rate-determining step; the initial sites are completely regenerated subsequent to the detachment step; provided that the concentrations of the reactants are kept constant, steady state conditions with regard to the oxide surface species are established (Table 5.1). If, furthermore, the system is far from dissolution equilibrium, the back reaction can be neglected, and constant dissolution rates occur.

The scheme of Fig. 5.5a corresponds to steady state conditions (Table 5.1). We can now apply the general rate law (Eqs. 5.7, 5.8), the rate of the ligand-promoted dissolution, R_L, is proportional to the concentration of surface sites occupied by L (metal-ligand complex, >ML) or to the surface concentration of ligands, C_L^s (mol m^{-2}):

$$R_L = k_L' (\equiv ML) = k_L' C_L^s \qquad (5.9)$$

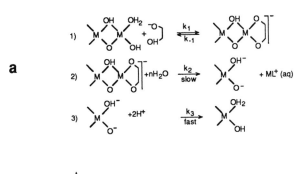

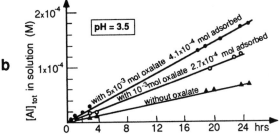

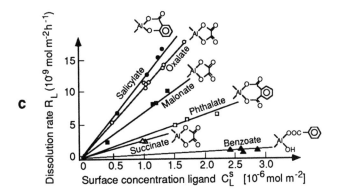

Figure 5.5

Promotion of the dissolution of an oxide by a ligand. The ligand illustrated here, in a short hand notation, is a bidentate ligand with two oxygen donor atoms (such as in oxalate, salicylate, citrate or diphenols).

a) The ligand-catalyzed dissolution reaction of a M_2O_3 can be described by three elementary steps:
    – a fast ligand adsorption step (equilibrium),
    – a slow detachment process, and
    – fast protonation subsequent to detachment restoring the incipient surface configuration.

b) The dissolution rate increases with increasing oxalate concentrations.

c) In accordance with the reaction scheme of a) the rate of ligand-catalyzed dissolution of δ-Al_2O_3 by the aliphatic ligands oxalate, citrate, and succinate, R_L (nmol m^{-2} h^{-1}), can be interpreted as a linear dependence on the surface concentrations of chelate complexes C_L^s.

(From Furrer and Stumm, 1986)

Table 5.1 Model Assumptions

① *Dissolution of slightly soluble hydrous oxides:*
   Surface process is rate-limiting
   Back reactions can be neglected if far away from equilibrium

② *The hydrous oxide surface, as a first approximation, is treated like a cross-linked polyhydroxo-oxo acid:*
   All functional groups are treated as if they were identical (mean field statistics)

③ *Steady state of surface phase:*
   Constancy of surface area
   Regeneration of active surface sites

④ *Surface defects, such as steps, kinks, and pits, establish surface sites of different activation energy, with different rates of reaction:*

$$\text{Acitve sites} \xrightarrow{\text{faster}} \text{Me(aq)} \qquad \text{(a)}$$

$$\text{Less active sites} \xrightarrow{\text{slower}} \text{Me(aq)} \qquad \text{(b)}$$

   Overall rate is given by (a):
   Steady-state condition can be maintained if a constant mole fraction, χ_a, of active sites to total (active and less active) sites is maintained, i.e., if active sites are continuously regenerated

⑤ *Precursor of activated complex:*
   Metal centers bound to surface chelate, or surrounded by n protonated functional groups
   $(C_H^s / S) \ll 1$

where k_L' is the reaction constant [time^{-1}]. C_L^s, the surface concentration of the ligand (oxalate) [mol m^{-2}], was determined analytically by determining the quantity of oxalate that was removed from the solution by adsorption and by considering the specific surface area of the Al_2O_3. The relationship between surface concentration and solute concentration is also obtained from the ligand exchange equilibrium constant. As shown in Fig. 5.5b, the experimental results are in accord with Eq. (5.9).

Fig. 5.5c illustrates the effects of various ligands upon the dissolution rate, and that a surface chelate (ring structure of the ligand bound to the metal center at the surface) is more efficient in enhancing the dissolution rate. Furthermore, the acceleration increases in the series Salicylate > oxalate > malonate > phtalate > succinate which indicates that a 5-ring chelate is more efficient than a 6-ring or 7-ring chelate.

One may note that at the same surface coverage, C_L^s, different dissolution rates are observed (Fig. 5.5c). At the same surface coverage, C_L^s, the extent of surface protonation is about the same; thus, the configuration and structure of the surface complex is of influence.

Mean Field Statistics and Steady State

However, we have to reflect on one of our model assumptions (Table 5.1). It is certainly not justified to assume a completely uniform oxide surface. The dissolution is favored at a few localized (active) sites where the reactions have lower activation energy. The overall reaction rate is the sum of the rates of the various types of sites. The reactions occurring at differently active sites are parallel reaction steps occurring at different rates (Table 5.1). In parallel reactions the fast reaction is rate determining. We can assume that the ratio (mol fraction, χ_a) of active sites to total (active plus less active) sites remains constant during the dissolution; that is the active sites are continuously regenerated after Al(III) detachment and thus steady state conditions are maintained, i.e., a mean field rate law can generalize the dissolution rate. The reaction constant k_L' in Eq. (5.9) includes χ_a, which is a function of the particular material used (see remark 4 in Table 5.1). In the activated complex theory the surface complex is the precursor of the activated complex (Fig. 5.4) and is in local equilibrium with it. The detachment corresponds to the desorption of the activated surface complex.

Fig 5.6 gives a simplified idea on the geometry of a square lattice model reflecting an octahedral coordination.

The postulate of steady state during dissolution reaction (Table 5.1) implies a continuous reconstitution of the surface with the maintenance of a constant distribution of the various surface sites and the steady state concentration of the surface complexes. Fig. 5.7 presents experimental evidence that the concentration of the surface ligand – in line with Fig. 5.5a – remains constant during the surface controlled dissolution reaction.

5.4 The Proton-Promoted Dissolution Reaction

The dissolution reaction under acid conditions requires protons, which may become bound to the surface oxide ions and weaken critical bonds; thus, detachment of the metal species into the solution results. Another part of the consumed protons replaces the metal ions, leaving the solid surface and thus maintaining the charge balance.

A scheme for the dissolution reaction of a trivalent oxide is given in Fig. 5.8. Al-

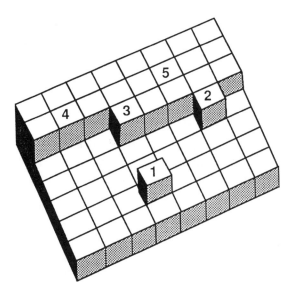

Figure 5.6

The geometry of a square lattice surface model

The five different surface sites are:

    1 = adatom   3 = kink

    2 = ledge   4 = step   5 = face

The five types of octahedral surface complexes are bonded to 1, 2, 3, 4, and 5 neighboring surface links (ligands). From the point of view of surface reactivity (e.g., dissolution rate), obviously the various surface sites have different activation energies, the adatom site (1) is most reactive, and the face site (5) (linked to five neighboring sites) is least reactive. The overall dissolution rate is based on the parallel dissolution reactions of all sites, but the overall dissolution kinetics is dictated by the fastest individual reaction rate. The latter is essentially given by the product of the first-order reaction rate specific for each type of site and the relative concentration of surface sites of each category. Monte Carlo methods, where individual activation energies were assigned to the distinct sites, were able to show that a steady state distribution of the various surface sites can be maintained during the dissolution and that one type of surface site essentially accounts for the overall dissolution rate. The Monte Carlo model (Wehrli, 1989) suggests that the kink sites (3), although reacting much slower than ledge and adatom sites but being present at much higher relative concentrations than the less linked surface sites, control the overall dissolution rate.

though this representation cannot account for individual crystallographic structures, it attempts to illustrate a typical sequence of the reaction steps which occur on the surface. The adsorption of protons at the surface is very fast (Hachiya et al., 1984); thus, surface protonation is faster than the detachment of the metal species, so it can be assumed that the concentration of protons at the surface is in equilibrium with the solution. Surface protonation may be assumed to occur at random. But the protons may move fast from one functional group to another and occupy terminal hydroxyl as well as bridging oxo or hydroxo groups. Tautomeric equilibria may be assumed. The detachment process (step 4 in Fig. 5.8) is (far from equilibrium) the slowest of the consecutive steps. If the steady state conditions are maintained, the following equation describes the overall reaction rate:

$$R_H = k_4 \langle D \rangle \tag{5.10}$$

Under steady state conditions, i.e., if the original surface sites are regenerated completely after the detachment step (Table 5.1) and if it is assumed that surface protonation equilibria are retained and kept constant by controlling the solution pH, one may write

$$d\langle A \rangle /dt = d\langle B \rangle /dt = d\langle C \rangle /dt = d\langle D \rangle /dt = 0 \tag{5.11}$$

If the fraction of surface that is covered with protons is smaller than 1 ($\chi_H \ll 1$), the surface density of singly, doubly, and triply protonated surface sites (B, C, and D, respectively (see Fig. 5.6)) can be described as probability functions of the surface protonation C_H^s.

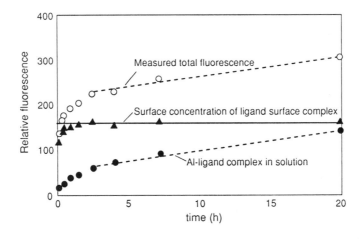

Figure 5.7

Ligand promoted dissolution of aluminum oxide

The ligand is hydroxyquinoline-sulfonate (HQS) which forms fluorescent Al-HQS complexes in solution. HQS forms surface complexes with the Al-centers of aluminum oxide surface; these surface complexes are also fluorescent. Fluorescence as a function of time during HQS-promoted dissolution of aluminum oxide. Surface-associated fluorescence was calculated from the difference between measured total and dissolved fluorescence.

The increase in total and dissolved fluorescence over time results from dissolution of the oxide. The rate of increase, i.e., the slope of the dashed lines, is proportional to the dissolution rate.

Conditions: 0.14 g/ℓ oxide, pH 7.5, [HQS]$_T$ = 500 µM added at t = 0.

(From Hering and Stumm, 1991)

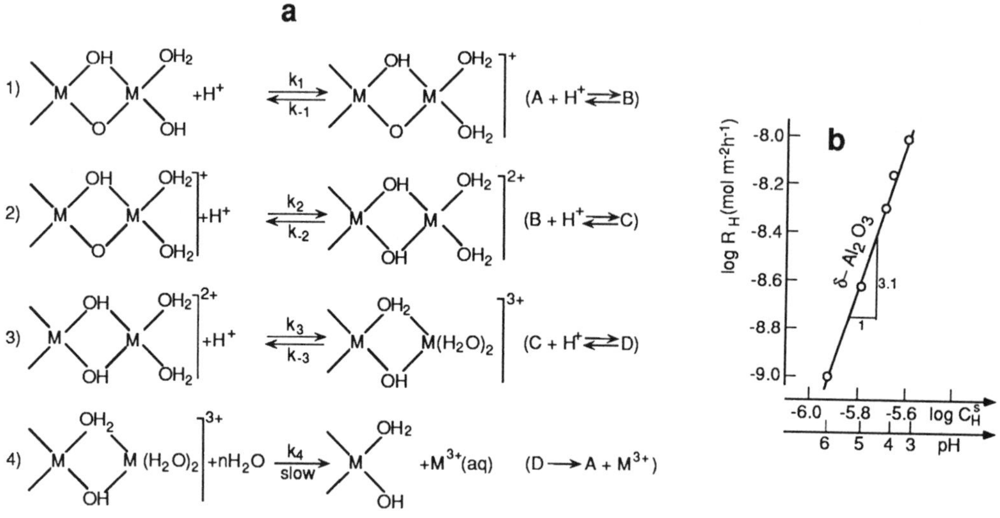

Figure 5.8

a) Schematic representation of the proton-promoted dissolution process at a M_2O_3 surface site. Three preceding fast protonation steps are followed by a slow detachment of the metal from the lattice surface.

b) The reaction rate derived from individual experiments with δ-Al_2O_3 (see Fig. 5.2a) is proportional to the surface protonation to the third power

(From Furrer and Stumm, 1986)

A weakening of the critical metal-oxygen bonds occurs as a consequence of the protonation of the oxide ions neighboring a surface metal center and imparting charge to the surface of the mineral lattice. The concentration (activity) of D should reflect that three of such oxide or hydroxide ions have to be protonated. If there is a certain numer of surface-adsorbed (bound) protons whose concentration C_H^s (mol m^{-2}) is much lower than the density of surface sites, S (mol^{-2}), the probability of finding a metal center surrounded with three protonated oxide or hydroxide ions is proportional to $(C_H^s/S)^3$. Thus, as has been derived from lattice statistics by Wieland et al. (1988), the activity of D is related to $(C_H^s)^3$, and the rate of proton-promoted dissolution, R_H (mol m^{-2} h^{-1}), is proportional to the third power of the surface protonation:

$$R_H \propto \langle D \rangle \propto \left(C_H^s\right)^3$$

that is,

$$R_H = k_H' \, (\equiv MeOH_2^+)^3 \tag{5.12a}$$

For another oxide, for which the dissolution mechanism requires only two preceding protonation steps, the rate would be proportional to $(C_H^s)^2$. Generally, for all oxides, which dissolve by an acid-promoted surface controlled process, the following rate equation may be postulated (see Fig. 5.9)

$$R_H = k\ (\equiv MeOH_2^+)^j \qquad ; R_H = k_H'\ (C_H^s)^j \qquad (5.12b)$$

where j is an integer if dissolution occurs by one mechanism only. In simple cases j corresponds to the charge of the central ion. If more than one mechanism occurs simultaneously, the exponent j will not be an integer.

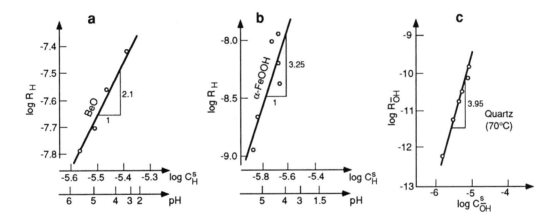

Figure 5.9

The dependence of the rate of dissolution, R_H or R_{OH} (mol m^{-2} h^{-1}), of
a) BeO (Furrer and Stumm, 1986),
b) α-FeOOH (Zinder et al., 1986),
c) and SiO$_2$ (quartz) (Guy and Schott, 1989; Knauss and Wolery, 1989) on the surface concentration of protons $(C_H^s)^3$, or hydroxide ions, C_{OH}^s (mol m^{-2}).

Enhancement of Dissolution by Deprotonation

The dissolution rate of most oxides increases both with increasing surface protonation and with decreasing surface protonation, equivalent to the binding of OH$^-$ ligands; thus, in the alkaline range the dissolution rate increases with increasing pH (Chou and Wollast, 1984; Schott, 1990; Brady and Walther, 1990); (see Fig. 5.9c).

$$R_{OH} = k_{OH}'\ (C_{OH}^s)^i \qquad (5.13)$$

Experimentally accessible Parameters

Surface protonation and deprotonation are experimentally directly accessible from alkalimetric or acidimetric surface tritrations. The surface concentrations $(\equiv MOH_2^+)$ or $(\equiv MO^-)$ are nonlinearly related to H^+ by surface complex formation equilibria or by semi-empirical relations; in other words,

$$(\equiv MOH_2^+) \propto [H^+]^n \qquad (\equiv MO^-) \propto [H^+]^{-m} \qquad (5.14)$$

Many authors have shown the empirical rate law

$$R_H = k_H [H^+]^n \qquad (5.15a)$$

where n is typically between 0 and 0.5. These observations are relevant for the assessment of the impact of acid rain on weathering rates. As Eq. (5.15) suggests, the rate of weathering does not increase linearly with the acidity of water. A tenfold increase in $[H^+]$ leads to an increase in the dissolution rate by a factor of about 2 – 3. Similarly, in the alkaline range the empirical rate law holds:

$$R_{OH} = k_{OH} [H^+]^{-m} \qquad (5.15b)$$

The Overall Rate of Dissolution

The surface charge of the mineral is an important factor in the polarization of the lattice bonds on the surface. Thus, one may generalize that the dissolution rate is related to the surface charge imparted to the surface by H^+ and/or OH^-; the rate increases both with increasing positive surface charge with decreasing pH values of the solution and with increasing negative surface charge with increasing pH values. The minimum dissolution rates are observed at the pH_{pzc}.

The overall rate of dissolution is given by

$$R = k_H' \left(C_H^s\right)^j + k_{OH}' \left(C_{OH}^s\right)^i + k_L' \left(C_L^s\right) + k_{H_2O}' \qquad (5.16)$$

the sum of the individual reaction rates, assuming that the dissolution occurs in parallel at different metal centers (Furrer and Stumm, 1986). The last term in Eq. (5.16) is due to the effect of hydration and reflects the pH-independent portion of the dissolution rate.

For the acidic range (pH < pH_{pzc}) we can propose tentatively – on the basis of the relationship for surface charge due to surface protonation given in Fig. 3.4 – (see Eqs. 3.10 and 5.12b) a general semiempirical rate law:

$$\log R_H = \log k_H' + j(\log K_F + m \, \Delta pH) \qquad (5.17a)$$

or

$$\log R_H = \log k_H' + j(0.14 \Delta pH - 6.5) \qquad (5.17b)$$

where $\Delta pH = pH_{pzc} - pH$; and R_H = rate of proton promoted dissolution [e.g., mol $m^{-2}\ h^{-1}$], j = an integer corresponding in ideal cases to the valency of the central ion of the dissolving oxide, $K_F = 10^{-6.5}$ (the Freundlich adsorption constant according to Fig. 3.4) and $m = 0.14$ (corresponding to the Freundlich slope in Fig. 3.4), and k_H' the rate constant.

But factors other than the surface charge can become important such as the effects of specific adsorption of cations and anions on the degree of surface protonation (see Example 5.1).

5.5 Some Case Studies

Most oxides show, in accordance with the general rate equation (5.16) the same trend with regard to the rate dependence on pH: a decrease in pH in the acid range and an increase in pH in the alkaline range.

Silica. Although the dissolution of $SiO_2(s)$:

$$SiO_2(s) + 2\,H_2O = H_4SiO_4(aq) \qquad (5.18)$$

does not involve H^+ or OH^- ions, the rate of solution is dependent on pH as shown in Fig. 5.10a).

The pH_{pznpc} is around pH $2-3$. As Fig. 5.10 illustrates, both the positive surface charge due to bound proton, and the negative surface charge, due to deprotonation (equivalent to bound OH^-) enhances the dissolution rate. The some kind of pH dependence is observed also with silicates (Fig. 5.10b). In Fig. 5.10b the rate data for some silicates (25° C) show a dependence of log rate on pH (between pH = 8 and pH = 12) of 0.3. (A similar value is observed for other silicates.) As has been pointed out by Brady and Walther (1989), the ubiquitous 0.3 slope at basic pH reflects the fact that Si–O bonds are the likely sites of precursors for the rate determining step, at least above pH 8. Furthermore, it appears that the silica contribution dominates overall surface charge. Mechanistically, according to Brady and Walther, this implies that SiO bonds in silicates are polarized and weakened by the presence of the charged surface species $\equiv$Si–O$^-$. This ultimately leads to the detachment of the silicon atom. That detachment of silica leads to stoichiometric dissolution at high pH can be explained using albite as an example. As protons are exchanged for surface alkali cations during the first hour of albite dissolution, the release of alkalis is not considered rate-limiting (Brady and Walther, 1989).

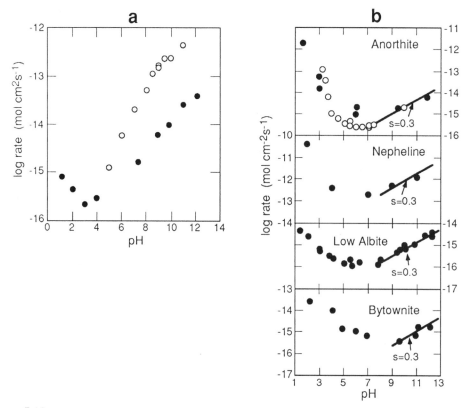

Figure 5.10

pH-dependent dissolution of silica and silicates
a) Rate of dissolution of silica as a function of pH:
   ○, vitreous silica in 0.7 M NaCl solution (Wirth and Gieskes, 1979);
   ●, Quartz in 0.2 M NaCl solution (Wollast and Chou, 1985).
b) 25° C dissolution rates of anorthite (Squares: Amrhein and Suarez,1988; Hexagons: Holdren and Speyer, 1987; Circles: Brady and Walther, in prep.), nepheline (Tole et al., 1986), low albite (Chou and Wollast, 1984), bytownite (Brady and Walther, in prep.). S denotes slope. (Modified from Brady and Walther, 1989.)

The dissolution of quartz is accelerated by bi- or multidentate ligands such as oxalate or citrate at neutral pH-values. The effect is due to surface complex formation of these ligands to the SiO_2-surface (Bennett, 1991). In the higher pH-range the dissolution of quartz is increased by alkali cations (Bennett, 1991). Most likely these cations can form inner-spheric complexes with the $\equiv SiO^-$ groups. Such a complex formation is accompanied by a deprotonation of the oxygen atoms in the surface lattice (see Examples 2.4 and 5.1). This increase in C_{OH}^s leads to an increase in dissolution rate (see Fig. 5.9c).

HCO_3^- enhances the Dissolution Rate of Hematite. Fe(III) in natural waters is present as hydroxo complexes, especially $Fe(OH)_2^+$, $Fe(OH)_3(aq)$, $Fe(OH)_4^-$. In addition a carbonato complex – $Fe(CO_3)_2^-$ – is present in seawater and at the surface of solid iron(III)(hydr)oxides. Fig. 5.11 shows the dependence of the dissolution rate as a function of the hydrogen carbonato surface complex

$$\text{Rate} \propto \langle \equiv Fe_2O_3 - HCO_3^- \rangle \tag{5.19}$$

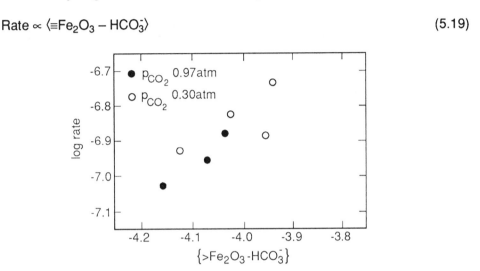

Figure 5.11

Dependence of the dissolution rate of hematite, α-Fe_2O_3 (mol m^{-2} h^{-1}) on the surface complex of the HCO_3^--Fe(III) complex (mol m^{-2}).

(From Bruno, Stumm, Wersin and Brandberg, 1991)

UO_2. Bruno et al. (1991) have studied the kinetics of dissolution of uranium dioxide, $UO_2(s)$, under strongly reducing conditions. The average uranium concentration in natural waters under reducing conditions are between 10^{-7} and 10^{-8} M. Even under reducing conditions O/U ratios are often larger than 2. The solubility of UO_{2+x} solid phases is normally higher than that of pure UO_2 under reducing conditions. As pointed out by Bruno et al., (1991), this indicates that under these conditions, uranium is initially dissolved from the primary sources and then it is reprecipitated as UO_2. This possible mobilization mechanism is relevant in the case of the disposal of UO_2 spent nuclear fuel.

The dissolution of pure UO_2 is surface controlled; it is proton promoted up to pH $\cong$ 7 (pH$_{pzc}$ = 6.8 $\pm$ 0.3) (see Fig. 5.12). In line with Eq. (5.12b) and (5.17) the dissolution rate has been found to be proportional to the activity of the protonated surface species according to the expression

$$R_H = k_H' \langle \equiv UOH_2^+ \rangle^4 \tag{5.20}$$

confirming the ideal correlation found between the oxidation state of the central ion and the order with respect to the activity of the protonated surface complex.

As the data in Fig. 5.12 show, and as was pointed out by Bruno et al. (1991), the half time for the dissolution reaction of UO_2 in the pH-range of most natural waters and under reducing conditions is in the order of days. If we compare this with typical residence times of undisturbed ground waters (years) we can conclude that the dissolution of $UO_2(s)$ and the mobility of uranium under these conditions is thermodynamically and not kinetically controlled.

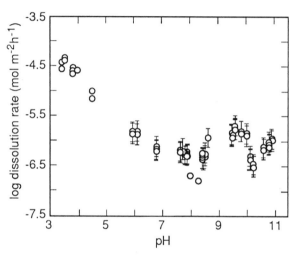

Figure 5.12

Rates of dissolution of $UO_2(s)$ obtained in a thin layer continuous flow-through reactor as a function of pH.

(From Bruno, Casas and Puigdomènech, 1991)

Dissolution Rates of various Al-Minerals and Fe-Minerals

Table 5.2 compares the dissolution rate of various Al-minerals. The differences are remarkable. At pH = 3, the half life of surface sites of different aluminum (hydr)-oxides varies from 2 years (corundum) to 20 hours (bayerite). The large difference in rates must be due to different coordinative arrangements of the active surface groups. Although no detailed theory is available, it is perhaps reasonable to assume, that the dissolution rate increases with the frequency of surface groups which be present as endstanding $\equiv$Al-OH groups.

Similar considerations apply to the allotropic forms of other minerals. For example acid oxalate (pH = 3) extraction (in absence of light) is used to distinguish between ferrihydrite and hematite and goethite. Within 2 hours the ferrihydrate is dissolved while leaving goethite and hematite essentially undissolved (Schwertmann and Cornell, 1991).

Table 5.2   Dissolution rates of various Al-minerals

Mineral	Rate pH = 3 (25° C) R_H [mol m^{-2} s^{-1}]	k [s^{-1}] [a]	$t_{1/2}$ [days] [b]
Corundum	2.0×10^{-13}	1.2×10^{-8}	669
Muscovite	3.9×10^{-13}	2.79×10^{-8}	288
Kaolinite	8.3×10^{-13}	5.93×10^{-8}	135
δ-Al$_2$O$_3$	3.0×10^{-12}	1.83×10^{-7}	44
Bayerite	1.5×10^{-10}	9.16×10^{-6}	0.88

[a] R_H divided by the density of surface sites (1.4×10^{-5} mol m^{-2} for muscovite and kaolinite and 1.64×10^{-5} mol m^{-2} for the other minerals; k [s^{-1}] reflects the mean reactivity of surface sites
[b] The half life of a surface site
(Modified from Wehrli, Wieland and Furrer, 1990)

5.6 Case Study: Kaolinite

The weathering of silicates has been investigated extensively in recent decades. It is more difficult to characterize the surface chemistry of crystalline mixed oxides. Furthermore, in many instances the dissolution of a silicate mineral is incipiently incongruent. This initial incongruent dissolution step is often followed by a congruent dissolution controlled surface reaction. The rate dependence of albite and olivine illustrates the typical enhancement of the dissolution rate by surface protonation and surface deprotonation. A zero order dependence on [H$^+$] has often been reported near the pH$_{pzc}$; this is generally interpreted in terms of a hydration reaction of the surface (last term in Eq. 5.16).

The dissolution of clay minerals and micas is very slow under natural conditions. Hence they are characteristic secondary weathering products occurring in soils and sediments. Here we exemplify the dissolution characteristics of *kaolinite*. Kaolinite is a 1:1 phyllosilicate. As Fig. 3.9 illustrates, the fundamental unit of its structure is an extended sheet of two basal layers: a silica-type layer (siloxane) of composition $(Si_4O_{10})^{4-}$ and a gibbsite type layer of composition $(OH)_6Al_4(OH)_2O_4$ and an edge that consists of the two constituents Al(OH)$_3$ and SiO$_2$. The siloxane layer may contain some isomorphic substitution of Si by Al. Thus, (Si–O–Al)-groups establish a permanent structural charge which in turn may be responsible for some ion exchange reactions with Na$^+$ and Al^{3+}.

a

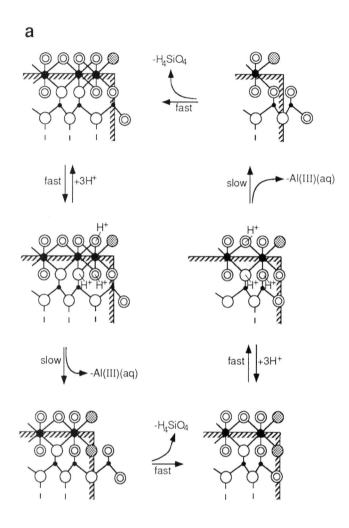

The sequential surface protonation of the kaolinite surface was illustrated in Fig. 3.11. As was explained in Chapter 3.6, the excess proton density can be interpreted as a successive protonation at the edge and of the gibbsite surface. The pH_{pzc} of the edge surface is about 7.5.

Fig. 5.13 displays the dissolution scenario of the gibbsite surface and of the edge surface. The dissolution reaction can be interpreted as a coupled release of Al and Si. The detachment of the Al center is the rate determining step. In a fast subsequent step Si is released from the same surface site. The Al(III) : H^+ stoichiometry of the precursor group (the group to be detached) is 1:3 at the gibbsite surface and 1:1 at the edge surface.

b

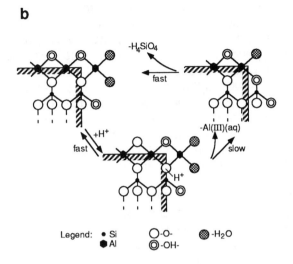

Legend: • Si   ◯-O-   ◉-H₂O
        ● Al   ◎-OH-

c

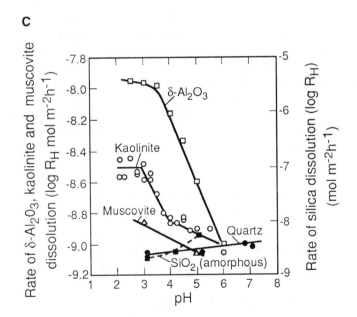

Figure 5.13

Dissolution of kaolinite; scenario for the gibbsite surface a) and the edge surface b). On the gibbsite surface three protons are needed to establish a detachable Al(III) group. In a subsequent fast step, Si is released. On the edge surface one proton is needed to cause the detachment of Al(III) and of Si. (From Wieland and Stumm, 1991.) c) The pH dependence of the proton-promoted dissolution rates of kaolinite, muscovite, and their constituent oxides of Al_2O_3 and amorphous SiO_2 or quartz, respectively. With increasing H^+ activity, the rate of Al detachment is promoted. (From Stumm and Wieland, 1990.)

Fig. 5.13 compares the proton promoted dissolution rate of kaolinite and muscovite with δ-Al$_2$O$_3$ and amorphous SiO$_2$ or quartz. These dissolution rates were obtained in batch dissolution experiments from zero order plots of Al(III)(aq) vs time and of Si(aq) vs time. The pH-dependence of the dissolution rate is the consequence of the sequential protonation of OH groups at the kaolinite surface. In the pH range above pH $\approx$ 4 the $\equiv$Al–OH groups at the edge surface are protonated (cf. Fig. 3.11) and the detachment of Al and Si from the edge face dominates the overall dissolution process. The protonation of the Al–OH–Al groups of the gibbiste surface predominates below pH = 5 and results in a steep increase with decreasing pH of the overall dissolution rate at pH < 4. Due to saturation of the gibbsite layer with protons the rates tend to reach a constant value at pH < 3.

In line with the stoichiometry Al : H = 1:1 at the edge surface and Al : H = 1:3 at the gibbsite surface and in accordance with Eqs. (5.12b) and (5.17b) the proton promoted rate law indicates a first order reaction (j = 1) with respect to protons bound to the edge surface and a third order dependence (j = 3) with respect to the mol fraction of protons bound to the gibbsite surface.

The dissolution in its initial phase is not fully congruent in the sense that the release of Si(aq) is slightly larger than that of Al(III)(aq), presumably because some of the Al(III) released becomes bound to the permanently charged Si–O–Al sites of the siloxane layer. The release of Si represents the "time" rate of kaolinite dissolution. Complex formers like oxalate or salicylate that form strong surface chelates with the Al(III) central ions (but not – at the concentration used – to any measurable extent on the Si-centers) enhance the dissolution rate significantly; furthermore, under these ligand promoted dissolution conditions, the dissolution reaction is fully congruent. This is another indirect evidence that the detachment of an Al-center is the rate determining elementary step of the dissolution reaction.

Example 5.1: Change in Surface Protonation as a Consequence of Metal Ion or Ligand Adsorption

In Fig. 3.5 we illustrated generally that an alkalimetric or acidimetric titration curve of a hydrous oxide dispersion becomes displaced by the adsorption of a metal ion or, – in opposite direction – by the adsorption of an anion (ligand).

$$\equiv\text{S-OH} + \text{Me}^{2+} \rightleftharpoons \equiv\text{S-OMe}^+ + \text{H}^+ \tag{i}$$

$$\equiv\text{S-OH} + \text{U}^{2-} \rightleftharpoons \equiv\text{SU}^- + \text{OH}^- \tag{ii}$$

In Figs. 2.10a and b experimental data on the shift in the titration curves were given and in Example 2.4 it was shown how these effects are quantified, and that the extent of adsorption can be determined from the displacement of the titration curve.

In this example we will calculate
1) the effect of ligand adsorption on surface protonation; and
2) the effect of a metal ion adsorption on surface deprotonation.
We choose a dilute hematite suspension.

1) *Effect of Ligand.* The hematite is characterized by the acidity constants

$$\equiv FeOH_2^+ = \equiv FeOH + H^+ \quad ; \; \log K_1^s = -7.25 \tag{iii}$$

$$\equiv FeOH = \equiv FeO^- + H^+ \quad ; \; \log K_2^s = -9.75 \tag{iv}$$

and by the following information:

$s = 4 \times 10^{-4}$ m² kg⁻¹; 3.2×10^{-1} mol surface sites per kg; 8.6×10^{-6} kg hematite is used per liter of solution. The ionic strength $I = 5 \times 10^{-3}$ M. The diffuse double layer model is used for the correction of electrostatic interaction.

The ligand, U^{2-}, is characterized by the following constants:

$$H_2U = H^+ + HU^- \quad ; \; \log K_1 = -5.0 \tag{v}$$

$$HU^- = H^+ + U^{2-} \quad ; \; \log K_2 = -9.0 \tag{vi}$$

and interacts with the Fe_2O_3 surface according to the following complex formation constant:

$$\equiv Fe\text{-}OH + H_2U = \equiv FeU^- + H^+ + H_2O \quad ; \; \log K^s = 2 \tag{vii}$$

The constants given for reactions (v), (vi) and (vii) are representative of those of a ligand containing a carboxylic acid and a phenolic hydroxo group.

With the equilibria defined in (iii) to (vii) and the mass balance for

$$[H_2U_T] = [H_2U] + [Hu^-] + [U^{2-}] + [\equiv FeU^-] \tag{viii}$$

and for

$$[\equiv Fe\text{-}OHTOT] = [\equiv FeOH_2^+] + [\equiv FeOH] + [FeO^-] + [\equiv FeU^-] \tag{ix}$$

we have 7 equations and calculate all solute and surface species present at a given pH.

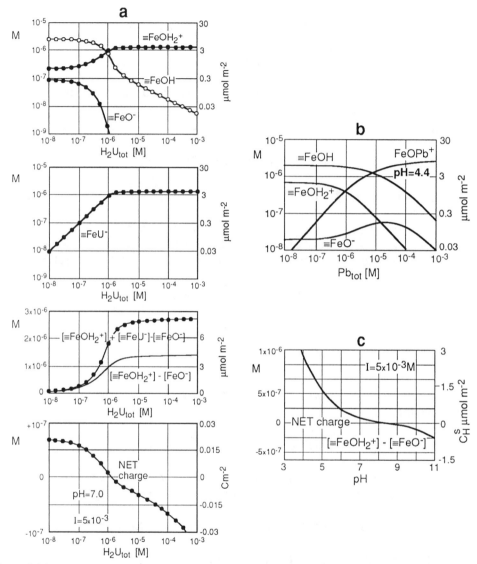

Figure 5.14

Effect of ligands and metal ions on surface protonation of a hydrous oxide. Specific Adsorption of cations and anions is accompanied by a displacement of alkalimetric and acidimetric titration curve (see Figs. 2.10 and 3.5). This reflects a change in surface protonation as a consequence of adsorption. This is illustrated by two examples:

a) Binding of a ligand at pH = 7 to hematite which increases surface protonation
b) Adsorption of Pb^{2+} to hematite at pH = 4.4 which reduces surface protonation
c) Surface protonation of hematite alone as a function of pH (for comparison). All data were calculated with the information given in Example 5.1. $I = 5 \times 10^{-3}$ M.

Points given are calculated.

The results are given for pH = 7 in Fig. 5.14a for various total concentrations of H_2U added to the system. Surface complex formation progressively increases with $[H_2U_T]$ and reaches saturation at concentrations above 10^{-6} M.

Two consequences are important:
1. the net surface charge of the hydrous oxide decreases (it becomes negative when the total ligand added exceeds 10^{-6} M);
2. the surface protonation increases.

Expressed as C_H^s [mol m^{-2}], $\equiv FeOH_2^+$ increases from ca. 6×10^{-7} mol m^{-2} to ca. 4×10^{-6} mol m^{-2}. The latter proton density corresponds to that of a hematite suspension at pH = 3.5 (in absence of H_2U_T).

2) *Effect of Metal Ion.* To the hematite suspension already characterized one adds Pb(II). The surface complex formation is characterized by

$$\equiv FeOH + Pb^{2+} = \equiv FeOPb^+ + H^+ \qquad ; \log K^s = 4.7 \qquad (x)$$

The calculations are similar and the result is displayed in Fig. 5.14b for pH = 4.4. Obviously Pb-adsorption is accompanied by an increase in net charge and a marked decrease in surface protonation $[\equiv FeOH_2^+]$. Plausibly, this reduction in C_H^s can decrease the dissolution rate.

5.7 Experimental Apparatus

Various devices can be used to determine the kinetics and rates of chemical weathering. In addition to the batch pH-stats, flow through columns, fluidized bed reactors and recirculating columns have been used (Schnoor, 1990). Fig. 5.15a illustrates the fluidized bed reactor pioneered by Chou and Wollast (1984) and further developed by Mast and Drever (1987). The principle is to achieve a steady state solute concentration in the reactor (unlike the batch pH-stat, where solute concentrations gradually build up). Recycle is necessary to achieve the flow rate to suspend the bed and to allow solute concentrations to build to a steady state. With the fluidized bed apparatus, Chou and Wollast (1984) could control the Al(III) concentration (which can inhibit the dissolution rate) to a low level at steady state by withdrawing sample at a high rate.

Fig. 5.15b shows a thin-film continuous flow reactor used by Bruno et al. (1991) for determining the dissolution rate of UO_2 under reducing conditions. A known weight of $UO_2(s)$ was enclosed into the reactor between two membrane filters (0.22 μm). The reducing conditions of the feed solution were obtained by bubbling $H_2(g)$ in the presence of a palladium catalyst. The dissolution rates determined using continu-

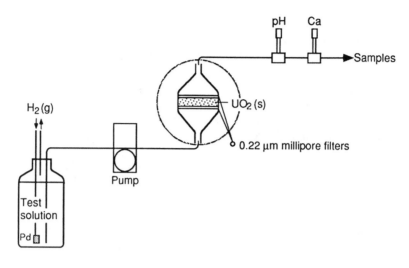

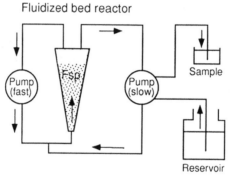

Figure 5.15

Continuous flow-type reactors to measure dissolution rates

a) experimental scheme of the thin-film continuous flow reactor used for example by Bruno, Casas and Puigdomènech (1991) to determine dissolution rate of UO_2 under reducing conditions;
b) schematic diagram of the fluidized-bed reactor by Chou and Wollast (1984), and developed further by Mast and Drever (1987).

ous flow reactors are based on the U(IV)-concentration of the effluent at steady state. The amount of U(IV) dissolved depends on the reaction time which is related to the residence time of the test solution in the reactor given by

$$t = V/Q$$

where V is the volume of solution in contact with the solid phase, and Q is the flow rate. The dissolution rate values are calculated using the equation

$$r_{diss} = Q\,(\ell\,s^{-1}) \times [U(IV)]\,(M) = mol\;s^{-1}$$

where $[U(IV)]$ is the concentration of U of the effluent. The values can then be normalized with respect to the total surface area of the solid phase.

5.8 Incongruent Dissolution as a Transient to Congruent Dissolution

As was mentioned in the introduction to this chapter "diffusion-controlled dissolution" may occur because a *thin layer* either in the liquid film surrounding the mineral or on the surface of the solid phase (that is depleted in certain cations) limits transport; as a consequence of this, the dissolution reaction becomes incongruent (i.e., the constituents released are characterized by stoichiometric relations different from those of the mineral. The objective of this section is to illustrate briefly, that even if the dissolution reaction of a mineral is initially incongruent, it is often a surface reaction which will *eventually* control the overall dissolution rate of this mineral. This has been shown by Chou and Wollast (1984). On the basis of these arguments we may conclude that in natural environments, the steady-state surface-controlled dissolution step is the main process controlling the weathering of most oxides and silicates.

Initial Incongruent Dissolution. We essentially follow the explanation given by Schnoor (1990) for a representative example of an initial incongruent dissolution.

Fig. 5.16 illustrates the schematic building of a cation depleted layer of thickness y from a hypothetical mineral with constituents A and B (stoichiometry 1:1). Initially incongruent dissolution of AB results in the rapid migration (diffusion) of constituent B from the core of the mineral grain through a layer that is depleted in B (Eq. 5.21)

$$\frac{dM_B}{dt} = -D_{BA}\,S\,\frac{dB_{AB}}{dr} \approx -D_{BA}\,S\,\frac{(B_{AB} - B_{surf})}{y} \tag{5.21}$$

$$\frac{dM_A}{dt} = kS\,(\equiv Me-H^+)^j \tag{5.22}$$

Eq. (5.22) results from the rate determining step of the proton-promoted detachment of a metal ion (Eq. 5.9). In Eqs. (5.21) and (5.22), M_B and M_A are, respectively the mass of B and A dissolving [mol]. D_{BA} is the diffusion coefficient [$m^2\;s^{-1}$] of B diffusing through A; S is the surface area [m^2]; B_{AB} is the concentration of ion B in solid AB [$mol\;m^{-3}$]; B_{surf} is the concentration of ion B at the surface of solid AB [mol^{-3}]; y is the thickness [m] of the cation depleted layer; k is the rate constant, j = exponent, corresponding, in ideal cases, to the valence of the ion; r is the radial distance [m] and $(\equiv Me-H^+)$ is the surface concentration of protons [$mol\;m^{-2}$].

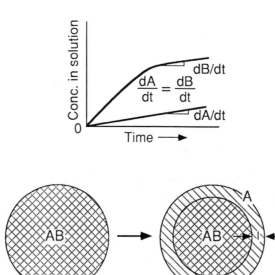

Figure 5.16

Schematic illustration of initial incongruent dissolution. In the initial stages, a mineral grain may develop a cation depleted layer ℓ, but eventually congruent dissolution is observed and the rate of dissolution of A (dA/dt) must equal B (dB/dt) for a 1:1 stoichiometry.

(From Schnoor, 1990)

The concentration of constituent B becomes negligible at the surface of the mineral grain. Gradually, the rate of mass diffusion of B (Eq. 5.21) through an increasing depleted layer (y) becomes slower and is equal to the rate of surface-controlled dissolution of A (Eq. 5.22). Thus, a pseudosteady state is attained and the depleted layer thickness stabilizes. The rates of reaction of solid layer diffusion (Eq. 5.21) and of surface controlled dissolution become equal:

$$D_{AB} \, S \, (B_{AB}/y) \; = \; kS \, \langle \equiv Me-H^+ \rangle^j \tag{5.23}$$

and at steady state

$$y \; = \; \frac{D_{AB} \, B_{AB}}{k \, \langle \equiv Me-H^+ \rangle^j} \tag{5.24}$$

The thickness, y, of the cation depleted layer at steady state with decrease with increasing [H⁺] in the solution. Therefore congruent dissolution will result immediately from low pH solution.

Congruent surface-controlled dissolution follows after the initial incongruent period.

5.9 Weathering and the Environmental Proton Balance

The classical geochemical material balance (Goldschmidt, 1933) assumes that the H^+ balance in our environment has been established globally by the interaction of primary (igneous) rocks with volatile substances:

$$\begin{matrix} \text{igneous} \\ \text{rocks} \end{matrix} + \begin{matrix} \text{volatile} \\ \text{substances} \end{matrix} \rightleftharpoons \text{seawater} + \text{atmosphere} + \text{sediments} \quad (5.25)$$

silicates	CO_2	pH $= 8$	$p_{O_2} = 0.2$	carbonates
carbonates	H_2O	pε $= 12$	$p_{CO_2} = 0.0003$	silicates
oxides	SO_2			
	HCl			

In an oversimplified way one can say that acids of the volcanoes have reacted with the bases of the rocks; the composition of the ocean (which is at the first endpoint (pH = 8) of the titration of a strong acid with a carbonate) and the atmosphere (which with its $p_{CO_2} = 10^{-3.5}$ atm is nearly in equilibrium with the ocean) reflect the proton balance of reaction (5.25). Oxidation and reduction are accompanied by proton release and proton consumption, respectively. (In order to maintain charge balance, the production of e^- will eventually be balanced by the production of H^+.) Furthermore, the dissolution of rocks and the precipitation of minerals are accompanied by H^+-consumption and H^+-release, respectively. Thus, as shown by Broecker (1971), the pε and pH of the surface of our global environment reflect the levels where the oxidation states and the H^+ ion reservoirs of the weathering sources equal those of the sedimentary products.

Weathering is a major H^+ consuming process and pH-buffering mechanism, not only globally and regionally but it also plays a major role in local watersheds in soil processes, in nutrient uptake by plants and in epidiagenetic reactions in sediments.

Acid atmospheric deposition causes acidification of waters and soils if the neutralization of the acids by weathering is too slow. In order to understand pH regulation one needs to consider the major H^+ yielding and H^+ consuming processes. Unifying definitions of acid and base neutralizing capacity (ANC, BNC or alkalinity and acidity) are helpful (see Appendix at the end of this chapter). Biologically mediated redox processes are important in affecting the H^+-balance. Among the redox processes that have a major impact on H^+ production and consumption are the synthesis and mineralization of biomass. Any uncoupling between photosynthesis and respiration affects acidity and alkalinity in terrestrial and aquatic ecosystems.

As Fig. 5.17 illustrates aggrading vegetation (forests and intensive crop production) produces acidity (see Eq. (viii) of the Appendix) since more cations are taken up by the plants (trees) H^+ is released through the roots. The protons released react with

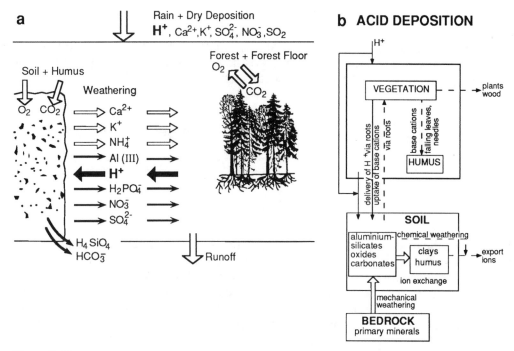

Figure 5.17

Effect of plants (trees) upon H^+-balance in soils

a) Competition between H^+ release by the roots of growing trees (aggrading forest) and H^+ ion consumption by chemical weathering of soil minerals. The delicate proton balance can be disturbed by acid deposition.

b) Processes affecting the acid-neutralizing capacity of soils (including the exchangeable bases, cation exchange and mineral bases). H^+ ions from acid precipitation and from the release by the roots react by weathering carbonates, aluminum silicates and oxides and by surface complexation and ion exchange on clays and humus. Mechanical weathering resupplies weatherable minerals. Lines drawn out indicate flux of protons; dashed lines show flux of base cations (alkalinity). The trees (plants) act like a base pump.

(From Schnoor and Stumm, 1986)

the weatherable minerals to produce some of the cations needed by the plants. The H^+ balance in soils (production through the roots vs consumption by weathering) is delicate and can be disturbed by acid deposition.

If the weathering rate equals or exceeds the rate of H^+ release by the biota, such as would be the case in a calcareous soil, the soil will maintain a buffer in base cations and residual alkalinity. On the other hand, in noncalcareous "acid" soils, the rate of H^+ release by the biomass may exceed the rate of H^+ consumption by weathering and cause a progressive acidification of the soil. In some instances, the acidic atmospheric deposition may be sufficient to disturb an existing H^+ balance between aggrading vegetation and weathering reactions. A quantitative example is provided by the carefully established acidity budget in the Hubbard Brook Eco-system, (Driscoll and Likens, 1982). Humus and peat can likewise become very acid and deliver some humic or fulvic acids to the water. In wetlands certain plants such as the bog moss sphagnum avidly taking up base cations and releasing H^+ ions act like cation exchangers.

The acidification of the environment is accompanied by an increase in base in the aggrading biomass (wood and vegetation). The ash of wood and of vegetation is alkaline; one speaks of "potash". Fig. 5.17b illustrates that forest trees act like a base pump. The "export" of alkalinity occurs by the harvesting of the wood and the release of base cations to the aqueous runoff; this is balanced by the production of base cations in the weathering process of minerals. Ion exchange processes of soil minerals and of humus represents an intermediate buffering pool of fast exchange-able bases. The large pool of mineral bases is characterized by relatively slow kinetics of chemical weathering. If chemical weathering did not replace exchange-able bases in acid soils of temporal regions receiving acidic deposition, the base exchange capacity of the soils would be completely diminished over a period of 50 – 100 years.

5.10 *Comparison between Laboratory and Field Weathering Results*

So far we have discussed weathering rates and rate laws from laboratory experi-ments on pure minerals. These laboratory studies are meant to provide insight for natural systems (rates and variables that affect these rates). We may first try to com-pare laboratory and field results.

Chemical weathering introduces solutes into the natural waters draining (exported) from the catchment area. The rate of chemical weathering [mol per m^2 catchment area per year] – one also speaks of chemical erosion or chemical denudation – can be estimated from the concentrations of the solutes in the waters running off (rivers):

$$W_i = \frac{Q\, c_i}{A} \tag{5.26}$$

where W_i is the chemical weathering (dissolution) rate with regard to the solute i [mol per m^2 catchment area per year), c_i is the concentration of the solute i [mol m^{-3}], Q is the runoff [m^3 y^{-1}] and A is the (geographic) catchment area [m^2].

Fig. 5.18 illustrates the difference in water chemistry of rivers draining $CaCO_3$ (cal-cite) or crystalline rocks.

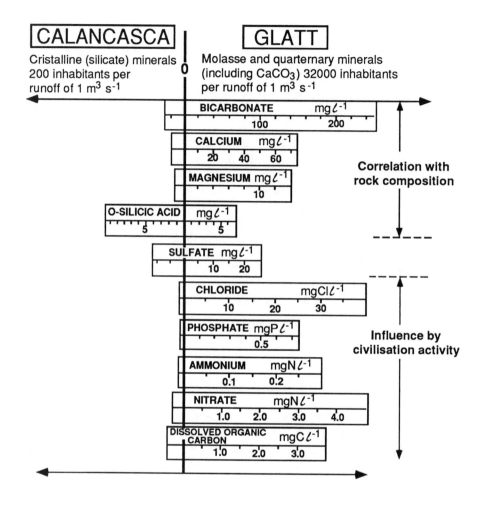

Figure 5.18

Comparison of the chemical composition of rivers Calancasca (southern Switzerland) draining silicate rocks and Glatt (northern Switzerland) draining $CaCO_3$ (calcite) bearing terrain. Because of the high population density anthropogenic effects are superimposed on the geological effects.

In Table 5.3 some laboratory dissolution rates are compared with field measurements on solute export in silicate terrain.

It may be noted that the field data on a per m² catchment area basis, from different geographical areas are remarkably similar ($10^{-2} - 10^{-1}$ equiv m^{-2} y^{-1}). In order to estimate from these rates, actual rates of dissolution of rocks on a per m² mineral surface area basis, we have to know, how many m² of effective (active) mineral surface is available per m² geographic area. This is not known, but the estimate of 10^5

Table 5.3   Laboratory and Field Weathering Rates

Mineral	Laboratory data dissolution rate pH = 4
	mol Si m^{-2} y^{-1}
Placioclase	10^{-3} [a]
Olivine	1.3 × 10^{-3} [a]
Biotite	1.2 × 10^{-3} [b]
Kaolinite	9 × 10^{-5} [c]
Muscovite	4.5 × 10^{-5} [c]
Quartz	2 × 10^{-5} [c]
δ-Al$_2$O$_3$	9 × 10^{-5} (mol Al^{3+}) [c] [d]
CaCO$_3$ (calcite)	10^{-2} (mol Ca^{2+})

Geographic area	Representative field measurements on cation export from silicate terrain
	equivalent cation m^{-2} (catchment) y^{-1}
Trnavka River Basin (CZ)	2 × 10^{-2} [e]
Coweeta Watersled NC (USA)	3.5 × 10^{-2} [f]
Filson Creek MN (USA)	1.5 – 3 × 10^{-2} [g]
Cristallina (CH)	2 – 3 × 10^{-2} [h]
Bear Brooks ME (USA)	9.6 × 10^{-2} [i]
Swiss Alps	4 – 10 × 10^{-2} [k]

[a]   Busenberg and Clemency (1976)
[b]   Grandstaff (1986)
[c]   Wieland and Stumm (1987)
[d]   mol Al m^{-2} y^{-1}
[e]   Paces (1983)
[f]   Velbel (1985)
[g]   Siegel and Pfannkuch (1984)
[h]   Giovanoli et al. (1989)
[i]   Schnoor (1990)
[k]   Zobrist and Stumm (1981)

m^2 surface area of mineral grains active in weathering available per m^2 of geographic area was made (Schnoor, 1990)[1]. If we assume such a conversion factor (10^5 m^2/m^2 geographic area) weathering rates in nature would be significantly slower than predicted from laboratory studies. There are various reasons for this apparent discrepancy. In addition to the possibility that our conversion factor is off – possibly by 1 or 2 orders of magnitude – the following reasons could be given: (1) unsaturated flow through soil macropores, limits the amount of weatherable minerals exposed to water; (2) dissolved Al(III) has been shown to inhibit the dissolution rates in soil macropores (hydrological control due to macropore flow through soils).

---

[1] Based on the following assumptions: 50 cm of saturated regolith (mantle rock), 0.5 m^2 g^{-1} surface area of mineral grains, 40 % of mineral grains active in weathering, density of minerals $\cong$ 2.5 g cm^{-3}.

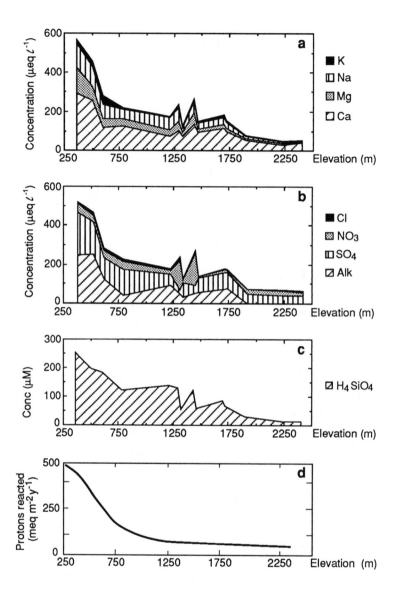

Figure 5.19

Effect of altitude of catchment area on water chemistry and weathering rates (Data from Zobrist and Drever, 1990)

The weathering rate in d) corresponds to a "smoothed" line reflecting the actual data. The change with altitude is primarily caused probably by an increase in effective surface area of weathered mineral grains. With decreasing elevation more soil is produced partially also due to increased biological activity.

Zobrist and Drever (1990) have estimated weathering rates in the southern Alps of Switzerland (gneiss with chlorite, mica, hydrobiotite and poorly crystalline mixed-layer material). They found that the weathering rates increase significantly with the decreasing catchment elevation. Fig. 5.19 shows that the concentrations of base cations, H_4SiO_4 and alkalinity decrease exponentially with increasing catchment elevation. This can be explained by decreasing rates of chemical weathering, from about 500 meq per m² (geographic) per year to ca. 20 meq m^{-2} y^{-1}. At high altitudes very little soil is available. Above the tree line biological activity is very limited; there is no or little production of organic matter. At lower elevations, photosynthesis and respiration activities are much higher than at high elevations. Plants and trees exude through their roots ligands that increase weathering rates of primary minerals and, in turn, the production of soil minerals and their specific surface areas.

Mechanical Weathering. Operationally one distinguishes between mechanical (physical) erosion and chemical weathering. The cracking of rocks by physical forces (e.g., melting and freezing of ice) produces suspended solids in the load of rivers. The rates of mechanical erosion are larger by about an order of magnitude than chemical erosion. The rates of mechanical erosion can be estimated from the sedimentation rates in underlying lakes (sediments deposited minus authigenic[1] solid products formed in the lake). As Fig. 5.17b suggests mechanical weathering, which produces materials of higher specific surface area, often proceeds chemical weathering.

Example 5.2: Chemical Erosion Rate

The Rhine River above Lake Constance (alpine and prealpine catchment area) averages the following composition:

Ca^{2+}	=	43	mg/ℓ
Mg^{2+}	=	9	mg/ℓ
Na^+	=	3.1	mg/ℓ
SO_4^{2-}	=	53	mg/ℓ
Cl^-	=	2.8	mg/ℓ
HCO_3^-	=	115	mg/ℓ
H_4SiO_4	=	6.5	mg/ℓ.

Annual precipitation amounts to 140 cm y^{-1} of which about 30 % is lost due to evaporation and evapo-transpiration.

What is the chemical weathering rate in this catchment area? Can it be subdivided to dissolution rates of individual minerals?

---

[1] authigenic = "formed on the spot"; the term is usually used to describe sedimentary materials formed in the ocean or in the lake.

The annual runoff approximates 1 m^3 per m^2 geographic area; and one m^3 of water contains the concentrations given. This amounts to

1.07	mol of Ca^{2+},
0.4	mol of Mg^{2+},
0.13	mol of Na^+,
0.08	mol of Cl^-,
1.9	mol of HCO_3^-,
0.55	mol of SO_4^{2-} and
7×10^{-2}	mol of H_4SiO_4

per m^3; thus, corresponding mol quantities have become dissolved per m^2.

We can try to represent these concentrations in terms of minerals that have dissolved. All the SO_4^{2-} minus SO_4^{2-} introduced by acid rain (on the average the rain contains 0.05 mol SO_4^{2-} m^{-3}) is "assigned" to the dissolution of $CaSO_4(s)$, all the Mg^{2+} to $MgCO_3(s)$ (component of dolomite, $CaMg(CO_3)_2(s)$) and the remaining Ca^{2+} to $CaCO_3(s)$, Cl^- to $NaCl$ and the H_4SiO_4 to a feldspar $NaAlSi_3O_8$.

Thus, we can estimate the erosion rates:

	mol m^{-2} y^{-1}	g m^{-2} y^{-1}
$CaSO_4$	0.5	68
$MgCO_3$	0.4	33
$CaCO_3$	0.6	60
$NaCl$	0.08	4.7
$Si(s)$	0.07 or	5.5
$NaAlSi_3O_8$	0.023	

Total chemical erosion: 3.1 eq m^{-2} y^{-1} or 170 g m^{-2} y^{-1}

These estimates are based on some crude assumptions; nevertheless, the chemical erosion rates obtained are believed to be within ± 20 %.

Note: Mechanical erosion has been estimated for the same catchment area to be 1150 g m^{-2} y^{-1}.

Estimate on the basis of results on experimental weathering rates and the number of functional groups (or central ions) per unit area, how long it takes to dissolve one "monomolecuar" layer of a representative silicate mineral.

There are about 10 functional groups per nm^2 (cf. Chapter 2.1). That corresponds to 10^{19} groups or $S_T = 1.7 \times 10^{-5}$ mol groups per m^2. A representative weathering rate (Table 5.3) is $10^{-3} - 10^{-4}$ mol m^{-2} y^{-1}. Converting this in a pseudo first order rate law

$$R'_H = \frac{\text{weathering rate (mol m}^{-2}\text{ y}^{-1})}{\text{density of surface groups (mol m}^{-2})} \qquad (i)$$

This rough calculation shows that only a few monolayers are dissolved per year.

(The idea to this example comes from an article by B. Wehrli in EAWAG News 28/29, 1990.)

5.11 Case Study: Chemical Weathering of Crystalline Rocks in the Catchment Area of Acidic Ticino Lakes, Switzerland

Fig. 5.20 gives the water composition of four lakes at the top of the Maggia valley in the southern alps of Switzerland. Although these lakes are less than 10 km apart, they differ markedly in their water composition as influenced by different bedrocks in their catchments. All lakes are at an elevation of 2100 – 2550 m. The small catchments are characterized by sparse vegetation (no trees), thin soils and steep slopes.

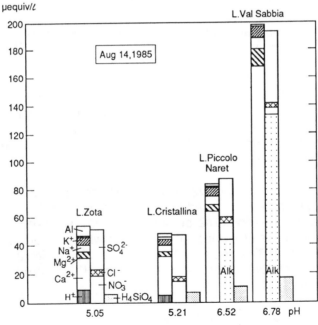

Figure 5.20

Comparison of water composition of four lakes influenced by different bedrocks in their catchments. Drainage areas of Lake Zota and Lake Cristallina contain only gneiss and granitic gneiss; that of Lake Piccolo Naret contains small amounts of calcareous schist; that of Lake Val Sabbia exhibits a higher proportion of schist.

(From Giovanoli, Schnoor, Sigg, Stumm, Zobrist; 1988)

The composition of Lakes Cristallina and Zota, situated within a drainage area characterized by the preponderance of gneissic rocks and the absence of calcite and dolomite, and Lake Val Sabbia, the catchment area of which contains dolomite, are markedly different. The waters of Lake Cristallina and Lake Zota exhibit mineral acidity (i.e., caused by mineral acids and HNO_3), their calcium concentrations are 10 – 15 µmol/liter and their pH is < 5.3. On the other hand, the water of Lake Val Sabbia is characterized by an alkalinity of 130 µmol/liter and a calcium concentration of 85 µmol/liter. The water of Lake Piccolo Naret is intermediate; its alkalinity is < 50 µmol/liter and appears to have been influenced by the presence of some calcite or dolomite in its catchment area.

Weathering Processes regulating the Chemical Water Composition. On the basis of simple mass-balance considerations plausible reconstructions were attempted for the contribution of the various weathering processes responsible for the residual water composition of the acidic Lake Cristallina. Stoichiometric reactions are presented in Table 5.3 for the interaction of acidic deposition (snow melt) with weatherable minerals in the drainage area. The composition of atmospheric deposition results generally from the interactions of strong acids (H_2SO_4, HNO_3), base (NH_3) and wind-blown dust and aerosols ($CaCO_3$, $MgCO_3$, NH_4NO_3, $NaCl$, KCl).

In Table 5.4 the contributions of the individual weathering reactions were assigned and combined in such a way as to yield the concentrations of Ca^{2+}, Mg^{2+}, Na^+, K^+, and H^+ measured in these lakes; the amounts of silicic acid and aluminum hydroxide produced and the hydrogen ions consumed were calculated stoichiometrically from the quantity of minerals assumed to have reacted. Corrections must be made for biological processes, such as ammonium assimilation and nitrification and the uptake of silicic acid by diatoms. Some of the H_4SiO_4 was apparently lost by adsorption on aluminum hydroxide and Fe(III)(hydr)oxides, but the extent of these reactions was difficult to assess.

Although an unequivocal quantitative mass balance could not be obtained, a plausible reaction sequence was deduced that accounts reasonably well for the residual chemical water composition. The amount of $CaCO_3$ that had to be dissolved to establish the residual water composition is about what can be accounted for by wind-blown calcite dust. The neutralization of the acidic precipitation by NH_3 was, subsequent to its deposition, largely annulled by the H^+ ions produced by nitrification and NH_4^+ assimilation.

$$NH_4^+ + 2\,O_2 \longrightarrow NO_3^- + 2\,H^+ + H_2O$$

According to Table 5.3, in the drainage area of Lake Cristallina, about 4 µmol of plagioclase, 2 µmole of epidote, 1 µmole of biotite, and 1 µmol of K-feldspar per liter of runoff water are weathered per year. This amounts to 19 meq of cations per m^2-year.

Figs. 5.21 a and b illustrate some of our observations (Giovanoli et al., 1988) on the effect of acid deposition on granitic gneiss. The surface of some white, macroscopic plagioclase grains were covered with etch pits. The presence of etch pits suggests that defects in the crystal structure are sites of strong preferential dissolution during geochemical processes (Blum and Lasaga, 1987). Such etch pits are also evidence for surface-controlled (rather than diffusion or transport-controlled) dissolution reactions. As shown in Fig. 5.21b, the etch pits were arranged along subgrain boundaries.

Although the elemental analysis indicates an enrichment in Al, no crystalline Al phase was observed in any of the sediment samples. Neither XRD nor electron diffraction showed a separate Al hydroxide phase.

5.12 Inhibition of Dissolution; Passivity

It is important to understand the factors that retard dissolution. The same question is especially relevant in technical systems, and in the corrosion of metals and building materials. Passivity is imparted to many metals by overlying oxides; the inhibition of the dissolution of these "passive" layers protects the underlying material.

Fig. 5.3 lists some of the factors that promote and inhibit dissolution. Obviously, substances which "block" surface functional groups or prevent the approach of dissolution promoting H^+, OH^-, ligands and reductants to the functional groups inhibit the dissolution. Very small concentrations of inhibitors can often be effective because it may suffice to block the functional groups of the solution active sites such as the ledge sites or the kink sites. Often competitive equilibria exist between the surface and solutes which block surface sites and, on the other hand, solutes which promote dissolution, e.g., H^+, ligands. The main role of many inhibitor is to prevent access of corrosion promoting agents to the surface.

Metal Ions. Metal ions, especially those which are specifically adsorbed in the pH region < 7, inhibit the proton promoted dissolution of oxides and silicates. This inhibition might be looked at as a competition in the binding of H^+ by the binding of metal ions, i.e., a surface metal center is no longer dissolution-active if bound to a metal ion. Alternatively, the inhibitory effect can be interpreted as being due, in part, to a lowering of surface protonation which occurs as a consequence of metal binding. As seen in Fig. 3.5 and exemplified in Example 5.1 and Fig. 5.14c, the addition at constant pH of a metal ion to a hydrous oxide reduces surface protonation, c_H^s, $(\Gamma_H - \Gamma_{OH^-})$, or σ_{H^+}. Because the dissolution rate, R_H, has been shown to be proportional to $(c_H^s)^i$ (Eq. 5.12) metal ion binding most likely decreases R_H (see Fig. 5.14. The effect of specifically adsorbable cations on the reduction of dissolution (weathering) rates of minerals is of importance in geochemistry. A documented case is the effect of cations other than uranium in the dissolution of uraninite (UO_2).

Table 5.4   *Weathering of Cristalline Rocks in the Catchment area of Lake Cristallina*

a) Possible reaction sequences coupled with stoichiometric calculations for establishing chemical composition of lake water (From Giovanoli, Schnoor, Sigg, Stumm and Zobrist, 1988)

Reactants	Substances produced or consumed (−)											
	H^+	Na^+	K^+	Ca^{2+}	Mg^{2+}	NH_4^+	$Al(OH)_{3-v}^y$	SO_4^{2-}	NO_3^-	Cl^-	H_4SiO_4	$[Al(OH)_3]_s$
	← µeq/liter →										← µmol/liter →	
Genesis of precipitation (wet and dry deposition)												
31 µeq H_2SO_4 } acids	31							31				
11 µeq NHO_3	11								11			
3 µeq NaCl } "dust"		3								3		
2 µeq KCl			2							2		
19 µeq $CaCO_3$ } bases	−19			19								
2 µeq $MgCO_3$	−2				2							
10 µeq NH_3	−10					10						
Resulting composition	11	3	2	19	2	10	< 1	31	11	5	< 1	< 1
After evaporation [1]	16	4	3	27	3	14	< 1	44	16	7	< 1	< 1
Lake Cristallina												
– Weathering reactions												
1 µmol calcite	−2			2								
4 µmol plagioclase (x = 0.25)	−5	3		2							13	5
2 µmol epidote (x = 0.5)	−8			8							6	5
1 µmol biotite (x = 0.5)	−2		1		1						3	1
1 µmol K-feldspar	−1		1								3	1
– Additional reactions												
NH_4 assimilation [2]	13					−13						
Dissolution of $[Al(OH)_3]_s$ [3]	−6						6					
Final composition												
Calculated	5	7	5	39	4	1	6	44	16	7	25 [4]	8
Measured September 1985	4	6	5	40	4	1	6	42	16	8	15	–

b) Dissolution reactions in approximate rank order for ease of weathering [5]

$CaCO_{3(s)} + 2 H^+ \longrightarrow Ca^{2+} + H_2CO_3$
calcite

$CaMg(CO_3)_{2(s)} + 2 H_2CO_3 \longrightarrow Ca^{2+} + Mg^{2+} + 4 HCO_3^-$
dolomite

$Na_{(1-x)}Ca_xAl_{(1+x)}Si_{(3-x)}O_{8(s)} + (1+x) H^+ + (7-x) H_2O \longrightarrow (1-x) Na^+ + x Ca^{2+} + (3-x) H_4SiO_4 + (1+x) Al(OH)_3$
plagioclase (x = mole fraction of anorthite) [6]

$Ca_2Al_{(3-x)}Fe_xSi_3O_{12}(OH)_{(s)} + 4 H^+ + 8 H_2O \longrightarrow 2 Ca^{2+} + 3 H_4SiO_4 + (3-x) Al(OH)_3 + x Fe(OH)_3$
epidote (x = mole fraction of iron)

$KMg_xFe_{(3-x)}AlSi_3O_{10}(OH)_{2(s)} + (1+2x) H^+ + (9-2x) H_2O \longrightarrow K^+ + x Mg^{2+} + 3 H_4SiO_4 + Al(OH)_3 + (3-x) Fe(OH)_2$
biotite (x = mole fraction of magnesium)

$KAlSi_3O_{8(s)} + H^+ + 7 H_2O \longrightarrow K^+ + 3 H_4SiO_4 + Al(OH)_3$
orthoclase (K-feldspar)

$KAl_3Si_3O_{10}(OH)_{2(s)} + H^+ + 9 H_2O \longrightarrow K^+ + 3 H_4SiO_4 + 3 Al(OH)_3$
muscovite

$Al(OH)_3 + 3 H^+ \longrightarrow Al^{3+} + 3 H_2O$
(noncrystalline)

$FeOOH_{(s)} + 3 H^+ \longrightarrow Fe^{3+} + 2 H_2O$
goethite

$SiO_{2(s)} + 2 H_2O \longrightarrow H_4SiO_4$
quartz

1) Evapoconcentration factor $= \dfrac{\text{precipitation}}{\text{runoff}} = \dfrac{1.7 \text{ m}}{1.2 \text{ m}} = 1.42$

2) NH_4 assimilation: $NH_4^+ \longrightarrow (NH_3)org + H^+$

3) $[Al(OH)_3]_s + vH^+ \longrightarrow Al(OH)_{3-v}^y$, v depends on the pH of the lake water

4) Loss of H_4SiO_4 due to incorporation in diatoms and adsorption on noncrystalline $Al(OH)_3$ not included

5) From greatest to least; these reactions could also be written, instead of with H^+ ions as reactants, with water and dissolved CO_2. In either scheme, the net result of the reaction is the same – cations (Ca^{2+}, Mg^{2+}, Na^+, K^+ are released, and alkalinity is produced via OH^- or HCO_3^- production or H^+ consumption. In all reactions, the equivalents of cations released are exactly balanced by the equivalents of acid consumed which corresponds to the alkalinity produced.

6) anorthite $= CaAl_2Si_2O_8$

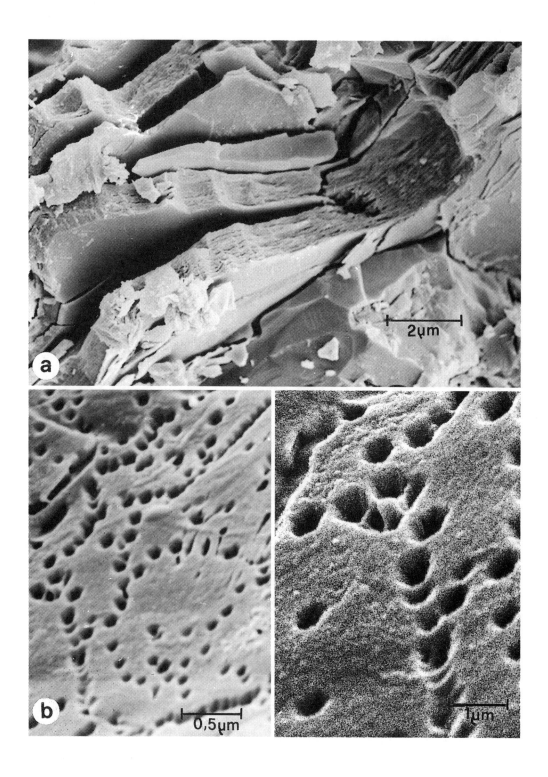

As shown by Grandstaff (1976) the cations, thorium, lead and the rare earths, associated with uraninite retard the dissolution of UO_2 significantly.

In the more alkaline pH-range surface-bound cations may increase the dissolution rate; the surface deprotonation increases c_{OH}^s (cf. Fig. 5.9c) and thus enhances the dissolution. This effect has been observed, above all, with silicates and glasses.

Ligands. Some ligands like oxalate (and other dicarboxylic acids), salicylate, F^-, EDTA (Ethylenediaminetetraacetate), NTA (Nitrilotriacetate) accelerate the dissolution while others, SO_4^{2-}, CrO_4^{2-}, benzoate tend to inhibit the dissolution. Some, like phosphate and arsenate, accelerate the dissolution at low pH and retard dissolution at pH > 4. Inert, not potential determining ions such as K^+, Na^+, Cl^-, NO_3^- have little effect. How can these different effects of ligands be explained? Why do similar kind of bidentates such as CrO_4^{2-} (inhibitor) and oxalate (dissolution promotor) act so differently? One hypothesis is that mononuclear complexes, especially if they are bidentate (and mononuclear), accelerate dissolution while *binuclear complexes* inhibit dissolution. The effect of binuclear complexes could plausibly be accounted for by inferring that much more energy is needed to remove simultaneously two center atoms from the crystalline lattice (Matjević, 1982). In case of phosphate and arsenate one could postulate (and there is some supporting experimental evidence) that at low pH mononuclear and at near neutral pH value bi- or tri-nuclear surface complexes are formed. In case of CrO_4^{2-} and oxalate($C_2O_4^{2-}$) one may imply that in the former case binuclear and in the latter case mononuclear complexes (both bidentate) are formed. One could also argue that the leaving groups formed at the surface, the chromato and the oxalato complex are, from a solution point of view, completely different. Furthermore, the electron donor properties of the (oxidizing) chromate and (potentially reducing) oxalate ligands are remarkable different. A further role of the CrO_4^{2-} is to maintain a high redox potential at the oxide-water interface and thus imparting passivity and preventing reductive dissolution. Any chromate that is reduced becomes adsorbed even at relatively low pH values, as Cr(III) to the oxide surface; the adsorbed Cr^3 or $CrOH^{2+}$ surface complex is an excellent inhibitor to the dissolution. In case of phosphate and silicate one needs also to consider that at higher concentrations of the solutes at near-neutral pH values solid iron silicate and iron phosphate surface films could be formed. Reductive dissolution and its inhibition will be discussed in Chapter 9.

Figure 5.21

Electron micrographs of weathered silicate minerals
  a) Scanning electron micrograph of weathered gneiss, showing packs of mica splitting into sheets. Attack was apparently from edges and propagated through interlayer cleavage. Sheet surfaces appear clean and unattacked.
  b) Scanning electron micrograph of corroded feldspar grain. Same region in two enlargements showing the etch pits aligned along subgrain boundaries. Etch pits have narrow size distribution.
(From Giovanoli, Schnoor, Sigg, Stumm and Zobrist, 1988)

Corrosion; Passive Films

To what extent is our know-how on the reactivity of oxides useful for the understanding of corrosion reactions and passivity. The corrosive behaviour of a few metals is essentially determined by the kinetics of the dissolution of the corrosion products. This seems to be the case for Zn in HCO_3^- solutions, for passive iron in acids and passive Al in alkaline solutions (Grauer and Stumm, 1982). The mechanism of the dissolution of iron and of the passivation of this dissolution is extremely complex. We may not know exactly the composition of the passive film; but it has been suggested that it consists of an oxide of $Fe_{3-x}O_4$ with a spinel structure. The passive layer seems to vary in composition from Fe_3O_4 (magnetite), in oxygen-free solutions, to $Fe_{2.67}O_4$ in presence of oxygen. It may also consist of a duplex layer consisting of an inner layer of Fe_3O_4 and an outer layer of γ-Fe_2O_3. The coulometric reduction of the passive layer gives two waves which are interpreted either by the reduction of two different layers, Fe_2O_3 and Fe_3O_4 or by successive reduction of Fe_3O_4 to lower valence oxides and its further reduction to metallic iron. Fig. 5.22 represents a schematic model of the hydrated passive film on iron as proposed by Bockris and collaborators (Pou et al., 1984). Obviously, the hydrated passive film on iron displays the coordinative properties of the surface hydroxyl groups.

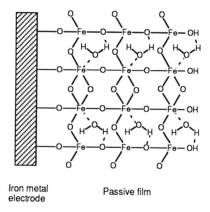

Iron metal electrode   Passive film

Figure 5.22

Schematic representation of the hydrated passive film on iron
(From Pou et al., 1984)

Since the dissolution rate of passive metals is apparently related to the dissolution rate of the passive film, some of our informations on the effect of solution variables on the dissolution reactivity of such type of oxides appear applicable to the interpretation of some of the factors that enhance or reduce passivity.

   i) *Protons*. Obviously, surface protonation will enhance dissolution. For the passivation of iron, critical current density must be exceeded which increases with decreasing pH;

ii) *Fluoride and Chloride.* These nucleophilic ions form surface complexes and may even permeate into the crystalline lattice; they destroy especially at low pH the passivity of iron oxides films;

iii) *Reductants* not only tend to lower the electrode potential and may reduce O_2 but also favor the reductive dissolution of passive iron oxide films (see Chapter 9); if these reductants are able to form surface complexes, especially surface chelates, they can facilitate the electron transfer to the oxide. Presence of Fe(II) in solution can catalyze this reductive dissolution. *Oxidants* added to the water as an inhibitor may prevent this reductive dissolution and restore high electrode potential;

iv) Surface complex forming *ligands* like oxalate, salicylate and phenols, especially at low pH, tend to dissolve the passive film;

v) *Cations* like Zn^{2+}, Al^{3+}, Cr^{3+}, block surface sites and thus stabilize the passive film; these metal ions reduce surface protonation;

vi) *Phosphates*, silicates, polyphosphates or polymeric ligands may have dissolution enhancing effects at low pH and inhibition effects at neutral or alkaline pH values. Electrode kinetic effects of ligands on the "passivity" of aluminum oxide have been shown by Zutić and Stumm (1984).

Some of the passive films have been characterized as semiconductors; in this case, corrosion of these oxides may imply transfer of holes (h^+) from the valence band to the reductant and of electrons (e^-) from the conduction band, in the case of iron(III) oxides as Fe(II).

Appendix

Alkalinity, a "Product of Weathering"

Most weathering reactions consume protons and produce alkalinity (Reactions (5.2) and (5.3), and Table 5.4). Here we briefly review the definitions and illustrate the value of the alkalinity-acidity concept in the context of the weathering reactions.

One has to distinguish between the H^+ concentration as an intensity factor and the availability of H^+, the H^+-ion reservoir as given by the base neutralizing capacity, BNC, or the H-acidity [H-Acy]. The BNC relates to the alkalinity [Alk] or acid neutralizing capacity, ANC, by

$$H\text{-Acy} \qquad\qquad = \quad -[Alk] \qquad\qquad\qquad (A.5.1)$$

$$\text{Base neutralizing capacity} \;=\; -\text{Acid neutralizing capacity}$$

The BNC can be defined by a net proton balance with regard to a reference level – the sum of the concentrations of all the species containing protons in excess of the reference level, less the concentrations of the species containing protons in deficiency of the proton reference level. For natural waters, a convenient reference level (corresponding to an equivalence point in alkalimetric titrations) includes H_2O and H_2CO_3:

$$[H\text{-Acy}] = [H^+] - [HCO_3^-] - 2\,[CO_3^{2-}] - [OH^-] \qquad\qquad (A.5.2)$$

The acid-neutralizing capacity, ANC, or alkalinity [Alk] is related to [H-Acy] by

$$-[H\text{-Acy}] = [Alk] = [HCO_3^-] + 2\,[CO_3^{2-}] + [OH^-] - [H^+] \qquad (A.5.3)$$

Considering a charge balance for a typical natural water

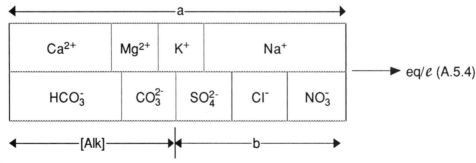

eq/ℓ (A.5.4)

we realize that [Alk] and [H-Acy] also can be expressed by a charge balance – the equivalent sum of conservative cations, less the sum of conservative anions ([Alk] = a – b). The conservative cations are the base cations of the strong bases $Ca(OH)_2$, KOH, and the like; the conservative anions are those that are the conjugate bases of strong acids (SO_4^{2-}, NO_3^-, and Cl^-).

$$[Alk] = [HCO_3^-] + 2\,[CO_3^{2-}] + [OH^-] - [H^+]$$
$$= [Na^+] + [K^+] + 2\,[Ca^{2+}] + 2\,[Mg^{2+}] - [Cl^-] - 2\,[SO_4^{2-}] - [NO_3^-] \qquad (A.5.6)$$

The [H-Acy] for this particular water, obviously negative, is defined ([H-Acy] = b – a) as

$$[\text{H-Acy}] = [Cl^-] + 2\,[SO_4^{2-}] + [NO_3^-]\ \ [Na^+] - [K^+] - 2\,[Ca^{2+}] - 2\,[Mg^{2+}] \qquad (A.5.6)$$

These definitions can be used to interpret interactions of acid precipitation with the environment.

A simple accounting can be made: Every base cationic charge unit (for example, Ca^{2+} or K^+) that is removed from the water by whatever process is equivalent to a proton added to the water, and every conservative anionic charge unit (from anions of strong acids – NO_3^-, SO_4^{2-}, or Cl^-) removed from the water corresponds to a proton removed from the water; or generally,

$$\sum\Delta\,[\text{base cations}] - \sum\Delta\,[\text{conservative anions}] = +\Delta[Alk] = -\Delta[\text{H-Acy}] \quad (A.5.7)$$

If the water under consideration contains other acid- or base-consuming species, the proton reference level must be extended to the other components. In addition to the species given, if a natural water contains organic molecules, such as a carboxylic acid (H-Org) and ammonium ions, the reference level is extended to these species.

One usually chooses the ammonium species NH_4^+ and H-org as a zero-level reference condition. Operationally, we wish to distinguish between the acidity caused by strong acids (mineral acids and organic acids with pK < 6) – typically called *mineral acidity* or *free acidity,* which often is nearly the same as the free-H^+ concentration – and the total acidity given by the BNC of the sum of strong and weak acids (Johnson and Sigg, 1985; Sigg and Stumm, 1991).

Disturbance of H^+ Balance from Temporal or Spatial Decoupling of the Production and Mineralization of the Biomass. In a most general way the synthesis (assimilation) and the decomposition (respiration) of biomass can be written stoichiometrically as

$$a\ CO_2(g) + b\ NO_3^- + c\ HPO_4^{2-} + d\ SO_4^{2-} + g\ Ca^{2+} + h\ Mg^{2+}$$

$$+ i\ K^+ + f\ Na^+ + x\ H_2O + (b + 2c + 2d + 2g - 2h - i - f)\ H^+$$

$$\rightleftharpoons \langle C_a\ N_b\ P_c\ S_d \cdots Ca_g Mg_h\ K_i\ Na_f\ H_2O_m \rangle_{biomass} + (a + 2b)\ O_2(g) \qquad (A.5.8)$$

Eq. (A.5.8) can perhaps be appreciated more simply by considering a "mean" stoichiometry of the terrestrial biomass: $C : N : P : S : Ca : Mg : Al : Fe \approx 800 : 8 : 1 : 1 : 4 : 1 : 1 : 1$.

Using our definition for alkalinity or ANC whereby any decrease (increase) in concentrations of base cations (e.g., K^+, Ca^{2+}, Fe^{2+}, etc.) or any increase (decrease) in concentrations of "acid anions" (e.g., NO_3^-, HPO_4^{2-}, SO_4^{2-}, etc.) is accompanied by a decrease (increase) in alkalinity. Thus, as illustrated in Fig. 5.17 net synthesis of terrestrial biomass (e.g., on the forest and forest floor, where more cations than anions are taken up by the plants (trees), is accompanied by a release of H^+ to the environment.

ANC of Humus. The acid neutralizing capacity of a solution containing humus is primarily due to the sum of concentrations of dissociated humus and of free OH^-:

$$ANC = (TFA - n_H)\ c_s + [OH^-] - [H^+] \qquad (A.5.9)$$

where TFA is the total functional group acidity in mol charge units per kilogram and c_s is the humus concentration in kg ℓ; n_H is the formation function [mol charge units kg^{-1}]

$$n_H = \frac{C_A - C_B + [OH^-] - [H^+]}{C_S} \qquad (A.5.10)$$

(This formation function was derived for an oxide in Eq. (ii) of Example 2.1.) (Sposito, 1989). Clearly ANC will increase with pH. Soil humus is important in the buffering of acid soil.

Reading Suggestions

Berner, R. A., and A. C. Lasaga (1989), "Modeling the Geochemical Carbon Cycle", *Scientific Amer.* **260**, 74-81.

Blum, A. E., and A. C. Lasaga (1987), "Monte Carlo Simulations of Surface Reaction Rate Laws", in W. Stumm, Ed., *Aquatic Surface Chemistry*, pp. 255-292.

Brady, P. V., and J. V. Walther (1989), "Controls on Silicate Dissolution Rates in Neutral and Basic pH Solutions at 25°C", *Geochim. Cosmochim. Acta* **53**, 2823-2830.

Chou, L., and R. Wollast (1985), "Steady-State Kinetics and Dissolution Mechanisms of Albite", *Am. J. Sci.* **285**, 963-993.

Ducker, W. A., T. J. Senden, and R. M. Pashley (1991), "Direct Measurement of Colloidal Forces Using an Atomic Force Microscope", *Nature* **353**, 239-241.

Furrer, G, and W. Stumm (1986), "The Coordination Chemistry of Weathering, I. Dissolution Kinetics of δ-Al_2O_3 and BeO," *Geochim. & Cosmochim. Acta* **50**, 1847-1860.

Giovanoli, R., J. L. Schnoor, L. Sigg, W. Stumm, and J. Zobrist (1989), "Chemical Weathering of Crystalline Rocks in the Catchment Area of Acidic Ticino Lakes, Switzerland," *Clays Clay Min.* **36**, 521-529.

Grandstaff, D. E. (1986), "The Dissolution Rate of Forsteritic Olivine from Hawaiian Beach Sand", in S. M. Colman and D. P. Dethier, Eds., *Rates of Chemical Weathering of Rocks and Minerals*, Academic Press, pp. 41-59.

Lasaga, A. C. (1981), "Rate Laws of Chemical Reactions", in A. C. Lasaga and R. J. Kirkpatrick, Eds., *Kinetics of Geochemical Processes*, Soc. Amer. Rev. in Mineral. 8, Washington D.C, pp. 1-68.

Stumm, W., and R. Wollast (1990), "Coordination Chemistry of Weathering: Kinetics of the Surface-Controlled Dissolution of Oxide Minerals", *Reviews of Geophysics* **28/1**, 53-69.

Stumm, W., Ed. (1990), *Aquatic Chemical Kinetics, Reaction Rates of Processes in Natural Waters,* W. Stumm, Ed., Wiley-Interscience, New York.

Sverdrup, H. A. (1990), *Release due to Chemical Weathering*, Lund University Press, Lund

Chapter 6

Precipitation and Nucleation [1]

6.1 Introduction
The Initiation and Production of the Solid Phase

The birth of a crystal and its growth provide an impressive example of nature's selectivity. In qualitative analytical chemistry inorganic solutes are distinguished from each other by a separation scheme based on the selectivity of precipitation reactions. In natural waters certain minerals are being dissolved, while others are being formed. Under suitable conditions a cluster of ions or molecules selects from a great variety of species the appropriate constituents required to form particular crystals.

Minerals formed in natural waters and in sediments provide a record of the physical chemical processes operating during the period of their formation; they also give us information on the environmental factors that regulate the composition of natural waters and on the processes by which elements are removed from the water. The memory record of the sediments allows us to reconstruct the environmental history of the processes that led to the deposition of minerals, in the past.

Pronounced discrepancies between observed composition and the calculated equilibrium composition illustrate that the formation of the solid phase, for example, the nucleation of dolomite and calcite in seawater, is often kinetically inhibited, and the formation of phosphates, hydrated clay and pyrite is kinetically controlled.

The Role of Organisms in the Precipitation of Inorganic Constituents

Organisms produce significant chemical differentiation in the formation of solid phases. The precipitation of carbonates, of opal, and of some phosphatic minerals by aquatic organisms has long been acknowledged. Within the last two decades, many different kinds of additional biological precipitates have been found. Life has succeeded in largely substituting for, or displacing to a varying extent, inorganic precipitation processes in the sea in the course of the last 6×10^8 years. It appears that metastable mineral phases and, more commonly, amorphous hydrous phases are the initial nucleation products of crystalline compounds in biological mineral precipitates. Amorphous hydrous substances have been shown to persist in the mature, mineralized hard parts of many aquatic organisms. (Mann, Webb and Williams (eds.), 1989.)

---

[1] In writing this chapter, the author enjoyed the assistance of Philippe Van Cappellen.

The Processes Involved in Nucleation and Crystal Growth

Various processes are involved in the formation of a solid phase from an over-saturated solution (Fig. 6.1). Usually three steps can be distinguished:

1) The interaction between ions or molecules leads to the formation of a critical cluster or nucleus.

$$X + X \rightleftharpoons X_2$$

$$X_2 + X \rightleftharpoons X_3$$

$$X_{j-1} + X \rightleftharpoons X_j \qquad \text{(critical cluster)}$$

Nucleation: $\quad X_j + X \longrightarrow X_{j+1}$ \hfill (6.1)

Nucleation corresponds to the formation of the new centers from which *spontaneous* growth can occur. The nucleation process determines the size and the size distribution of crystals produced.

2) Subsequently, material is deposited on these nuclei

$$X_{j+1} + X \longrightarrow \text{crystal growth}$$ \hfill (6.2)

and crystallites are being formed (crystal growth).

3) Large crystals may eventually be formed from fine crystallites by a process called ripening.

A better insight into the mechanisms of the individual steps in the formation of crystals would be of great help in explaining the creation and transformation of sedimentary deposits and biological precipitates. Valuable reviews are available on the principles of nucleation of crystals and the kinetics of precipitation and crystal growth (Zhang and Nancollas, 1990; Steefel and Van Cappellen, 1990; Van Cappellen, 1991). Only a few important considerations are summarized here to illustrate the wide scope of questions to be answered in order to predict rates and mechanisms of precipitation in natural systems.

Biomineralization. The processes controlling biomineralization are summarized in Fig. 6.1c. Organized biopolymers at the sites of mineralization are essential to these processes. In unicellular organisms these macromolecules act primarily as spatial boundaries through which ions are selectively transported to produce localized supersaturation within discrete cellular compartments. In many instances, particularity in organisms such as the diatoms that deposit shells of amorphous silica, the final shape of the mineral appears to be dictated by the ultrastrucure of the membrane-bound compartment. Thus, a diversity of mineral shapes can be biologi-

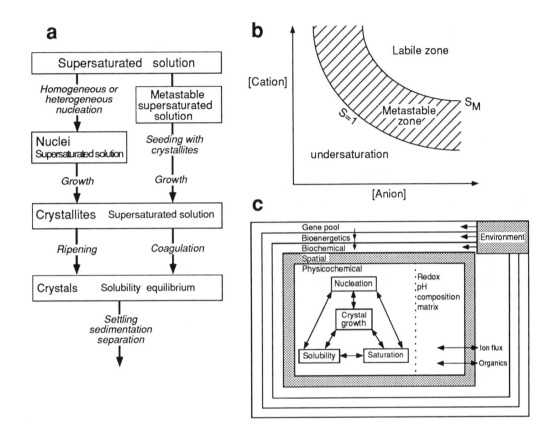

Figure 6.1

a) Simplified scheme of processes involved in nucleation and crystal growth (Nancollas and Reddy, 1974).

b) Schematic solubility isotherm of a solid electrolyte. Below a certain supersaturation, S_M, the nucleation rate is virtually zero and the solution under these conditions can be stable for long periods without precipitation. The range $1 < S < S_M$ is the metastable zone, within which crystal growth can be achieved without the complication of concommitant nucleation if the solution is seeded with crystallites. Foreign surfaces may also induce nucleation in the metastable zone (heterogeneous nucleation). (Modified from Zhang and Nancollas, 1990)

c) The control processes in biomineralization.
In organisms, there are several different interconnecting levels that regulate the physical chemical properties of mineralization (solubility, supersaturation, nucleation and crystal growth). An essential condition for controlled mineralization is spatial localization arising from the compartmentalization of biological space. This permits direct regulation of physicochemical and biochemical properties in the mineralization zone. Nucleation, in particular, can be mediated by organic polymeric substrates in or on the spatial boundary. At a higher level of organization, mineralization is under biochemical and bioenergetic constraints and, ultimately, under control at the gene level. The interplay of these control processes with the external environment is also of fundamental importance. (From Mann, 1988)

cally moulded by constraining the space available to the growing mineral. In many multicellular oganism crystallographic properties are related to the surface structure of the macromolecules, e.g., bone is made up of microscopic plate-like calcium phosphate crystals formed within and between fibrils of collagen (Mann, 1988).

6.2 Homogeneous Nucleation

If one gradually increases the concentration of a solution, exceeding the solubility product with respect to a solid phase, the new phase will not be formed within a specified amount of time until a certain degree of supersaturation has been achieved. Stable nuclei can only be formed after an activation energy barrier has been surmounted.

We first review the classical theory. The free energy of the formation of a nucleus, ΔG_j, consists essentially of energy gained (volume free energy) from making bonds and of work required to create a surface:

$$\Delta G_j = \Delta G_{bulk} + \Delta G_{surf} \tag{6.3}$$

For a nucleus, the first quantity (always negative for a supersaturated solution) can be expressed as

$$\Delta G_{bulk} = -jkT \ln \frac{a}{a_0} = -jkT \ln S \tag{6.4}$$

where j is the number of molecular units ("monomers") in the nucleus or, expressed in terms of volume for a spherical nucleus, $j = 4 \pi r^3/3 V$ where V is the "molecular" volume; where a is the actual concentration (activity) and a_0 the concentration at solubility equilibrium of the solutes that characterize the solubility; a/a_0 is the saturation ratio S

$$S = \frac{a}{a_0} = \left(\frac{IAP_0}{K_{s0}} \right)^{\frac{1}{\eta}} \tag{6.5}$$

where IAP_0 is the ion activity product (of the actual activities in the oversaturated solution) and K_{s0} is the solubility product; η is the number of ions in the formula unit of a mineral $A_\alpha B_\beta$ (i.e., $\eta = \alpha + \beta$). Because of the "normalization" by η, the saturation ratio S is independent of the way the formula is written, e.g., $Ca_5(PO_4)_3OH$ or $Ca_{10}(PO_4)_6(OH)_2$.

The second quantity in Eq. (6.3) is given (spherical nucleus) by

$$\Delta G_{surf} = 4 \pi r^2 \bar{\gamma} \tag{6.6}$$

where $\bar{\gamma}$ is the interfacial energy (assumed to be independent of cluster size). Hence the free energy of the formation of a spherical cluster can be written as

$$\Delta G_j = -\frac{4 \pi r^3}{3 V} kT \ln S + 4 \pi r^2 \bar{\gamma} \tag{6.7}$$

In Fig. 6.2a ΔG_j is plotted as a function of j for a few values of the saturation ratio $a/a_0 = S$. Obviously the activation energy ΔG^* decreases with increasing saturation ratio, as does the size of the critical nucleus, r^* or r_j.

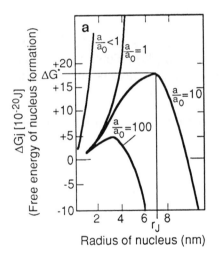
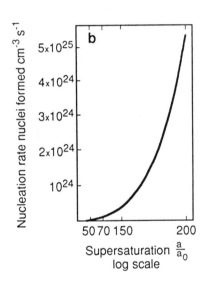

Figure 6.2

Nucleation. The energy barrier and the nucleation rate depend critically on the supersaturation.
a) Free energy of formation of a spherical nucleus as a function of its size, calculated for different saturation ratios (a/a_0). The height of the maximum, ΔG^* is the activation barrier to the nucleation process of nucleus of radius r_j.
b) Double logarithmic plot of nucleation rate versus saturation ratios (a/a_0) calculated with equations
   (6.10) and (6.11). The curves have been calculated for the following assumptions: $\bar{\gamma} = 100$ mJ
   m^{-2}; "molecular" volume $V = 3 \times 10^{-23}$ cm^3; collision frequency efficiency (Eq. 6.11) $\bar{A} = 10^{30}$ cm^{-3}
   s^{-1}.

With increasing size of the cluster the first term on the right hand side of Eq. (6.7) (increasing with r^3) outweighs the second term (increasing with r^2); for large crystals the second term becomes negligible.

Small crystallites are more soluble than large crystals; hence the energy barrier is related to the additional free energy needed to form the more soluble nuclei. Thermodynamically it can be shown (e.g., Stumm and Morgan, 1981) that $d \ln K_{s0}/d\bar{S} = 2/3 \, \bar{\gamma}/RT$, or

$$\log K_{s0_{(\bar{S})}} = \log K_{s0_{(\bar{S}=0)}} + \frac{2/3 \, \bar{\gamma}}{2.3 \, RT} \bar{S} \tag{6.8}$$

where $\bar{S}$ is the molar surface.

For an aggregate of spherical shape we can rewrite Eq. (6.8):

$$kT \ln \frac{a}{a_0} = \frac{2 \, V\bar{\gamma}}{r} \tag{6.9}$$

As before, the saturation ratio S can also be expressed as a/a_0, where a and a_0 are the actual and equilibrium activities, respectively, of the solutes that characterize the solubility. Once nuclei of critical size X_{j+1} (in Eq. 6.1) have been formed, crystallization is spontaneous.

The activation energy ΔG^* can be calculated by inserting into Eq. (6.8) r_j as obtained from Eq. (6.9)

$$r^* = \frac{2\bar{\gamma}V}{kT \ln(a/a_0)} \tag{6.10}$$

for which

$$\Delta G^* = \frac{16 \, \pi\bar{\gamma}^3 V^2}{3 \, [kT \ln(a/a_0)]^2} \tag{6.11}$$

The rate at which nuclei form, J, may be represented according to conventional rate theory as

$$J = \bar{A} \exp\left(\frac{-\Delta G^*}{kT}\right) \tag{6.12}$$

where $\bar{A}$ is a factor related to the efficiency of collisions of ions or molecules. Accordingly, the rate of nucleation is controlled by the interfacial energy, the supersaturation, the collision frequency efficiency, and the temperature. For given values of T, $\bar{A}$, and $\bar{\gamma}$, the nucleation rate J (nuclei formed cm^{-3} sec^{-1}) can be calculated as a function of IAP_0/K_{s0} (Fig. 6.2b). J is critically dependent upon the supersaturation. For example, using the conditions specified for Fig. 6.2a, one can calculate that nucleation is almost instantaneous at a supersaturation of 100, while (homogeneous) nucleation should not occur even within very long time spans for a 10-fold supersaturation.

At a high degree of supersaturation, the nucleation rate is so high that the precipitate formed consists mostly of extremely small crystallites. Incipiently formed crystallites might be of a different polymorphous form than the final crystals. If the nucleus is smaller than a one-unit cell, the growing crystallite produced initially is most likely to be amorphous; substances with a large unit cell tend to precipitate initially as an amorphous phase ("gels").

The most important "message" of this chapter is that there is a *critical supersaturation* that must be exceeded before homogeneous nucleation can occur. The background given is an essential preparation for the introduction of heterogeneous nucleation.

6.3 Heterogeneous Nucleation

Heterogeneous nucleation, however, is in many cases the predominant formation process for crystals in natural waters. In a similar way as catalysts reduces the activation energy of chemical reaction, foreign solids may catalyze the nucleation process by reducing the energy barrier. Qualitatively, if the surface of the solid substrate matches well with the crystal, the interfacial energy between the two solids is smaller than the interfacial energy between the crystal and the solution, and nucleation may take place at a lower saturation ratio on a solid substrate surface than in solution.

Phase changes in natural waters are almost invariably initiated by heterogeneous solid substrates. Inorganic crystals, skeletal particles, clays, sand, biological surfaces can serve as suitable substrate.

In writing on heterogeneous nucleation we have been influenced, especially by Steefel and Van Cappellen (1990) and by Van Cappellen (1991). More generally Eq. (6.3) can be rewritten as

$$\Delta G = -mkT \ln \frac{IAP_0}{K_{s0}} + \bar{\gamma} A \tag{6.3a}$$

where m is the number of formula units of the mineral in the crystal, i.e., $m = j/\eta$, and A is the surface area of the crystal; the latter can be expressed as

$$A = \alpha \left(\frac{V_M}{\eta N_A} \right)^{2/3} j^{2/3} \tag{6.13}$$

where V_M is the molar volume of the solid phase and N_A is the Avogadro number; α is a geometric factor that depends on the shape of the crystal, e.g., for a sphere $\alpha = (36\,\pi)^{1/3} = 4.84$; the mean ionic radius, $\bar{r}$, (typically $1 - 2$ Å)

$$\bar{r} = \tfrac{1}{2}\left(\frac{V_M}{\eta\,N_A}\right)^{1/3}$$

We can rewrite Eq. (6.13) once more in a most general form

$$\Delta G = -jkT \ln S + \alpha\bar{\gamma}\, 4\,\bar{r}^2\, j^{2/3} \tag{6.3b}$$

In the case of heterogeneous nucleation the interfacial energy needs some redefinition because the nucleus is now formed in part in contact with the solution and in part in contact with the surface of the solid substrate:

Homogeneous nucleation:

$$\Delta G_{surf} = \bar{\gamma}_{CW}\, A \tag{6.15}$$

Heterogeneous nucleation (Van Cappellen, 1991):

$$\Delta G_{interf} = \bar{\gamma}_{CW}\, A_{CW} + (\bar{\gamma}_{CS} - \bar{\gamma}_{SW})\, A_{CS} \tag{6.16}$$

The suffixes CW, CS, SW refer to cluster-water, cluster-substrate and substrate-water, respectively.

A surface-catalytic effect is observed, as mentioned above, when the surface of the solid substrate "matches well" with the crystal to be formed, i.e., when

$$\bar{\gamma}_{CS} < \bar{\gamma}_{CW} \tag{6.17a}$$

In ideal cases (epitaxial) $\bar{\gamma}_{CS}$ becomes very small

$$\bar{\gamma}_{CS} \longrightarrow 0 \tag{6.17b}$$

and the solid-solution interfacial energy of the substrate is similar to that of the cluster

$$\bar{\gamma}_{SW} \approx \bar{\gamma}_{CW} \tag{6.17c}$$

As a consequence, for a "good" substrate:

$$\Delta G_{interf} \approx \gamma_{CW}(A_{CW} - A_{CS}) \tag{6.18}$$

When the attachment of the substrate to the precipitate to be formed is strong, the clusters tend to spread themselves out on the substrate and form thin surface islands. A special limiting case is the formation of a surface nucleus on a seed crystal of the same mineral (as in surface nucleation crystal growth). As the cohesive bonding within the cluster becomes stronger relative to the bonding between the cluster and the substrate, the cluster will tend to grow three-dimensionally (Steefel and Van Cappellen, 1990).

On the other hand, if $\bar{\gamma}_{SW} \gg \bar{\gamma}_{CW}$, the precipitate tends to form a structurally continuous coating on the substrate grain. The interfacial energy (Eq. 6.16) may even become negative and the activation barrier vanishes. An example reflecting this condition is the growth of amorphous silica on the surface of quartz (Wollast, 1974).

6.4 The Interfacial Energy and the Ostwald Step Rule

The above considerations show that the interfacial energy is of utmost importance in determining the thermodynamics and kinetics of the nucleation process. Unfortunately, however, there are considerable uncertainities on the values of interfacial free energies. Values determined from contact angle measurements are significantly lower than those determined from the dependence of solubility upon molar surface of the crystallites. Furthermore, reliable data on $\bar{\gamma}_{CS}$ are lacking.

It is useful to know, that for a given type of crystals (oxides, sulfates, carbonates), the interfacial mineral-aqueous solution free energy, $\bar{\gamma}$ (or $\bar{\gamma}_{CW}$), increases with decreasing solubility (Schindler, 1967). Nielsen (1986) cites the following empirical relationship

$$\frac{4 \, \bar{r}^2 \, \bar{\gamma}}{kT} = 4.70 - 0.272 \ln C_{sat} \tag{6.19}$$

where C_{sat} is the solubility in moles of formula units of mineral per liter and, $\bar{r} =$ the mean ionic radius [m], the units of $\bar{\gamma}$ are Joules m^{-2}. Table 6.1 (from Van Cappellen, 1991) lists interfacial energies, the solubilities (for oxides and hydroxides C_{sat} values were calculated for pH = 7) and values for the mean ionic radius.

The Ostwald Step Rule, or the rule of stages postulates that the precipitate with the highest solubility, i.e., the least stable solid phase will form first in a consecutive precipitation reaction. This rule is very well documented; mineral formation via precursors and intermediates can be explained by the kinetics of the nucleation process. The precipitation sequence results because the nucleation of a more soluble phase is kinetically favored over that of a less soluble phase because the more soluble phase has the lower solid-solution interfacial tension ($\overline{\gamma}_{CW}$) than the less soluble phase (Eq. 6.19). In other words, a supersaturated solution will nucleate first the least stable (often an amorphous solid phase) because its nucleation rate is larger than that of the more stable phase (Fig. 6.4). While the Ostwald Step Rule can be explained on the basis of nucleation kinetics, there is no thermodynamic contradiction in the initial formation of a finely divided precursor.

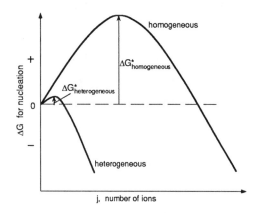

Figure 6.3

Schematic representation of the effect of a solid substrate to catalyze (for a given saturation ratio S) the nucleation.
It is assumed that $\overline{\gamma}_{CS} < \overline{\gamma}_{CW}$. The exact curves depend also on the geometry of the crystals formed.

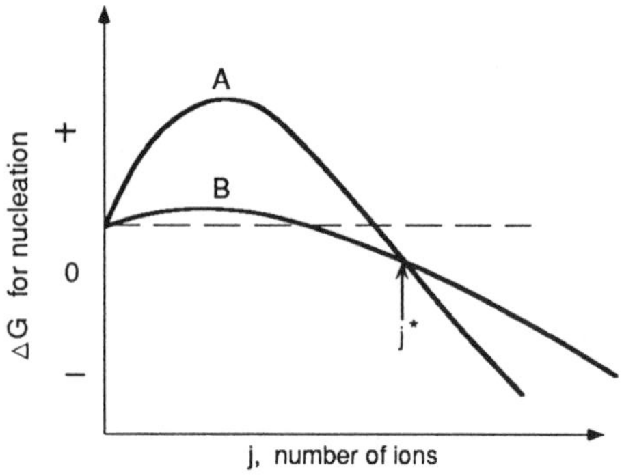

Figure 6.4

Schematic plot of free energy of formation of clusters from solution as a function of size (number of ions in the cluster). Curve A corresponds to the (thermodynamically) ultimately formed more stable phase, while curve B corresponds to the precursor phase.
(Modified from Van Cappellen, 1991). The particle size, j*, is critical for the inversion from one polymorphous form to another.

The Size Dependence of the Solubility has also a thermodynamic base. As an example we follow arguments presented by Schindler (1967) and consider as an example the reaction

$$Cu(OH)_2(s) = CuO(s) + H_2O$$

Schindler and his coworkers determined the solubility constants and the influence of molar surface, $\overline{S}$, upon solubility (25° C)

$Cu(OH)_2$: $\log *K_{s0} = 8.92 + 4.8 \times 10^{-5} \overline{S}$; $\overline{\gamma} = 410 \pm 130$ mJ m^{-2}

CuO: $\log *K_{s0} = 7.89 + 8 \times 10^{-5} \overline{S}$; $\overline{\gamma} = 690 \pm 150$ mJ m^{-2}

(log $*K_{s0}$ is defined by the reaction $Cu(OH)_2(s) + 2\,H^+ = Cu^{2+} + 2\,H_2O$ and in an analogous way for $CuO(s)$.) (Fig. 6.5a.)

Table 6.1	Surface free energies			
Mineral	Formula	σ (mJ/m²)	C_{sat} (M)	$2\,\bar{r}$ ($\times 10^{10}$ m)
Fluorite	CaF_2	120 [2]	2×10^{-4}	2.39
Calcite	$CaCO_3$	94 [2]	6×10^{-5}	3.13
Witherite	$BaCO_3$	115 [1]	1×10^{-4}	3.36
Cerussite	$PbCO_3$	125 [1]	4×10^{-6}	3.23
Gypsum	$CaSO_4 \times 2\,H_2O$	26 [11]	1.5×10^{-2}	3.14
Celestite	$SrSO_4$	85 [1]	4×10^{-4}	3.37
Barite	$BaSO_4$	135 [1]	1×10^{-5}	3.51
F-apatite	$Ca_5(PO_4)_3F$	289 [3]	6×10^{-9}	3.08
OH-apatite	$Ca_5(PO_4)_3OH$	87 [4]	7×10^{-6}	3.08
OCPp	$Ca_4H(PO_4)_3 \times H_2O$	26 [3]	2×10^{-4}	3.15
Portlandite	$Ca(OH)_2$	66 [1]	6×10^{-5}	2.64
Brucite	$Mg(OH)_2$	123 [1]	1.5×10^{-4}	2.39
Goethite	$FeOOH$	1600 [6]	1×10^{-12}	2.26
Hematite	Fe_2O_3	1200 [6]	1×10^{-12}	2.16
Zincite	ZnO	770 [7]	2×10^{-5}	2.28
Tenorite	CuO	690 [7]	3×10^{-7}	2.16
Gibbsite (001)	$Al(OH)_3$	140 [5]	7×10^{-8}	2.37
Gibbsite (100)		483 [5]		
Quartz	SiO_2	350 [8]	1×10^{-4}	2.32
Amorph. Silica	SiO_2	46 [9]	2×10^{-3}	2.32
Kaolinite	$Al_2Si_2O_5(OH)_4$	200 [10]	1×10^{-6}	2.33

OCPp = Octacalcium phosphate (subscript p = precursor)

References:
[1] Nielsen and Söhnel (1971)
[2] Christoffersen et al. (1988)
[3] Van Cappellen (1991)
[4] Arends et al. (1987)
[5] Smith and Hem (1972)
[6] Berner (1980)
[7] Schindler (1967)
[8] Parks (1984)
[9] Alexander et al. (1954)
[10] Steefel and Van Cappellen (1990)
[11] Chiang et al. (1988)

Solubilities are calculated from a variety of sources compiled by the author. The mean ionic diameter is calculated by Eq. (6.13) using molar volumes in CRC Handbook of Chemistry and Physics. The value of a for OCPp is calculated from the crystallographic data on OCP in Brown et al. (1962).

Thus, the solubility of $Cu(OH)_2$ is ca 10 times greater than that of CuO. The inversion of $Cu(OH)_2$ into CuO should occur exergonically. However, if CuO is very finely divided, it becomes less stable than coarse $Cu(OH)_2$. Fig. 6.5c shows the variations of the reaction free energy with particle size for the conversion of hematite to goethite.

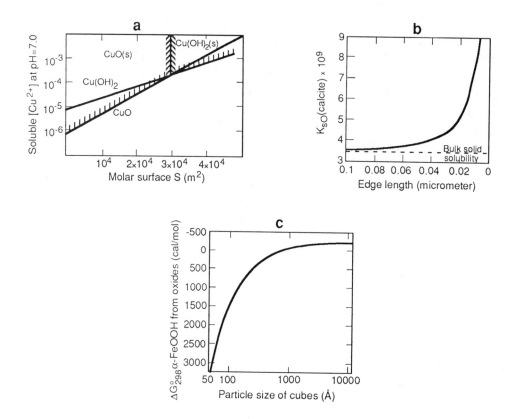

Figure 6.5

a) Influence of molar surface upon solubility of CuO and of $Cu(OH)_2$ at pH = 7.0. (From data on solubility constants and surface tensions by Schindler (1967) (The relations depicted have been validated experimentally only for $\overline{S} < 10^4$ m².) The figure suggests that $Cu(OH)_2(s)$ becomes more stable than CuO(s) for very finely divided Cu0 crystals ($\overline{S} > 3 \times 10^4$ m², d < 40 Å). Plausibly, in precipitating Cu(II), $Cu(OH)_2(s)$ may be precipitated (d = very small), but CuO(s) becomes more stable than $Cu(OH)_2$ upon growth of the crystals, and an inversion of $Cu(OH)_2$ into the more stable phase becomes possible.

b) Change in calcite solubility with particle size, assuming cubic shape and $\overline{\gamma}$ = 85 mJ m⁻². (From Morse and Mackenzie, 1990)

c) $\Delta G°$ for the reaction ½ α-Fe_2O_3 + ½ H_2O = α-FeOOH is plotted as a function of particle size assuming equal particle size for goethite and hematite. For equal-sized hematite and goethite crystals, goethite is more stable than hematite when the particle size exceeds 760 Å but less stable than hematite at smaller particle sizes. (From D. Langmuir and D.O. Whittemore, 1971)

The Precursor as a Substrate (Template) for the Formation of the More Stable Phase

As the precursor, e.g., an amorphous phase, precipitates and brings down the supersaturation of the solution, the more stable phase to be precipitated is using the precursor phase as a substrate for its own precipitation (Steefel and Van Cappellen, 1990). A classical example that documents this principle is the precipitation of calcium phosphates, where a metastable calcium phosphate precursor phase is nucleated initially and is then replaced, in some instances via an intermediate phase, by apatite. (Nancollas, 1990; Steefel and Van Cappellen, 1990).

Biomineralization. In biomineralization, inorganic elements are extracted from the environment and selectively precipitated by organisms. Usually, templates consisting of suitable macro-molecules serve as a substrate for the heterogeneous nucleation of bulk mineralized structures such as bone, teeth and shells. Biological control mechanisms are reflected not only in the type of the mineral phase formed but also in its morphology and crystallographic orientation (Mann et al., 1989; Lowenstamm and Weiner, 1989). Two examples (perhaps oversimplified) may illustrate the principle (Ochial, 1991):

1) *Silica Gel Formation in Diatoms.* The protein in the cell wall of several diatom species contain a relatively large proportion of serine (H_2N-$CH(CH_2OH)COOH$) which contains aliphatic hydroxo groups which can undergo condensation or ligand exchange reactions with salicic acid

The small aggregates of SiO_n so formed are considered to be the template for the nucleation and growth of silica.

2) *CaCO$_3$ Shells in Molluscs.* The major component of the organic matrix protein is a glycoprotein with a predominance of aspartic acid ($H_2NCH(CH_2COOH)COOH$) and glycine ($H_2N\ CH_2COOH$). The sequence apsartate $-$ X $-$ aspartate (where X is mostly glycine) in the protein is thought to nucleate $CaCO_3$.

$$
\begin{array}{ccc}
CO_3 & & CO_3 \\
| & & | \\
Ca & & Ca \\
\diagup \quad \diagdown & \diagup \quad \diagdown \\
O & O & O \\
| & | & | \\
C=0 & C=0 & C=0 \\
| & | & | \\
\end{array}
$$

$$- \; Asp - X - Asp - X - Asp - X -$$

There are various other factors which determine the crystal structure (calcite vs aragonite).

Several crystals, such as vaterite and calcite forms of $CaCO_3$, or α-glycine, have been nucleated *(induced oriented crystallization)* at the water surface covered with a monolayer film of carboxylic acids or aliphatic alcohols (compressed to "suitable" distances of the hydrophilic groups with a Langmuir balance) (Mann et al., 1988).

Molecular Recognition at Crystal Interfaces. As we have seen, the key concept for the understanding the different processes of the nucleation and crystal growth is "molecular" recognition at the interface; the surfaces of templates, nuclei and crystals can be thought of as beeing composed of "active sites" that interact stereospecifically with ions or molecules in solution, in a manner similar to the interactions of enzymes and substrates or antibodies and antigens. Crystals can be engineered with desired morphologies (Weissbuch et al., 1991).

6.5 Enhancement of Heterogeneous Nucleation by Specific Adsorption of Mineral Constituents

The "classical" theory of nucleation concentrates primarily on calculating the nucleation free energy barrier, ΔG^*. Chemical interactions are included under the form of thermodynamic quantities, such as the surface tension. A link with chemistry is made by relating the surface tension to the solubility which provides a kinetic explanation of the Ostwald Step Rule and the often observed disequilibrium conditions in natural systems. Can the chemical model be complemented and expanded by considering specific chemical interactions (surface complex formation) of the components of the cluster with the surface?

In addition to the matching of the structures of the surfaces of the mineral to be nucleated and the substrate, adsorption or chemical bonding of nucleus constituents to the surface of the substrate can be expected to enhance the nucleation. Surface complex formation and ligand exchange of crystal forming ions with the surface sites of the substrate, their partial or full dehydration and structural realignment, and perhaps the formation of ternary surface complexes, are essential, at least partially rate determining, steps in the heterogeneous nucleation process. One might argue that a critical surface concentration of surface constituent ions; i.e., a two-dimensional solubility product, must be exceeded before a nucleus is being formed.

In the nucleation of an ion lattice $\langle AB \rangle_n$ the following steps may occur

$$A(H_2O)_n^+ \quad + \; -SOH \quad \overset{k_1}{\rightleftharpoons} \quad -SOA(H_2O)_m + (n-m)\,H_2O + H^+$$

$$B(H_2O)_x^- \quad + \; -SOH \quad \overset{k_2}{\rightleftharpoons} \quad -SB(H_2O)_y \quad + \quad (x-y)\,H_2O \; + \; OH^-$$

$$-SOA(H_2O)_m + \; -SB(H_2O)_y \overset{k_3}{\rightleftharpoons} \langle AB \rangle \quad + \; 2\,-SOH + (m+y-1)\,H_2O$$

where $-SOH$ is a surface site with an OH functional group. The surface binding is accompanied with at least a partial dehydration of the ions A and B. If the last step is rate determining the rate of AB nucleus formation is given by

$$\frac{d\langle AB \rangle}{dt} = k_3\,\theta_A\,\theta_B \tag{6.20a}$$

i.e., the rate of nucleus formation is proportional to the product of the relative coverage of sites of A and B where

$$\theta_A = \frac{\text{sites occupied by A}}{\text{all sites available}}$$

The product of $\theta_A\,\theta_B$ relates to the probability to find neighboring sites occupied by A and B.

Eq. (6.20a) is an elaboration of the classic rate law (Eq. 6.12)

$$J = \bar{A}\,\exp\!\left(\frac{-\Delta G^*}{kT}\right) \tag{6.12}$$

Chemical adsorption or surface complexation as given in Eq. (6.21) attempts to relate to the "collision" factor $\bar{A}$ in Eq. (6.12) to the surface concentrations of adsorbed ions. By analogy to the treatment of activated processes the following "general" rate law for the rate of nucleation of the mineral $\langle AB \rangle$ on a foreign surface could then be proposed

$$J = k^* \, \theta_A \, \theta_B \, \exp\!\left(\frac{-\Delta G^*}{kT}\right)$$

(6.21)

where k^* includes the activation energy for the actual elementary step that controls the assembly of the nucleus.

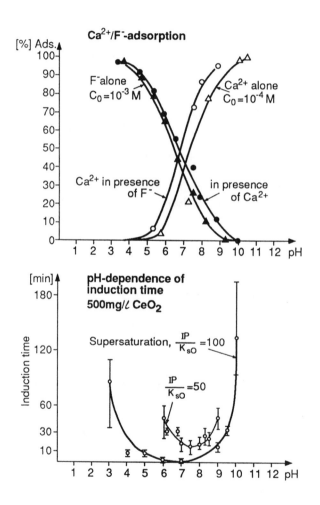

Figure 6.6

Heterogeneous nucleation of CaF_2 on CeO_2. It occurs only in pH range where Ca^{2+} and F^- are specifically bound to the CeO_2 surface. This surface coordination, accompanied by partial dehydration of the ions, appears to be a prerequisite for the nucleation. IP = Ion Product. (Data: H. Hohl)
(From Stumm, Furrer, Kunz, 1983)

An instructive example is given by the nucleation of calcium fluoride (CaF_2) on the substrate of CeO_2 (Fig. 6.6). CeO_2 and CaF_2 have the same crystalline structure and the same lattice distances. The CeO_2 substrate enhances the nucleation of CaF_2 only within the pH-range where both Ca^{2+} and F^- are specifically bound. As shown, induction times (operationally defined as the time necessary until the formation of the precipitate becomes apparent) are reduced significantly by the presence of a heterogeneous surface. Similar results have been obtained by Hohl et al. (1982) for the nucleation of MgF_2 on TiO_2. The heterogeneous nucleation rate can be retarded by cations and ligands that are competitively bound to the oxide surface.

In line with Eq. (6.20a), one could postulate for the rate of heterogeneous nucleation of CaF_2:

$$\frac{d\langle CaF_2 \rangle_n}{dt} = k\, \theta_{Ca}\, \theta_F^2 \tag{6.20b}$$

In natural waters $CaCO_3$ (calcite) nucleation occurs primarily heterogeneously. Many surfaces such as algae, biomass, aluminum silicates, aluminum oxides serve

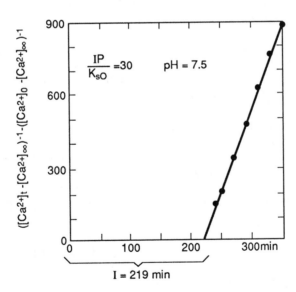

Figure 6.7

Heterogeneous nucleation of $CaCO_3$ on δ–Al_2O_3. Example for the sequence of nucleation and subsequent crystal growth. The latter is plotted as a 2nd order reaction (as is typical for screw dislocation catalysis).

(From Stumm, Furrer, Kunz, 1983)

as substrates. $\gamma\text{-}Al_2O_3$ is an excellent substrate for the nucleation of calcite (Fig. 6.7). The Ca^{2+} binding to the $\gamma\text{-}Al_2O_3$ increases with pH and becomes measurable above pH = 7. HCO_3^- and CO_3^{2-} bind also specifically to the Al_2O_3 substrate surface; correspondingly the tendency to form $CaCO_3$ nuclei is favored by a slightly alkaline pH-range.

As is well known (Morse and Berner, 1972) even very small concentrations of HPO_4^{2-} (and some organic solutes) inhibit nucleation of $CaCO_3$, most likely because these adsorbates block essential surface sites on the substrate or on the mineral clusters. Mg^{2+} is known to inhibit many nucleation processes, especially also the nucleation of Mg bearing minerals. The water exchange rate of Mg^{2+} is slower than that of many cations, such as Pb^{2+}, Cu^{2+}, Zn^{2+}, Cd^{2+}, Ca^{2+}. The inhibition effect of Mg^{2+} may be due to its sluggishness to (partial) dehydration. (Mg^{2+} has among the bivalent ions a very large enthalpy $(-\Delta H_H^o)$ of hydration.)

Certain acids with hydroxylic and carboxylic groups have been shown (Schwertmann and Cornell, 1991) to induce in Fe(III) solutions the formation of hematite because these acids may act as templates for the nucleation of hematite. These examples illustrate that a complete understanding and quantitative description of the rate of heterogeneous nucleation will have to include surface complexation and other adsorption processes.

The overall *kinetics* of crystal precipitation has to consider that the process consists of a series of consecutive processes; in simple cases, the slowest is the rate determining step. Assuming the volume diffusion is not the rate determining step, we still have at least the following reaction sequences:
  i) *Adsorption:*
      adsorption of constituent ions onto the substrate;
  ii) *Surface Nucleation:*
      diffusion of adsorbed ions; partial dehydration; formation of a two-dimensional nucleus; growth to three-dimensional nucleus;
  iii) *Crystal growth:*
      Each one of these sequential processes consist of more than one reaction step.

In a simplified scheme the reaction sequence could be depicted as in Fig. 6.8. The activation energy is for many minerals too large; they are not nucleated within the time of observation, although ΔG for the precipitation is favorable (driving force: S >> 1). Examples are dolomite in seawater and $CaCO_3$ (calcite) in seawater. While calcite readily nucleates from oversaturated fresh water, it is usually not precipitated from seawater outside of organisms. In both cases Mg^{2+} ion may be responsible for inhibiting the surface nucleation process. Organisms, e.g., foraminifera, can provide suitable templates for the heterogeneous nucleation of calcite or aragonite.

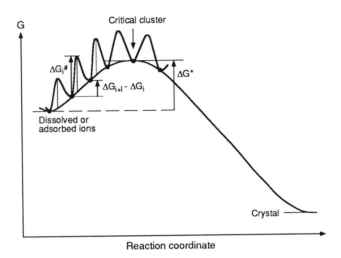

Figure 6.8

This figure illustrates schematically the relationship between the free energy of formation of clusters (ΔG_i) and the free energies of activation $\Delta G_i^\#$ of the elementary reactions $X_i + X \rightleftharpoons X_{i+1}$.

6.6 Surface Precipitation

In surface precipitation cations (or anions) which adsorb to the surface of a mineral may form at high surface coverage a precipitate of the cation (anion) with the constituent ions of the mineral. Fig. 6.9 shows schematically the surface precipitation of a cation M^{2+} to hydrous ferric oxide. This model, suggested by Farley et al. (1985), allows for a continuum between surface complex formation and bulk solution precipitation of the sorbing ion, i.e., as the cation is complexed at the surface, a new hydroxide surface is formed. In the model cations at the solid (oxide) water interface are treated as surface species, while those not in contact with the solution phase are treated as solid species forming a *solid solution* (see Appendix 6.2). The formation of a solid solution implies isomorphic substitution. At low sorbate cation concentrations, surface complexation is the dominant mechanism. As the sorbate concentration increases, the surface complex concentration and the mole fraction of the surface precipitate both increase until the surface sites become saturated. Surface precipitation then becomes the dominant "sorption" (= metal ion incorporation) mechanism. As bulk solution precipitation is approached, the mol fraction of the surface precipitate becomes large.

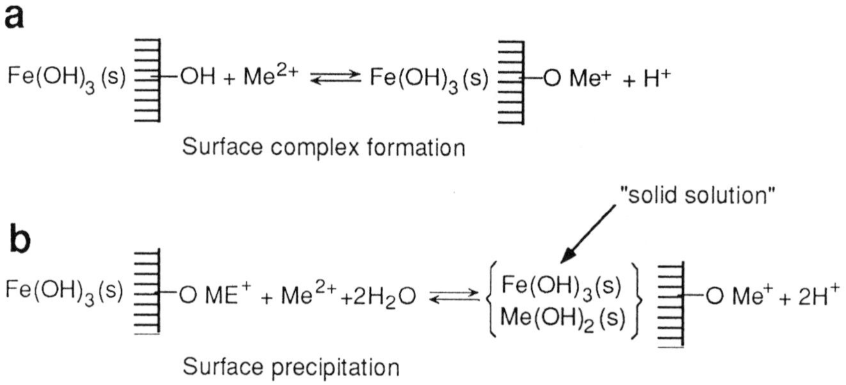

a

$$Fe(OH)_3 (s) \equiv\!\!\!-OH + Me^{2+} \rightleftharpoons Fe(OH)_3(s) \equiv\!\!\!-O\,Me^+ + H^+$$

Surface complex formation

"solid solution"

b

$$Fe(OH)_3(s) \equiv\!\!\!-O\,ME^+ + Me^{2+} + 2H_2O \rightleftharpoons \left\{ \begin{array}{l} Fe(OH)_3(s) \\ Me(OH)_2(s) \end{array} \right\} \equiv\!\!\!-O\,Me^+ + 2H^+$$

Surface precipitation

Figure 6.9

Schematic representation of surface precipitation on hydrous ferric oxide ($Fe(OH)_3(s)$)
a) At low surface coverage with Me^{2+}, surface complex formation dominates. Instead of the usual short-hand notation ($\equiv Fe–OH + Me^{2+} \rightleftharpoons \equiv FeOMe^+ + H^+$) we use one that shows the presence of $Fe(OH)_3(s)$.
b) With progressive surface coverage, surface precipitation may occur. The surface precipitate is looked at as a solid solution of $Fe(OH)_3(s)$ and $Me(OH)_2(s)$; some isomorphic substitution of Me(II) in Fe(III) occurs.
The model has been proposed by Farley, Dzombak and Morel (1985).

Fig. 6.10 shows idealized isotherms (at constant pH) for cation binding to an oxide surface. In the case of cation binding, onto a solid hydrous oxide, a metal hydroxide may precipitate and may form at the surface prior to their formation in bulk solution and thus contribute to the total apparent "sorption". The contribution of surface precipitation to the overall sorption increases as the sorbate/sorbent ratio is increased. At very high ratios, surface precipitation may become the dominant "apparent" sorption mechanism. Isotherms showing reversals as shown by e have been observed in studies of phosphate sorption by calcite (Freeman and Rowell, 1981).

Does Surface Precipitation occur at Concentrations lower than those calculated from the Solubility Product? As the theory of solid solutions (see Appendix 6.2) explains, the solubility of a constituent is greatly reduced when it becomes a minor constituent of a solid solution phase (curve b in Fig. 6.10).Thus, a solid species, e.g., $M(OH)_2$ can precipitate at lower pH values in the presence of a hydrous oxide (as a solid solvent), than in its absence.

Adsorption and Precipitation vs heterogeneous Nucleation and Surface Precipitation. There is not only a continuum between surface complexation (adsorption) and precipitation, but there is also obiously a continuum from heterogeneous nucleation to surface precipitation. The two models are two limiting cases for the initiation of precipitation. In the heterogeneous nucleation model, the interface is fixed and no mixing of ions occurs across the interface. As a consequence precipi-

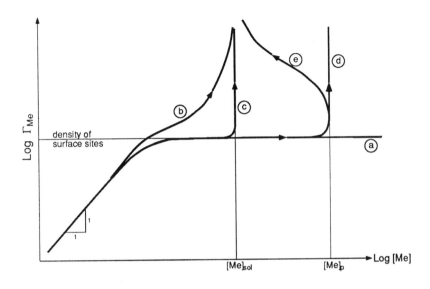

Figure 6.10

Schematic sorption isotherms of a metal ion (Me) on an oxide (XO_n) at constant pH:
a) adsorption only;
b) adsorption and surface precipitation via ideal solid solution;
c) adsorption and heterogeneous nucleation in the absence of a free energy nucleation barrier ($\Delta G^* \rightarrow 0$);
d) adsorption and heterogeneous nucleation of a metastable precursor;
e) same as in d but with transformation of the precursor into the stable phase.
The arrows show the isotherm evolution for continual addition of dissolved Me. The initial isotherm with the slope of 1 (in the double logaritmic plot) corresponds to a Langmuir isotherm (surface complex formation equilibrium). $[Me]_{sol}$ = solubility concentration of Me for the stable metal oxide; $[Me]_p$ = solubility concentration of Me for a metastable precursor (e.g., a hydrated Me oxide phase).
(From Van Cappellen; Personal Communication, 1991)

tation of the new solid phase does not occur until the solution becomes supersaturated. In the surface precipitation model, the interface is a mixing zone for the ions of the new solid phase and those of the substrate. The surface phase is treated as an ideal solid solution. This allows precipitation to start from solutions undersaturated with the pure phase. Furthermore, the composition of the surface phase can vary continuously from that of the pure substrate to that of the new phase. Whether in an actual case precipitation of a new phase approaches one or the other limiting model will depend, as has been pointed out by Van Cappellen (Personal communications, 1991), on the mixing energies of the pure new phase and the substrate. If the mixing enthalpy is small ($\Delta H_{mix} = 0$: ideal solid solution) solid solution formation should be favoured; when it is large nucleation should be favored. In other words, it should be possible to correlate the occurence of one or the other mechanism (or some combination of both) with the mismatch strain energy between the lattices of the substrate and the new phase, and with "chemical" differences in electronegativity and polarization between the substituting ions.

A Case Study on the Coprecipitation of Mn^{2+} on $FeCO_3(s)$ (siderite)

Natural carbonate minerals are rarely pure; most contain significant quantities of "impurities" (most commonly metal cations) in their structure. This assemblage is the final product of different processes occurring at the solid-water interface, which determines the availability and mobility of a given metal, and therefore records geochemical conditions during carbonate precipitation. The uptake of a cation onto a surface – similar to the ideas expressed on surface precipitation of metal ions on hydrous oxides – is thought to proceed via adsorption, surface precipitation leading to a solid solution end product (Sposito, 1986).

The solid solution formation can be characterized by

$$FeCO_3(s) + Mn^{2+} \rightleftharpoons MnCO_3(s) + Fe^{2+} \; ; \; K^{ss}$$

where K^{ss} is given by $(K_{s0(FeCO_3)} / K_{s0(MnCO_3)}) = 10^{-0.3}$ (I = 0, 25° C).

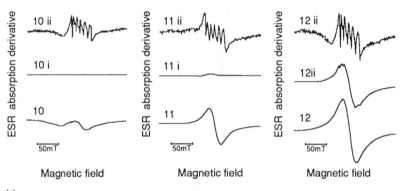

Figure 6.11

ESR spectra of Mn^{2+} sorbed on $FeCO_3(s)$; spectrum 10 with 0.4, spectrum 11 with 1.1, spectrum 12 with 3.7 mol % $MnCO_3$. The samples indexed with ii were prepared after two minutes of addition of Mn(II)-solution to the $FeCO_3(s)$ suspension; those indexed with i were prepared after one day of reaction time; and the samples without index are equilibrium samples and were prepared after about a week of equilibration. Spectra indexed ii were recorded at a gain 100 times higher. After the short equilibration time the solid-state ESR show the typical six-line spectrum of Mn^{2+}. (These six lines arise from the interaction of the electronic spin with the nuclear moment.) This is characteristic of the aquo Mn^{2+} ion and the spectrum may be attributed to Mn^{2+} loosely bound to the surface. After several hours of equilibration time a decrease of the symmetrical six-line spectrum is detected; this may be attributed to the formation of a surface complex (outer-sphere or inner-sphere). After a long (one week) equilibration time, the appearance of the broad signal is attributed to a $MnCO_3(s)$ phase.

(From Wersin, Charlet, Karthein and Stumm, 1989)

Studies on the kinetics of adsorption with electron spin resonance (ESR spectra are compatible with this mechanism, Fig. 6.11), i.e., a rapid initial adsorption is followed by an intermediate surface complexation process before the coprecipitation reaction itself occurred, giving rise possibly to a solid solution. Wersin et al. (1989) have attempted to establish a quantitative equilibrium model for this process. However, the ESR data of Wersin do not unambiguously support a solid solution formation, since the signal of the 3.7 % $MnCO_3$ is already very close to that of pure $MnCO_3$. Thus, it is possible that surface nuclei of relatively pure $MnCO_3$ are forming. Further studies are obviously necessary to distinguish in such a case between surface precipitation via ideal solid solution and heterogeneous nucleation of an additional phase.

Solid solutions of carbonates, especially also the problem of magnesian calcites, are discussed in Chapter 8.

6.7 Crystal Growth

The classical crystal growth theory goes back to Burton, Cabrera and Frank (BCF) (1951). The BCF theory presents a physical picture of the interface (Fig. 6.9c) where at kinks on a surface step – at the outcrop of a screw dislocation-adsorbed crystal constituents are sequentially incorporated into the growing lattice.

Different rate laws for crystal growth have been proposed. The empirical law, often used is

$$V = k(S - 1)^n \tag{6.22}$$

where V is the linear growth rate [length time^{-1}].

Often this law fits the experimental data well, especially at high degrees of supersaturation. Often an exponent n = 2 is found. Nielsen (1981) has explained this observation by assuming that for these solids the rate determining step is the integration of ions at kink sites of surface spirals (see Fig. 6.12).

Blum and Lasaga (1987) and Lasaga (1990) propose the very general rate equation

$$V = A(\ln S)^n \tag{6.23}$$

where A and n are adjustable parameters. In other words, the precipitation (or dissolution rate, for S < 1) depends on ΔG of the reaction (and on other parameters).

As Fig. 6.13 illustrates growth and dissolution are not symmetric with respect to the saturation state. At very high undersaturation, the rate of dissolution becomes independent of S and converges to the value of the apparent rate constant. This is why studies of dissolution far from equilibrium allow to study the influence of inhibition/ catalysis on the apparent rate constant, independently from the effect of S. The same is not true for crystal growth.

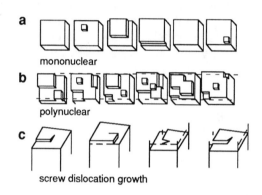

Figure 6.12

Surface models for crystal growth (figures from Nielsen, 1964)
a)   mononuclear growth
b)   polynuclear growth
c)   screw dislocation growth
Along the step a kink site is shown. Adsorbed ions diffuse along the surface and become preferentially incorporated into the crystal lattice at kink sites. As growth proceeds, the surface step winds up in a surface spiral. Often the growth reaction observed occurs in the sequence c, a, b.

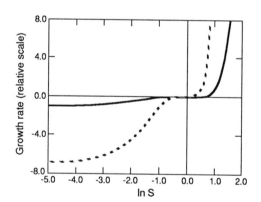

Figure 6.13

Surface reaction rate laws for dislocation-free surfaces. No surface diffusion allowed. Crystal growth for lnS > 0, dissolution for lnS < 0. Solid line, $\Phi/kT = 3.5$; dashed line, $\Phi/kT = 3.0$.
(From Blum and Lasaga, 1987)

Appendix

A.6.1 Solubility of Fine Particles

Finely divided solids have a greater solubility than large crystals. As a consequence small crystals are thermodynamically less stable and should recrystallize into large ones. For particles smaller than about 1 μm or of specific surface area greater than a few square meters per gram, surface energy may become sufficiently large to influence surface properties. Similarly the free energy of a solid may be influenced by lattice defects such as dislocations and other surface heterogeneities.

Precipitation and Dissolution

The change in the free energy ΔG involved in subdividing a coarse solid suspended in aqueous solution into a finely divided one of molar surface $\bar{S}$ is given by

$$\Delta G = \tfrac{2}{3} \bar{\gamma} \bar{S} \tag{A.6.1}$$

where $\bar{\gamma}$ is the mean free surface energy (interfacial tension) of the solid-liquid interface.

Eq. (A.6.1) can be derived (Schindler, 1967) from the thermodynamic statement that at constant temperature and pressure and assuming only one "mean" type of surface

$$dG = \mu_0 dn + \bar{\gamma} ds \tag{A.6.2}$$

$$\mu = \mu_0 + \frac{\bar{\gamma} ds}{dn} \tag{A.6.3}$$

This can be rewritten as

$$\mu = \mu_0 + \frac{M}{\rho} \bar{\gamma} \frac{ds}{dv} \tag{A.6.4}$$

where M = formula weight, ρ = density, n = number of moles, and s = surface area of a single particle.

Because the surface and the volume of a single particle of given shape are $s = kd^2$ and $v = ld^3$,

$$\frac{ds}{dv} = \frac{2s}{3v} \tag{A.6.5}$$

Since the molar surface is $\bar{S} = Ns$ and the molar volume $V = Nv = M/\rho$, where N is particles per mole, Eq. (A.6.4) can be rewritten as

$$\mu = \mu_0 + \tfrac{2}{3}\bar{\gamma}\,\bar{S} \tag{A.6.6}$$

or, in terms of Eq. (A.6.1)

$$\frac{d \ln K_{s0}}{d\bar{S}} = \tfrac{2}{3}\frac{\bar{\gamma}}{RT} \tag{A.6.7}$$

or

$$\log K_{s0(\bar{S})} = \log K_{s0(\bar{S}=0)} + \frac{\tfrac{2}{3}\bar{\gamma}}{2.3\ RT}\bar{S} \tag{A.6.8}$$

The specific surface effect can also be expressed by substituting $\bar{S} = M\alpha/\rho d$ where α is a geometric factor which depends on the shape of the crystals ($\alpha = k/l$) (compare Fig. 6.5).

A.6.2   *Solid Solution Formation* [1]

The solids occurring in nature are seldom pure solid phases. Isomorphous replacement by a foreign constituent in the crystalline lattice is an important factor by which the activity of the solid phase may be decreased. If the solids are homogeneous, that is, contain no concentration gradient, one speaks of homogeneous solid solutions. The thermodynamics of solid solution formation has been discussed by Vaslow and Boyd (1952) for solid solutions formed by AgCl(s) and AgBr(s).

To express theoretically the relationship involved we consider a two-phase system where AgBr(s) as solute becomes dissolved in solid AgCl as solvent. This corresponds to the reaction that takes place if AgCl(s) is shaken with a solution containing Br^-. The reaction might formally be characterized by the equilibrium

$$AgCl(s) + Br^- = AgBr(s) + Cl^-$$

---

[1] We follow here essentially the discussion presented in Stumm and Morgan (1981).

The equilibrium constant for this reaction, that is, the distribution constant D is given by

$$D = \frac{\dfrac{(AgBr(s))}{(AgCl(s))}}{\dfrac{(Br^-)}{(Cl^-)}} \approx \frac{\left(\dfrac{[Br^-]}{[Cl^-]}\right)_{solid}}{\left(\dfrac{[Br^-]}{[Cl^-]}\right)_{liquid}} \tag{A.6.9}$$

corresponds to the quotient of the solubility product constants of AgCl(s) and AgBr(s), where ( ) denotes activity and [ ] the concentration in a phase.

$$\frac{(Ag^+)\,(Cl^-)}{(AgCl(s))} = K_{s0_{AgCl}} \qquad\qquad \frac{(Ag^+)\,(Br^-)}{(AgBr(s))} = K_{s0_{AgBr}} \tag{A.6.10}$$

$$\frac{(Cl^-)\,(AgBr(s))}{(Br^-)\,(AgCl(s))} = \frac{K_{s0_{AgCl}}}{K_{s0_{AgBr}}} = D \tag{A.6.11}$$

The activity ratio of the solids may be replaced by the ratio of the mole fractions $(X_{AgCl} = n_{AgCl} / (n_{AgCl} + n_{AgBr})$ multiplied by activity coefficients:

$$\frac{(Cl^-)}{(Br^-)} \frac{X_{AgBr}}{X_{AgCl}} \frac{f_{AgBr}}{f_{AgCl}} = \frac{K_{s0_{AgCl}}}{K_{s0_{AgBr}}}$$

or
$$\tag{A.6.12}$$

$$\frac{X_{AgBr}}{X_{AgCl}} = \frac{K_{s0_{AgCl}}}{K_{s0_{AgBr}}} \frac{(Br^-)}{(Cl^-)} \frac{f_{AgCl}}{f_{AgBr}}$$

According to this equation the extent of dissolution of Br^- in solid AgCl (X_{AgBr}/X_{AgCl}) is a function of (a) the solubility product ratio of AgCl to AgBr; (b) the solution composition, that is, the activity ratio of Br^- to Cl^-; and (c) a solid solution factor, given by the ratio of the activity coefficients of the solid solution components (f_{AgCl}/f_{AgBr}).

As a first approximation, we may assume that f_{AgCl}/f_{AgBr} is equal to unity (ideal solid solution) and that the activity ratio of the species in the fluid may be replaced by the concentration ratio

$$\frac{[Cl^-]}{[Br^-]} \frac{X_{AgBr}}{X_{AgCl}} \approx \frac{K_{s0_{AgCl}}}{K_{s0_{AgBr}}} = D \tag{A.6.13}$$

The qualitative significance of solid solution formation can be demonstrated with the help of this simplified equation, using the following numerical example.

Consider a solid solution of 10 % AgBr in AgCl (90 %) which is in equilibrium with Cl^- and Br^-; the composition of the suspension is:

Aqueous Phase	Solid Phase
$[Cl^-]$ = $10^{-4.9}$ M	X_{AgCl} = 0.9
$[Br^-]$ = $10^{-8.4}$ M	X_{AgBr} = 0.1
$[Ag^+]$ = $10^{-4.9}$ M	

In accordance with Eq. (A.6.13), D = 400 ($\simeq$ quotient of the K_{s0} values at 25° C; $pK_{s0_{AgCl}}$ = 9.7 and $pK_{s0_{AgBr}}$ = 12.3).

The composition of the equilibrium mixture shows that Br^- has been enriched significantly in the solid phase in comparison to the liquid phase (D > 1). If one considered the concentrations of aqueous $[Br^-]$ and $[Ag^+]$, one would infer, by neglecting to consider the presence of a solid solution phase, that the solution is undersaturated with respect to AgBr ($[Ag^+]$ $[Br^-]/K_{s0_{AgBr}}$ = 0.1). Because the aqueous solution is in equilibrium with a solid solution, however, the aqueous solution is saturated with Br^-. Although the solubility of the salt that represents the major component of the solid phase is only slightly affected by the formation of solid solutions, the solubility of the minor component is appreciably reduced. The observed occurrence of certain metal ions in sediments formed from solutions that appear to be formally (in the absence of any consideration of solid solution formation) unsaturated with respect to the impurity can, in many cases, be explained by solid solution formation.

Usually, however, the distribution coefficients determined experimentally are not equal to the ratios of the solubility product because the ratio of the activity coefficients of the constituents in the solid phase cannot be assumed to be equal to 1. Actually observed D values show that activity coefficients in the solid phase may differ markedly from 1. Let us consider, for example, the coprecipitation of $MnCO_3$ in calcite. Assuming that the ratio of the activity coefficients in the aqueous solution is close to unity, the equilibrium distribution may be formulated as (cf. Eq. A.6.11)

$$\frac{K_{s0_{CaCO_3}}}{K_{s0_{MnCO_3}}} = D = \frac{[Ca^{2+}]}{[Mn^{2+}]} \frac{X_{MnCO_3}}{X_{CaCO_3}} \frac{f_{MnCO_3}}{f_{CaCO_3}} \tag{A.6.14}$$

$$D = D_{obs} \frac{f_{MnCO_3}}{f_{CaCO_3}} \tag{A.6.15}$$

The solubility product quotient at 25° C ($pK_{s0_{MnCO_3}}$ = 11.09, $pK_{s0_{CaCO_3}}$ = 8.37) can now be compared with an experimental value of D. The data of Bodine, Holland and Borcsik (1965) give D_{obs} = 17.4 (25° C). Because D is smaller than the ratio of the K_{s0} values, the solid solution factor acts to lower the solution of $MnCO_3$ in $CaCO_3$ significantly from that expected if an ideal mixture had been formed. If it is assumed that in dilute solid solutions (X_{MnCO_3} very small) the activity coefficient of the "solvent" is close to unity ($X_{CaCO_3} f_{CaCO_3} \simeq 1$), an activity coefficient for the "solute" is calculated to be f_{MnCO_3} = 31. Qualitatively, such a high activity coefficient reflects a condition similar to that of a gas dissolved in a concentrated electrolyte solution where the gas, also characterized by an activity coefficient larger than unity is "salted out" from the solution.

Solid Solution of Sr^{2+} in Calcite

It has been proposed (e.g., Schindler, 1967) that a solid solution of Sr^{2+} in calcite might control the solubility of Sr^{2+} in the ocean. We may estimate the composition of the solid solution phase (X_{SrCO_3}). The following information is available: The solubilities of $CaCO_3$(s) calcite and $SrCO_3$(s) in seawater at 25° C are characterized by p^cK_{s0} = 6.1 and 6.8, respectively. The equilibrium concentration of CO_3^{2-} is $[CO_3^{2-}]$ = $10^{-3.6}$ M. The actuel concentration of Sr^{2+} in seawater is $[Sr^{2+}] \simeq 10^{-4}$ M. For the distribution the following has been found:

$$\left(\frac{[Ca^{2+}]}{[Sr^{2+}]}\right)\left(\frac{X_{SrCO_3}}{X_{CaCO_3}}\right) = 0.14 \ (25° C)$$

Assuming a saturation equilibrium of seawater with Sr^{2+} – calcite, the equilibrium concentrations would be $[Ca^{2+}] \simeq 10^{-2.5}$ (= $K_{s0}/[CO_3^{2-}]$) and $[Sr^{2+}]$ = 10^{-4} M (= actual concentration). Thus, X_{SrCO_3}/X_{CaCO_3} = 0.004 and X_{CaCO_3} = 0.996.

It may be noted that, since the distribution coefficient is smaller than unity, the solid phase becomes depleted in strontium relative to the concentration in the aqueous solution. The small value of D may be interpreted in terms of a high activity coefficient of strontium in the solid phase, $f_{SrCO_3} \approx 38$. If the strontium were in equilibrium with strontianite, $[Sr^{2+}] \simeq 10^{-3.2}$ M, that is, its concentration would be more than six times larger than at saturation with $Ca_{0.996}Sr_{0.004}CO_3$(s). This is an illustration of the consequence of solid solution formation where with $X_{CaCO_3} f_{CaCO_3} \simeq 1$:

$$[Sr^{2+}] = \frac{X_{SrCO_3} f_{SrCO_3} {}^cK_{s0SrCO_3}}{[CO_3^{2-}]}$$

that is, *the solubility of a constituent is greatly reduced when it becomes a minor constituent of a solid solution phase.*

Heterogeneous Solid Solutions

Besides homogeneous solid solutions, *heterogeneous arrangement* of foreign ions within the lattice is possible. While homogeneous solid solutions represent a state of true thermodynamic equilibrium, heterogeneous solid solutions can persist in metastable equilibrium with the aqueous solution. Heterogeneous solid solutions may form in such a way that each crystal layer as it forms is in distribution equilibrium with the particular concentration of the aqueous solution existing at that time (Doerner and Hoskins, 1925; Gordon, Salutsky and Willard, 1959). Correspondingly, there will be a concentration gradient in the solid phase from the center to the periphery. Such a gradient results from very slow diffusion within the solid phase. Following the treatment given by Doerner and Hoskins, the distribution equilibrium for the reaction

$$CaCO_3(s) + Mn^{2+} = MnCO_3(s) + Ca^{2+} \tag{A.6.16}$$

is written as in Eq. (A.6.11), but we consider that the crystal surface is in equilibrium with the solution:

$$\left(\frac{[MnCO_3]}{[CaCO_3]}\right)_{crystal\ surface} \times \frac{[Ca^{2+}]}{[Mn^{2+}]} = D' \tag{A.6.17}$$

If $d[MnCO_3]$ and $d[CaCO_3]$, the increments of $MnCO_3$ and $CaCO_3$ deposited in the crystal surface layer are proportional to their respective solution concentrations, Eq. (A.6.18) is obtained

$$\frac{d[MnCO_3]}{d[CaCO_3]} = D' \frac{[Mn^{2+}]_0 - [Mn^{2+}]}{[Ca^{2+}]_0 - [Ca^{2+}]} \tag{A.6.18}$$

or, after rearrangement,

$$\frac{d[MnCO_3]}{[Mn^{2+}]_0 - [Mn^{2+}]} = D' \frac{d[CaCO_3]}{[Ca^{2+}]_0 - [Ca^{2+}]} \tag{A.6.19}$$

where $[Ca^{2+}]_0$ and $[Mn^{2+}]_0$ represent initial concentrations in the aqueous solution. Integration of Eq. (A.6.19) leads to

$$\log \frac{[Mn^{2+}]_0}{[Mn^{2+}]_f} = D' \log \frac{[Ca^{2+}]_0}{[Ca^{2+}]_f} \tag{A.6.20}$$

where $[Mn^{2+}]_f$ and $[Ca^{2+}]_f$ represent final concentrations in the aqueous solutions.

Most of the distribution coefficients measured to date for a variety of relatively insoluble solids are characterized by the Doerner-Hoskins relation. This relationship is usually obeyed for crystals that have been precipitated from homogeneous solution (Gordon, Salutsky and Willard). If the precipitation occurs in such a way that the aqueous phase remains as homogeneous as possible and the precipitant ion is generated gradually throughout the solution, large, well formed crystals likened to the structure of an onion are obtained. Each infinitesimal crystal layer is equivalent to a shell of an onion. As each layer is deposited, there is insufficient time for reaction between solution and crystal surface before the solid becomes coated with succeeding layers. Kinetic factors make the metastable persistence of such compounds possible for relatively long – often for geological – time spans.

Reading Suggestions

Berner, R. A. (1980), *Early Diagenensis, a Theoretical Approach*, Princeton University Press, Princeton N.Y., 241 p.

Christofferson, J., M. R. Christoffersen, W. Kibalczyc, and W. G. Perdok (1988), "Kinetics of Dissolution and Growth of Calcium Fluoride and Effects of Phosphate", *Acta Odontol Scand.* **46**, 325-336.

Lowenstamm, H. A., and S. Weiner (1985), *On Biomineralization*, Oxford University Press, New York.

Mann, S., J. Webb, and R. J. P. William (1989), *Biomineralization*, VCH Verlag, Weinheim.

Nielsen, A. E. (1964), *Kinetics of Precipitation*, Pergamon Press, Oxford, 151 p.

Steefel, C. I., and Ph. Van Cappellen (1990), "A New Kinetic Approach to Modelling Water Rock Interaction: The Role of Nucleation, Precursors, and Ostwald Ripening", *Geochim. Cosmochim. Acta* **54**, 2657.

Zhang, J. W., and G. H. Nancollas (1990), "Mechanism of Growth and Dissolution of Sparingly Soluble Salts", in M. F. Hochella Jr. and A. F. White, Eds., *Reviews in Mineralogy,* **23**, Washington D.C, pp. 365-396.

Chapter 7

Particle – Particle Interaction [1]

7.1 Introduction
Aquatic Particles as Adsorbents and Reactants

Solid particles are ubiquitous in all natural water and soil systems. High specific surface areas make such small particles efficient adsorbents for metals and other trace elements and for pollutants. The solid-water interface, mostly established by the particles, plays a commanding role in transporting and regulating most reactive elements in soil and water systems (Table 1.1). Aquatic particles are characterized by an extreme complexity and an extreme diversity – being organisms, biological debris, organic macromolecules, clays, various minerals and oxides – partially coated with organic matter and other solutes and mixtures of all of these (Table 7.1).

Particles in natural systems are usually characterized by a continuous particle size distribution. The distinction between particulate and dissolved compounds, conventionally made in the past by membrane filtration (0.45 µm) did not consider the small organic and inorganic *colloids* present in water. But a significant portion of organic matter and of inorganic matter, especially iron(III) and manganese(III, IV) oxides, sulfur and sulfides can be present as submicron particles that may not be retained by membrane filters. Recent measurements in the ocean led to the conclusion that a significant portion of the operationally defined "dissolved" organic carbon is in fact present in the form of colloidal particles.

Fig. 7.1 gives a size spectrum of water-borne particles. Particles with diameters less than 10 µm have been called *colloids*. In soils, the clay-sized and fine silt-sized particles are classified as colloids. Colloids do not dissolve, but instead remain as a solid phase in suspension. Colloids usually remain suspended because their gravitational settling is less than 10^{-2} cm s^{-1}. Under simplifying conditions (spherical particles, low Reynolds numbers), Stokes' law gives for the settling velocity, v_s

$$v_s = \frac{g}{18} \frac{\rho_s - \rho}{\eta} d^2 \qquad (7.1)$$

where g is the gravity acceleration (9.81 m s^{-2}), ρ_s and ρ are the mass density [g cm^{-3}] of the particle and of water, respectively, η is the absolute viscosity (at 20° C,

---

[1] In revising this chapter, the author profited from discussions with Rolf Grauer.

1.0×10^{-2} kg m^{-1} s^{-1}) and d is the diameter of the particle (cm). Note that v_s is proportional to the square of the particle diameter. Eq. (7.1) applies also to *flotation* (gravitational rising of suspended particles that are lighter than water; when $\rho_s < \rho$, $v_s < 0$).

The colloids are removed from the water either by settling if they aggregate or by filtration if they attach to the grains of the medium through which the solution passes (soils, ground water carriers, technological filters). *Aggregation of particles* (clays, hydrous oxides, humus, microorganisms, phytoplankton) refers in a general sense to the agglomeration of particles to larger aggregates. The process by which a colloidal suspension becomes unstable and undergoes gravitational settling is called *coagulation*. Sometimes the term *flocculation* is used to describe aggregation of colloids by bridging polymers, but all these terms are often used interchangeably.

Size spectrum of waterborne particles

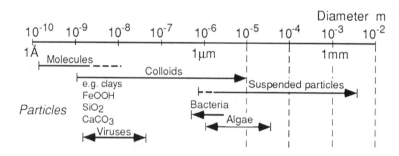

Figure 7.1

Suspended particles in natural and wastewaters vary in diameter from 0.001 to about 100 μm (1×10^{-9} to 10^{-4} m). For particles smaller than 10 μm, terminal gravitational settling will be less than about 10^{-2} cm sec^{-1}. The smaller particles (colloids) can become separated either by settling if they aggregate or by filtration if they attach to filter grains.

(From Stumm, 1977)

Aggregation of particles is important in natural water and soil systems in groundwater infiltration and groundwater transport and in water technology (coagulation and flocculation in water supply and waste water treatment, bioflocculation (aggregation of microorganisms and other suspended solids) in biological treatment processes; sludge conditioning (dewatering, filtration); filtration). Flotation is used both in water technology and in the separation of a specific mineral component from a mixture. Oceans and lakes are settling basins for particles; coagulation in these basins can be sufficiently rapid and extensive to affect suspended particle concentrations and sedimenting fluxes significantly. A significant fraction of riverborne colloids and suspended matter is coagulated and settled in the estuaries.

Table 7.1   Type of Colloids present in Natural Systems

① *River-borne Particles*

– Products of weathering and soil colloids, e.g., aluminum silicates, kaolinite, gibbsite, SiO_2
– Iron(III) and manganese(III,IV) oxides
– Phytoplankton, biological debris, humus colloids (colloidal humic acid), fibrils[1]
– So-called "dissolved" iron(III) consists mainly of colloidal Fe(III) oxides stabilized by humic or fulvic acids

② *Soil Colloids*

– Kaolinite particles. Typically about 50 unit layers of hexagonal plates are stacked irregularly and interconnected through H-bonding between the OH-groups of the octahedral sheet and the oxygens of the tetrahedral sheet (Fig. 3.9) (Sposito, 1989)
– Illite and other 2:1 layer type clay minerals. Platelike particles stacked irregularly
– Smectites and vermiculites have a lesser tendency to agglomerate because their layer charge is smaller than that of illite
– Humus, colloidal humic acids, fibrils
– Iron hydrous oxides
– Polymeric coatings of soil particles by humus, by hydrous iron(III) oxides and hydroxo-Al(III) compounds

③ *Sediment Colloids*

in addition to the colloids listed above:
– Sulfide and polysulfide colloids in anoxic sediments

④ *Biological Colloids*

– Microorganisms, virus, biocolloids, fibrils

---

1) fibrils = elongate organic colloids with a diameter of 2 – 10 nm and composed in part of polysaccharide

In dealing with colloids, the term *stability* has an entirely different meaning than in thermodynamics. A system containing colloidal particles is said to be stable if during the period of observation it is slow in changing its state of dispersion. The times for which sols[1] are stable may be years or fractions of a second. The large interface present in these systems represents a substantial free energy which by recrystallization or agglomeration tends to reach a lower value; hence, thermodynamically, the lowest energy state is attained when the sol particles have been united into aggregates. The term *stability* is also used for particles having sizes larger than those of colloids; thus, the stability of sols and suspensions can often be interpreted by the same concepts.

Historically, two classes of colloidal systems have been recognized; hydrophobic and hydrophilic colloids. In colloids of the second kind there is a strong affinity between the particles and water; in colloids of the first kind this affinity is negligible. There exists a gradual transition between hydrophobic and hydrophilic colloids. Gold sols, silver halogenides, and nonhydrated metal oxides are typical hydrophobic colloid systems. Gelatine, starch, gums, proteins, and so on, as well as all biocolloids (viruses, bacteria), are hydrophilic. Hydrophobic and hydrophilic colloids have a different stability in the same electrolyte solution. Macromolecular colloids and many biocolloids are often quite stable. Many colloid surfaces relevant in water systems contain bound H_2O molecules at their surface. Adsorption of suitable polymers may for steric reasons impart stability.

As we shall see colloid stability can be affected by electrolytes and by adsorbates that affect the surface charge of the colloids and by polymers that can affect particle interaction by forming bridges between them, or by sterically stabilizing them.

Colloids are ubiquitous, they are found everywhere in concentrations of above 10^7 to 10^8 particles per liter of water. Natural colloids are also found in subsurface and groundwaters. All these colloids are efficient adsorbents. Relatively little is known how these colloids are generated and how they are dissolved again.

Actinide Colloids. Actinide cations undergo hydrolysis in water. Hydrolysis is a step to polynucleation and thus to the generation of actinide colloids; the polynuclear hydrolysis species become readily adsorbed to the surface of natural colloids. A great part of M(III), M(IV) and M(VI) present in groundwater are colloid bound. Because of the potential possibility that such colloids migrate in an aquifer system, actinide colloids in groundwaters are presently a subject of various investigations (see Kim, 1991). Size distribution has been estimated by ultrafiltration, scanning electron microscopy and photocorrelation spectroscopy. Ultrafiltration facilitates the characterization of a number of size groups, down to ca. 1 nm diameter (see Chapter 7.10), while scanning electron microscopy determines the particle number down to ca. 50 nm. The number counting of colloids can also be performed by photo-

---

[1]   A colloid dispersed in a liquid is known as a sol.

acoustic detection of light scattering. The presence of actinide colloids can be veryfied by Laser-induced Photoacoustic Spectroscopy (Kim, 1991).

7.2 Kinetics of Particle Agglomeration [1]

The rate of particle agglomeration depends on the frequency of collisions and on the efficiency of particle contacts (as measured experimentally, for example, by the fraction of collisions leading to permanent agglomeration). We address ourselves first to a discussion of the frequency of particle collision.

Frequency of Collisions between Particles. Particles in suspension collide with each other as a consequence of at least three mechanisms of particle transport:
1) Particles move because of their thermal energy (Brownian motion). Coagulation resulting from this mode of transport is referred to as *perikinetic.*
2) If colloids are sufficiently large or the fluid shear rate high, the relative motion from velocity gradients exceeds that caused by Brownian (thermal) effects (*orthokinetic* agglomeration).
3) In settling, particles of different gravitational settling velocities may collide (agglomeration by differential settling).

The time-dependent decrease in the concentration of particles (N = number of particles per cubic centimeter) in a *monodisperse* suspension due to *collisions by Brownian motion* can be represented by a second-order rate law

$$-\frac{dN}{dt} = k_p N^2 \tag{7.2}$$

or

$$\frac{1}{N} - \frac{1}{N_0} = k_p t \tag{7.3}$$

where k_p is the rate constant [cm^3 s^{-1}]

As given by Von Smoluchowski, k_p can be expressed as

$$k_p = 4D\pi d \tag{7.4}$$

where D is the Brownian diffusion coefficient and d is the diameter of the particle. In practice, the right hand side of Eqs. (7.2) and (7.3) are multiplied by α_p. α_p is the fraction of collisions leading to permanent agglomeration and is an operational parameter for the stability ratio. For example, $\alpha = 10^{-2}$ means that one out of hundred collisions leads to mutual attachment. The diffusion coefficient in Eq. (7.4) can

---

[1] This section is largely based on Stumm and Morgan (1981).

be expressed as (Einstein-Stokes) $D = kT/3\pi\eta d$, where η is the absolute viscosity. With this substitution we obtain

$$-\frac{dN}{dt} = \alpha_p \frac{4kT}{3\eta} N^2 \qquad (7.5)$$

The rate constant k_p is on the order of 5×10^{-12} cm^3 sec^{-1} for water at 20° C and for $\alpha_p = 1$. Thus, for example, a turbid water containing 10^6 particles cm^{-3} will reduce its particle concentration by half within a period of ca. 2.5 days (2×10^5 sec) provided that all particles are completely destabilized and that the particles are sufficiently small (e.g., d < 1 μm) so that collisions result from Brownian motion only.

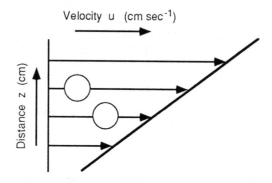

Figure 7.2
Particle collision in an idealized shear field of velocity gradient du/dz.

Agitation may accelerate the aggregation of larger particles. The velocity of the fluid may vary both spatially and temporally. The spatial changes in velocity are characterized by a velocity gradient (see Fig. 7.2). Since particles that follow the fluid motion will also have different velocities, opportunities exist for interparticle contact. The rate of decrease in particles due to agglomeration of particles (having uniform particle size) under the influence of a mean velocity gradient G [time^{-1}] can be described by

$$-\frac{dN}{dt} = \frac{2}{3} \alpha_0 G d^3 N^2 \qquad (7.6)$$

where α_0 (defined in the same way as α_p) is the fraction of collisions leading to permanent agglomeration and d is the particle diameter. It is useful to consider the ratio of the rate at which contacts occur by orthokinetic agglomeration to the rate at which contacts occur by perikinetic agglomeration (Eqs. (7.5), (7.6)); assuming

$$\alpha_0 = \alpha_p \tag{7.7}$$

$$\frac{(dN/dt)_{ortho}}{(dN/dt)_{peri}} = \frac{Gd^3}{2kT}\,\eta \tag{7.8}$$

In water at 25° C containing colloidal particles having a diameter of 1 μm, this ratio is unity when the velocity gradient is 10 sec^{-1}.

If the volume of solid particles is conserved during agglomeration, the volume fraction of colloidal particles, the volume of colloids per unit volume of suspension, ϕ, can be expressed as

$$\phi = \frac{\pi}{6}\,d_0^3\,N_0 \tag{7.9}$$

where N_0 is the initial number of particles and d_0 is the initial particle diameter; Eq. (7.6), that is, the reaction rate for a homogeneous colloid, may then be expressed as a pseudo-first-order reaction:

$$-\frac{dN}{dt} = \frac{4}{\pi}\,\alpha_0\,\phi GN \tag{7.10}$$

A numerical example might again illustrate the meaning of this rate law. For 10^6 particles of diameter d = 1 μm, ϕ becomes approximately 5×10^{-7} cm^3 cm^{-3}. For $\alpha_0 = 1$ and for a turbulence characterized by a velocity gradient G = 5 sec^{-1} (this corresponds to slow stirring in a beaker – about one revolution per second), the first-order constant $[(4/\pi)\phi G]$ is on the order of 3×10^{-6} sec^{-1}. Hence, a period of ca. 2.7 days would elapse until the concentration of particles is halved as a result of orthokinetic agglomeration.

The overall rate of decrease in concentration of particles of any size is given by Eqs. (7.6) and (7.10) by assuming additivity of the separate mechanisms

$$-\frac{dN}{dt} = \alpha_p\,\frac{4kT}{3\eta}\,N^2 + \alpha_0\,\frac{4\phi G}{\pi}\,N \tag{7.11}$$

The first term usually becomes negligible for particles with a diameter d > 1 μm, whereas the second term is less important than the first term, at least incipiently, for particles with a diameter d < 1 μm.

Heterodisperse Suspensions. The rate laws given above apply to monodisperse colloids. In polydisperse systems the particle size and the distribution of particle sizes have pronounced effects on the kinetics of agglomeration (O'Melia, 1978). For the various transport mechanisms (Brownian diffusion, fluid shear, and differential settling), the rates at which particles come into contact are given in Table 7.2.

These rate constants are compared for two cases in Fig. 7.3. It follows that heterogeneity in particle size can significantly increase agglomeration rates.

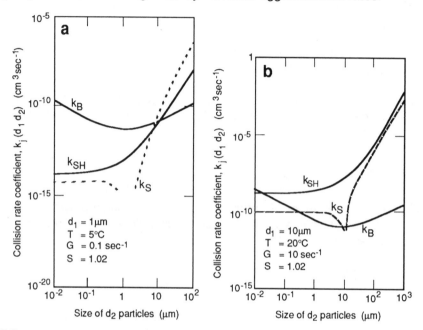

Figure 7.3

Effects of particle size on collision rate constants for agglomeration.
Left: $d_1 = 1\,\mu m$;
right: $d_1 = 10\,\mu m$.

Example 7.1:   Effects of Particle Size on Agglomeration Rate

Compare the agglomeration rate of an aqueous suspension containing 10^4 virus particles per cubic centimeter ($d = 0.01\,\mu m$) with that of a suspension containing, in addition to the virus particles, 10 mg liter^{-1} bentonite (number conc. $= 7.35 \times 10^6$ cm^{-3}; $d = 1\,\mu m$). The mixuture is stirred, $G = 10$ sec^{-1}, and the temperature is 25° C. Complete destabilization, $\alpha = 1$, may be assumed. (This example is from O'Melia, 1978.)

Let us neglect bentonite–bentonite particle interactions. We calculate from the equations given in Table 7.2 the following rate constants:

$$
\begin{aligned}
k_p &= 2 \times 10^{-12} \text{ cm}^3 \text{ sec}^{-1} & \text{(i)}\\
k_b &= 3.1 \times 10^{-10} \text{ cm}^3 \text{ sec}^{-1} & \text{(ii)}\\
k_{sh} &= 1.7 \times 10^{-12} \text{ cm}^3 \text{ sec}^{-1} & \text{(iii)}\\
k_s &= 7.8 \times 10^{-13} \text{ cm}^3 \text{ sec}^{-1} & \text{(iv)}
\end{aligned}
$$

According to Eq. (4) (Table 7.2), the time required to halve the concentration of the virus particles in the suspension containing the virus particles only would be almost 200 days. In the presence of bentonite ($k_b = 3.1 \times 10^{-10}$ cm^3 sec^{-1} and $N_{d_2} = 7.35 \times 10^6$ cm^{-3}) we find after integrating that the free virus concentration after 1 hour of contact is only 2.6 particles cm^{-3}. This example illustrates that the presence of larger particles may aid significantly in the removal of smaller ones, even when Brownian diffusion is the predominant transport mechanism.

Table 7.2   Agglomeration Kinetics of Colloidal Suspensions[a]

Transport mechanism	Rate constant for heterodisperse suspensions		Rate constant If $d_1 = d_2$ [b]	
Brownian diffusion	$k_b = \dfrac{2}{3}\dfrac{kT}{\eta}\dfrac{(d_1 + d_2)^2}{d_1 d_2}$	(1)	$k_p = \dfrac{4kT}{3\eta}$	(4)
Laminar shear	$k_{sh} = \dfrac{(d_1 + d_2)^3}{6} G$	(2)	$k_o = \dfrac{2}{3} d_p^3 G$	(5)
Differential settling	$k_s = \dfrac{\pi g (p - 1)}{72\eta}(d_1 + d_2)^3 (d_1 - d_2)$	(3)	$k_s = 0$	(6)

[a] The rate at which particles of sizes d_1 and d_2 come into contact by the jth transport mechanism is given by $F_j = k_j N_{d_1} N_{d_2}$.

F_j = collision rate in collisions per unit volume (cm^{-3} sec^{-1});
k_j = bimolecular rate constant (cm^3 sec^{-1}) for the jth mechanism;
N_{d_1} and N_{d_2} = number concentrations of particles of size d_1 and d_2, respectively (cm^{-3});
k = Boltzmann constant (1.38×10^{-23} J K^{-1});
η = absolute viscosity (g cm^{-1} sec^{-1}, or kg m^{-1} sec^{-1});
p = specific gravity of the solids (g cm^{-3});
g = gravity acceleration (cm sec^{-2});
G = mean velocity gradient (sec^{-1});
T = absolute temperature (K).

[b] A factor of 2 is applied so that the collisions are not counted double.

7.3 *Colloid-Stability; Qualitative Considerations*

In a qualitative way, colloids are stable when they are electrically charged (we will not consider here the stability of hydrophilic colloids – gelatine, starch, proteins, macromolecules, biocolloids – where stability may be enhanced by steric arrangements and the affinity of organic functional groups to water). In a physical model of colloid stability particle repulsion due to electrostatic interaction is counteracted by attraction due to van der Waal interaction. The repulsion energy depends on the surface potential and its decrease in the diffuse part of the double layer; the decay of the potential with distance is a function of the ionic strength (Fig. 3.2c and Fig.

3.3). The van der Waal attraction energy, in a first approximation is inversely proportional to the square of the intercolloid distance.

Fig. 7.4 illustrates that at the repulsion energy is affected by ionic strength while attraction energy is not. At small separations attraction outweighs repulsion and at intermediate separations repulsion predominates. With increasing ionic strength, the attraction preponders over larger interparticle distances. As Fig. 7.4 shows, a shallow so-called secondary minimum in the net interaction energy may lead to particle adhesion at larger distances between the particles. The adhesion of bacteria to one another and to surfaces as discussed in Chapter 7.9 may occur in this secondary minimum.

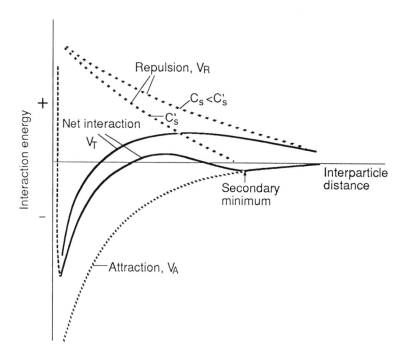

Figure 7.4

Physical model for colloid stability

Schematic forms of the curves of interaction energies (electrostatic repulsion V_R, van der Waals attraction V_A, and total (net) interaction V_T) as a function of the distance of surface separation. Summing up repulsive (conventionally considered positive) and attractive energies (considered negative) gives the total energy of interaction. Electrolyte concentration c_s is smaller than c'_s. At very small distances a repulsion between the electronic clouds (Born repulsion) becomes effective. Thus, at the distance of closest approach, a deep potential energy minimum reflecting particle aggregation occurs. A shallow so-called secondary minimum may cause a kind of aggregation that is easily counteracted by stirring.

However, chemical factors in addition to those of the electric double layer need to be considered. The specific adsorption of H^+, OH^-, metal ions and ligands (as well as the attachment of polymers) to the colloid surface affects the surface charge and the surface potential and, in turn, the colloid stability. In order to estimate colloid stability we need to know the surface potential. This cannot be measured, but – as we have seen in Chapter 3.2 – it can be calculated from the surface charge by using Gouy Chapman theory. Surface charge can, at least in principle, be determined experimentally by measuring all the charged species that have become adsorbed. In simple systems, e.g., hydrous oxides the surface charge can be predicted with the help of the surface complex formation model.

We will first illustrate how chemical variables affect surface charge and surface potential. Then we will discuss more quantitatively a physical stability model which depends on the surface potential (which, in turn, has been determined by chemical factors).

7.4 Colloid Stability; Effects of Surface Chemistry on Surface Potential

The relationship between surface charge and surface potential was discussed in Chapter 3.3 and specifically in Eq. (3.8). Here we illustrate first how the surface charge and colloid stability are affected by chemical interactions. We use as an introductory example the pH dependence of the surface charge of hematite (a case already encountered in Chapter 3; see Fig. 3.3) and its effect on the coagulation rate (Liang and Morgan, 1990). In Fig. 7.5 the potentiometrically determined (alkalimetric-acidimetric titration) surface charge is given as a function of pH for different ionic strengths. The surface charge may be compared with the measured electrophoretic mobility (Fig. 7.5b). (From the latter the zeta potential, ζ, can be calculated which is however – with exception of pH_{pzc} where $\zeta = \psi = 0$ – smaller than ψ.) The colloid stability in Fig. 7.5c is expressed as the stability ratio, W, (this is equal to α^{-1} in Eqs. (7.5) and (7.6)) and is measured experimentally by comparing the actual coagulation rate with that predicted theoretically. W is the ratio of the "fast" coagulation rate to the actual rate. Relative colloid stability is obtained from measurements of coagulation rates. In these the particle concentrations are measured as a function of time and the rate is calculated from fitting Eqs. (7.5) or (7.10). In the latter case (orthokinetic coagulation) the slope of the semilogaritmic plot of W vs time gives the coagulation rate. The concentration of the particles as a function of time can under suitable conditions be determinded spectrophotometrically. The theoretical basis of such measurements go back to Troelstra and Kruyt (1943). At pH = 8.5, the $pH_{pzc}$of the colloidal hematite particles the stability ratio is unity for all ionic strengths. At pH $\approx$ 5 and low ionic strength, coagulation is slowed down thousand-fold relative to that of pH_{pzc}, a consequence of the high positive charge density from $\equiv FeOH_2^+$ groups on the surface (compare σ and mobility data with W). At higher ionic strength-values (I = 0.05 – 0.1) the pH range of minimum stability is widened.

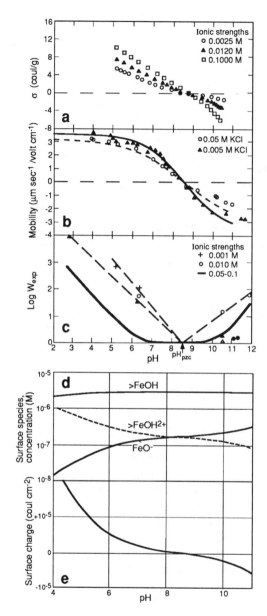

Figure 7.5

Comparison of hematite surface charge, [coul g^{-1}], electrophoretic mobility, and stability ratio, W_{exp}, as a function of pH. Note that at pH$_{pzc}$ the net surface charge and mobility are both zero, and the stability is a minimum.

The experimental stability ratio (W), the potentiometrically-determined surface charge, and the electro-kinetic mobility of 70 nm particles over the pH range from 3 to 11 are shown. The drawn-out line in Fig. c summarizes experiments obtained with I = 0.05 – 0.1. (Modified from Liang and Morgan, 1990.)

Figs. d) and e) give the results of simple equilibrium calcuations that have been made with equilibrium constants and surface characteristics of α-Fe$_2$O$_3$ given by Liang and Morgan (see Example 7.2).

Example 7.2: Surface Charge as a Function of pH for Hematite

Calculate the concentration of surface species and the surface charge as a function of pH for a hematite suspension which has the same characteristic as that used in the experiments of Liang and Morgan.

Specific surface area 40 $m^2 g^{-1}$, acidity constants of $\equiv FeOH_2^+$: $pK_{a1}^s(int) = 7.25$, $K_{a2}^s = 9.75$, site density = 4.8 nm^{-2}, hematite conc = 10 mg/ℓ. Ionic strength: 0.005. For the calculation the diffuse double layer model shall be used.

The results are given in Figs. 7.5 d) and e). The semiquantitative agreement between experimental data and calculated data is obvious. The surface charge estimated can be converted into a surface potential on the basis of the diffuse double layer model from which a stability could be calculated.

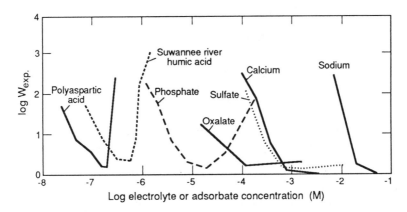

Figure 7.6

Summary plot of experimentally derived stability ratios, W_{exp}, of hematite suspensions, as a function of added electrolyte or adsorbate concentration at pH around 6.5 (pH = 10.5 for Ca^{2+} and Na^+). Hematite concentration is about 10 – 20 mg/ℓ. The stability ratio, W_{exp}, was determined from measurements on the coagulation rate; it is the reciprocal of the experimentally determined collision efficiency factor, α.

(From Liang and Morgan, 1990)

The surface charge of metal oxides (due to surface protonation) as a function of pH can be predicted if their pH_{pzc} are known with the help of the relationship given in Fig. 3.4. Fig. 7.6 exemplifies the effect of various solutes on the colloid stability of hematite at pH around 6.5 (pH = 10.5 for Ca^{2+} and Na^+) (Liang and Morgan, 1990).

Simple electrolyte ions like Cl^-, Na^+, SO_4^{2-}, Mg^{2+} and Ca^{2+} destabilize the iron(III) oxide colloids by compressing the electric double layer, i.e., by balancing the surface charge of the hematite with "counter ions" in the diffuse part of the double

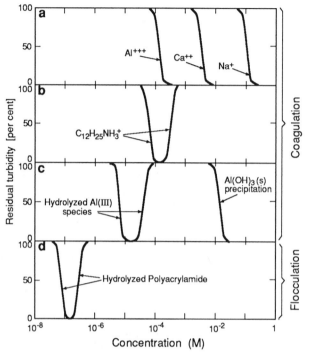

Figure 7.7

Schematic agglomeration curves for several different destabilizing agents

a) "Compaction" of the double layer by counter ions in accordance with the Schulze Hardy Rule;
b) Coagulation by specifically adsorbable organic cation;
c) Coagulation and restabilization by Al-hydroxo polymers; at higher dosage $Al(OH)_3$ precipitates and enmeshes dispersed colloids;
d) Destabilization by strongly sorbable polymers of charge equal to that of the colloid. Coagulation results from "bridging" of the colloids by polymer chains.

(From O'Melia, 1972)

layer. Usually these "simple" electrolyte ions are not – to any large extent – specifically bound[1] to the oxide surface and are thus not able to cause a charge reversal. In contrast, as Fig. 7.6 illustrates, oxalate, phosphate, humic and fulvic acids, as well as polyelectrolytes such as polyaspartic acid are specifically bound (ligand exchange) and able, at higher concentrations, to cause a charge reversal; i.e., the surface complex bound inorganic and organic species alter the metal oxide surface charge (charge or potential determining ions). The surface complex formation constants of different ligands can be used to predict the surface speciation and the surface charge from which the surface potential can be calculated. The effectiveness

---

[1] Specifically adsorbed species are those that are bound by interactions other than electrostatic ones. To what extent SO_4^{2-} and Ca^{2+} can form inner-sphere complexes is not yet well established. SO_4^{2-} is able to shift the point of zero proton condition of many oxides.

of ions in effecting the surface charge of iron(III) oxide surfaces and in coagulating or stabilizing colloidal Fe_2O_3 under comparable pH conditions with respect to pH_{pzc} (+ or −) is given by the decreasing sequence of the surface complex forming constants, e.g., phosphate > oxalate > phthalate.

The same kind of information in a more general way is given in Fig. 7.7 where schematic curves of residual turbidity as a function of coagulant dosage for a natural water treated with Na^+, Ca^{2+}, and Al^{3+}, with hydrolyzed metal ions, or with polymeric species are given. It is evident that there are dramatic differences in the coagulating abilities of simple ions (Na^+, Ca^{2+}, Al^{3+}), and hydrolyzed metal ions (charged multimeric hydroxo metal species), and other species that interact chemically with the colloid; that includes species of large ionic or molecular size.

Obviously, *two different mechanisms* occur in coagulation.
1) Inert electrolytes, i.e., ions which are not specifically adsorbed, compress the double layer and thus reduce the stability of the colloids (Fig. 7.4). A critical coagulation concentration, C_s or ccc, can be defined (see Eqs. (4) and (5) in Table 7.3) which is independent of the concentration of the colloids (Schulze-Hardy Rule).
2) Specifically adsorbable species (e.g., species that form surface complexes) affect the surface charge of the colloids and therefore their surface potentials, thereby affecting colloid stability. The critical coagulation concentration, ccc, for a given colloid surface area concentration, decreases with increasing affinity of the sorbable species to the colloid. ccc is no longer independent of the colloid concentration; there is a "stoichiometry" between the concentration necessary to just coagulate and the concentration of the colloids (Stumm and O'Melia, 1968). This is illustrated in a simple example of Fig. 7.8.

In *seawater* the "thickness" of the double layer as given by κ^{-1} (Eq. 3.9) is a few Ångstrøms, equal approximately to a hydrated ion. In other words, the double layer is compressed and hydrophobic colloids, unless stabilized by specific adsorption or by polymers, should coagulate. Some of this coagulation is observed in the estuaries where river water becomes progressively enriched with electrolytes (Fig. 7.14a). That these colloids exist in seawater for reasonable time periods is caused
1) by the small concentrations of colloids (relatively small contact opportunities),
2) the stabilization of colloids by specific adsorption, and
3) the presence of at least partially hydrophobic biological particles which dominate the detrital phases.

Specifically sorbable species that coagulate colloids at low concentrations may restabilize these dispersions at higher concentrations. When the destabilization agent and the colloid are of opposite charge, this restabilization is accompanied by a reversal of the charge of the colloidal particles. Purely coulombic attraction would not permit an attraction of counter ions in excess of the original surface charge of the colloid.

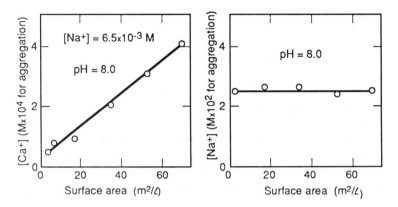

Figure 7.8

Relationship between MnO_2 colloid surface area concentration and ccc of Ca^{2+}; a stoichiometric relationship exists between ccc and the surface area concentration; in case of Na^+, however, this interaction is weaker, so that primarily compaction of the diffuse part of the double layer causes destabilization.

(From Stumm, Huang and Jenkins, 1970)

Polymers

The adsorption of polymers on colloids can
1) enhance colloid stability or
2) induce flocculation[1].
In steric stabilization the colloids are covered with a polymer sheath stabilizing the sol against coagulation by electrolytes. In sensitization or adsorption flocculation, the addition of very small concentrations of polymers or polyelectrolytes leads to destabilization (Lyklema, 1985).

As shown in Fig. 7.7d polymers can destabilize colloids even if they are of equal charge as the colloids. In polymer adsorption (cf. Fig. 4.16) chemical adsorption interaction may outweigh electrostatic repulsion. Coagulation is then achieved by *bridging* of the polymers attached to the particles. LaMer and coworkers have developed a chemical bridging theory which proposes that the extended segments attached to one of the particles can interact with vacant sites on another colloidal particle.

Flocculation of microorganisms is thought to be affected by a bridging mechanism of polymers excreeted by the microorganisms or exposed to the microbial surface under suitable physiological conditions. More recently new work on adhesion of bacteria have been carried out (Van Loosdrecht et al., 1990): see Chapter 7.9.

---

[1] Some authors distinguish between coagulation and flocculation and use the latter term for colloid agglomeration by bridging of polymer chains.

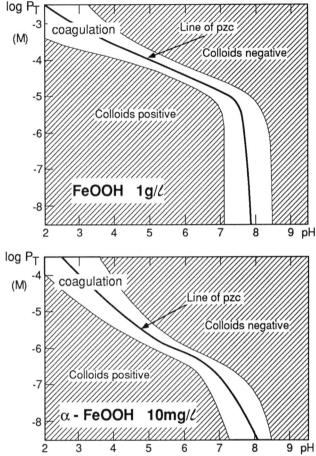

Figure 7.9

Colloid stability in the Fe(III)(hydr)oxide-phosphate system. Surface complexation equilibria were used to calculate the concentration domains of positively charged and negatively charged colloids and of nearly uncharged phosphate surface complexes on FeOOH.
(From Stumm and Sigg, 1979)

How to Model?

Which model should be used to calculate surface charge density and surface potential? Although various physical descriptions of the interface are equivalent in satisfying the charge material balance of adsorbed ions at the interface, the potential is somewhat model-dependent. Use of the diffuse double layer model provides consistency in colloid stability modelling (Liang and Morgan, 1990). To model colloid stability, it is assumed that the charge (potential) determining ions are equilibrated with the surface and a constant potential is maintained. The charge and potential distribution in the diffuse layer is governed by the Gouy Chapman theory.

Stumm and Sigg (1979) predicted the coagulation stability domains of goethite based on surface chemistry and surface potential estimates (Fig. 7.9).

Example 7.3:   Reversal of Surface Charge of Hematite by the Interaction with a Ligand

Estimate the variation of surface charge of a hematite suspension (same characteristics as that used in Example 7.2) to which various concentrations of a ligand H_2U (that forms bidentate surface complexes with the Fe(III) surface groups, $\equiv FeU^-$; such a ligand could be oxalate, phtalate, salicylate or serve as a simplified model for a humic acid; we assume acidity constants and surface complex formation constants representative for such ligands. The problem is essentially the same as that discussed in Example 5.1. We recalculate here for pH = 6.5.

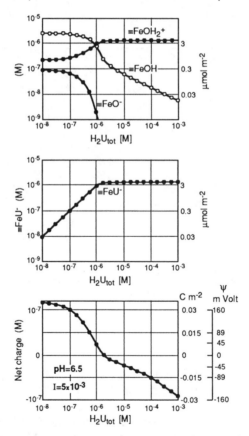

Figure 7.10

Interaction of hematite with a bidentate ligand H_2U. The relative concentrations of surface species, expressed as M, are given as a function of H_2U_T (added to the system). Coagulation is expected to occur at concentrations near the charge reversal. Conditions are given in Example 7.3 (pH = 6.5, I = 5×10^{-3}). Individual points refer to computed data.

The following equilibrium constants characterize the system:

$$H_2U \qquad\quad = H^+ + HU^-; \qquad\qquad \log K_{a1} = -5 \qquad (i)$$
$$HU^- \qquad\qquad = H^+ + U^{2-}; \qquad\qquad \log K_{a2} = -9 \qquad (ii)$$
$$\equiv FeOH_2^+ \qquad = \equiv FeOH + H^+; \qquad\quad \log K_1^s = -7.25 \quad (iii)$$
$$\equiv FeOH \qquad\quad = \equiv FeO^- + H^+; \qquad\quad \log K_2^s = -9.75 \quad (iv)$$
$$\equiv FeOH + H_2U = \equiv FeU^- + H^+ + H_2O; \qquad \log K^s = 2 \qquad (v)$$
$$H_2U_T \qquad\qquad = [H_2U] + [HU^-] + [U^{2-}] + [\equiv FeU] \qquad\qquad (vi)$$
$$[\equiv FeOH_T] \qquad = [\equiv FeOH_2^+] + [\equiv FeOH] + [\equiv FeO^-] + [\equiv FeU^-] \qquad (vii)$$

The diffuse double layer model for $[I = 0.005\ M]$ is used to correct for electrostatic effects. The calculation is made with a MICROQL program adapted for surface chemical interactions and adapted for a personal computer. Fig. 7.10 gives the calculations for increasing concentrations of H_2U_T. The surface charge is reversed at a concentration of ca. $2 \times 10^{-6}\ M$. This is in qualitative agreement of the experimentally obtained curve for humic acids, given in Fig. 7.6. With humic acids the experimentally determined concentration for charge reversal is somewhat smaller than that calculated in Fig. 7.10); this can be accounted for perhaps by inferring a polymeric effect (association of humate anions at the surface).

Adsorption of and Coagulation by Fatty Acids

Fig. 7.11 gives results on coagulation of hematite suspension by fatty acids. As concentration is increased, an influence on hematite coagulation rate becomes notice-

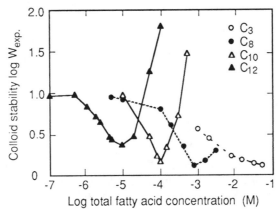

Figure 7.11

Experimentally derived stability ratio, W_{exp}, of hematite suspensions, plotted as a function of fatty acid concentration at pH 5.2. The ionic strength is 50 milimolar NaCl and hematite concentration is 34.0 mg/ℓ. Lauric acid is denoted by C_{12}, capric acid by C_{10}, caprylic acid by C_8 and propionic acid by C_3. (From Liang and Morgan, 1990)

able. Increase in fatty acid concentration first makes hematite less stable, and a minimum stability ratio is reached. The stability ratio then increases sharply as the fatty acid concentration is increased beyond the critical coagulation concentration. Plots of relative coagulation rate versus fatty acid concentration for C_8, C_{10} and C_{12} are similar to one another, but successive critical coagulation concentrations differ by a factor of about 10. For C_3, the stability ratio merely flattens out as the concentration is increased beyond a critical coagulation concentration. That a charge reversal occurs can only be accounted for by two-dimensional association of the adsorbed anions of the fatty acids (hemicelle formation). The situation is comparable to that described for the adsorption of sodium dodecyl sulfate on alumina, described in Chapter 4.5 and in Fig. 4.11.

7.5 A Physical Model for Colloid Stability [1]

To what extent can theory predict the collision efficiency factor? Two groups of researchers, *Derjagin* and *Landau*, and *Verwey* and *Overbeek*, independently of each other, have developed such a theory (the DLVO theory) (1948) by quantitatively evaluating the balance of repulsive and attractive forces that interact most effective tool in the interpretation of many empirical facts in colloid chemistry.

The DLVO theory considers van der Waals' attraction and diffuse double-layer repulsion as the sole operative factors. It calculates the interaction energy (as a function of interparticle distance) as the reversible isothermal work (i.e., Gibbs free energy) required to bring two particles from distance ∞ to distance d. Physically the requirement is that at any instant during interaction the two double layers are fully equilibrated (Lyklema, 1978). The mathematics of the interaction are different for the interaction of constant-potential surfaces than for the interaction of constant-charge surfaces. As long as the interaction is not very strong, that is, as long as the surfaces do not come too close, it does not make too much difference (Lyklema, 1978). Table 7.3 gives the approximate equations for the constant-potential case. In order to illustrate the use of units, we calculate the interaction energy of two flat plates in Example 7.3.

Example 7.4: Calculating Interaction Energy

Calculate the interaction energy of two flat Gouy plates of 25-mV surface potential (assumed to be constant), at 25° C in a 10^{-3} M NaCl solution, at a distance of 10 nm. A Hamaker constant $A_{11(2)} = 10^{-19}$ J may be used.

Using (1) to (3) of Table 7.3, and considering that, for a 10^{-3} M NaCl solution $\kappa^{-1} = 9.5 \times 10^{-9}$ m, we calculate

---

[1] Modified from Stumm and Morgen (1981)

an exponential fashion with increasing separation. V_R increases roughly in proportion with Ψ_d^2 (for small Ψ_d, tanh $u \approx u$). The distance characterizing the repulsive interaction is similar in magnitude to the thickness of a single double layer (κ^{-1}). Thus, the range of repulsion depends primarily on the ionic strength. The energy of attraction due to van der Waals attraction dispersion forces was plotted in the lower part of Fig. 7.12 as a function of separation This curve varies little for a given value of the Hamaker constant A which depends on the density and polarizability of the dispersed phase but is essentially independent of the ionic makeup of the solution. Often values for A between 10^{-19} and 10^{-20} J are adopted. Summing up repulsive and attractive energies gives the total energy of interaction. Conventionally, the repulsive potential is considered positive and the attractive potential negative. At small separations, attraction outweighs repulsion, and at intermediate separations repulsion predominates. This energy barrier is usually characterized by the maximum (net repulsion energy) of the total potential energy curve, V_{max} (Fig. 7.12). The potential energy curve shows, under certain conditions, a secondary minimum at larger interparticle distances ($d \approx 10^{-6}$ cm). This secondary minimum depends among other things on the choice of the Hamaker constant and on the dimensions of the particles involved; it is seldom deep enough to cause instability but might help in explaining certain loose forms of adhesion or agglomeration.

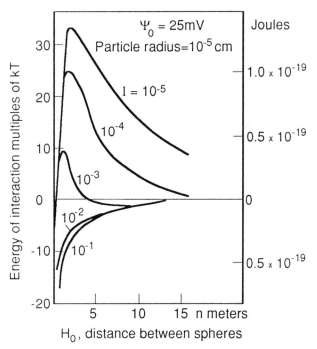

Figure 7.12

Physical model for colloid stability. Net energy of interaction for spheres of constant potential surface for various ionic strengths (1:1 electrolyte) (cf. Verwey and Overbeck).

Table 7.3   Colloid Stability as Calculated from van der Waals Attraction and Electrostatic Diffuse Double-Layer Repulsion [a) b)]

Additive interactions of repulsive interaction energy V_R and attraction energy V_A:

$$V_T = V_R + V_A \tag{1}$$

Repulsive interaction per unit area between flat plates:

$$V_R = \frac{64 n_s kT}{\kappa} \left[\tanh \left(\frac{ze\psi_d}{4kT} \right) \right]^2 e^{-\kappa d} \tag{2a}$$

for spherical particles:

$$V_R = \frac{64 \pi n_s kT}{\kappa^2} \frac{(a + \delta)^2}{R} \left[\tanh \left(\frac{ze\psi_d}{4kT} \right) \right]^2 e^{-\kappa(H - 2\delta)} \tag{2b}$$

Van der Waals attraction per unit area[c)] for flat plates:

$$V_A = - \frac{A_{11(2)}}{12 \pi d^2} \tag{3a}$$

for spherical particles:

$$V_A = - \frac{A_{11(2)}}{6} \left(\frac{2}{s^2 - 4} + \frac{2}{s^2} + \ln \frac{s^2 - 4}{s^2} \right) \tag{3b}$$

where

$$s = \frac{R}{a}$$

For very short particle distances (3b) may be replaced by

$$V_A = \frac{A_{11(2)} a}{12 H} \tag{3c}$$

Electrolyte concentration required to just coagulate the colloids (25° C): [d)]

$$c_s = 8 \times 10^{-36} \frac{[\tanh(ze\Psi/4kT)]^4}{A_{11(2)}^2 z^6} \tag{4}$$

Valence effect on stability for small Ψ:

$$c_s = \frac{3.125 \times 10^{-38}}{A_{11(2)}^2 z^2} \left(\frac{e\Psi}{kT} \right)^4 \tag{5}$$

Stability Ratio:

$$W = 2a \int_{2a}^{\infty} \exp \left(\frac{V_T}{kT} \right) R^{-2} \, dR \tag{6}$$

a) a = particle radius (cm);
    A = Hamaker constant (J);
    c_S = concentration of salt, mM;
    d = interaction distance between two surfaces (cm);
    e = elementary charge, 1.6×10^{-19} C;
    H = shortest interaction distance between two spherical particles (cm);
    k = Boltzmann constant, 1.38×10^{-23} J K^{-1};
    n_S = number of "molecules" or ion pairs (cm^{-3});
    R = distance between centers of two spheres (cm);
    W = stability ratio $(1/\alpha)$ (in the kinetic equations, k_j can be replaced by k_j/W);
    T = absolute temperature (K);
    V = interaction energy (J m^{-2});
    z = charge of ion (valence);
    κ = reciprocal thickness of double layer (cm^{-1});
    Ψ = potential (mV) at the plane where the diffuse double-layer begins.

b) Cf. J. Lyklema (1978). Most equations below stem from DLVO theory. As exact solutions do not exist, recourse must be made to approximations.

c) $A_{11(2)}$ is the Hamaker constant [dimension (energy)] that applies for the interaction between particles 1 in medium 2. This quantity can be related to the corresponding constants for attration in vacuum between particles 1 or between particles 2 by $A_{11(2)} = \left(\sqrt{A_{11}} - \sqrt{A_{22}}\right)^2$.

d) Calculated for the condition that $V_T \leq 0$ and $\delta V_T/\delta d = 0$.

$$V_R = \frac{64 \text{ mol m}^{-3} \times 6.02 \times 10^{23} \times 1.38 \times 10^{-23} \text{ J K}^{-1} \times 298.13 \text{ K}}{1.05 \times 10^8 \text{ m}^{-1}}$$

$$\left[\times \tanh \left(\frac{1.6 \times 10^{-19} \text{ C} \times 25 \times 10^{-3} \text{ V}}{4 \times 1.38 \times 10^{-23} \text{ J K}^{-1} \times 298.13 \text{ K}}\right)\right]^2$$

$$\times \exp - (1.05 \times 10^8 \text{ m}^{-1} \times 10^{-8} \text{ m})$$

$$= 1.5 \times 10^{-3} \text{ J m}^{-2} \times 5.7 \times 10^{-2} \times 0.35$$

$$= 3.0 \times 10^{-5} \text{ J m}^{-2}$$

$$V_A = -\frac{10^{-19} \text{ J}}{12 \times 3.14 \times 10^{-16} \text{ m}^2}$$

$$= -2.65 \times 10^{-5} \text{ J m}^{-2}$$

$$V_T = 3.5 \times 10^{-6} \text{ J m}^{-2}$$

In Fig. 7.4 the energies of interaction (double-layer repulsion, V_R, and van der Waals attraction, V_A, and net total interaction, V_T) were plotted as a function of distance of the separation of the surfaces. As (2a) of Table 7.3 shows, V_R decreases in

The Stability Ratio. We have already defined the stability ratio operationally. A stable colloidal dispersion is characterized by a high-energy barrier, that is, by a net repulsive interaction energy. Fuchs has defined a stability ratio W that is related to the area enclosed by the resultant curve of the energy of interaction versus separation distance. W is the factor by which agglomeration is slower than in the absence of an energy barrier. The conceptually defined W should correspond to the operationally determined α (W = α^{-1}). In a first approximation the stability ratio W is related to the height of the potential energy barrier V_{max}. The latter is conveniently expressed in units of kT. If V_{max} exceeds the value of a few kT, relatively stable colloids will be found. For example, if $V_{max} \simeq$ 15 kT, only 1 out of 10^6 collisions will be successful ($\alpha = 10^{-6}$). Fig. 7.12 shows how V_{max} typically decreases with an increase in electrolyte concentration.

Some of the pertinent interactions that affect colloid stability are readily apparent from Figs. 7.4 and 7.12. The main effect of electrolytes is a more rapid decay of the repulsion energy with distance and to compact the double layer (reducing κ^{-1}). Experimentally it is known that the charge of the counterion plays an important role. The critical electrolyte concentration required just to agglomerate the colloids is proportional to $z^{-6} A_{11(2)}^{-2}$ for high surface potential, and to $z^{-2} A_{11(2)}^{-2}$ at low potentials [(4) and (5) in Table 7.3]. This is the theoretical basis for the qualitative *valency rule of Schulze and Hardy.*

Limitations of the DLVO Theory

The DLVO theory is a theoretical construct that has been able to explain many experimental data in at least a semiquantitative manner; it illustrates plausibly that at least two types of interactins (attraction and repulsion) are needed to account for the overall interaction energy as a function of distance between the particles.

Measurements on the interaction force between two mica plates (Israelachvili and Adams, 1978) have shown good agreement with the theory in dilute KNO_3 solution at distances larger than ca. 5 nm. Similar results have recently been obtained with measurements between SiO_2 with the help of atomic forme microscopic measurements (Ducker et al., 1991). In solutions containing bivalent ions the agreement between theory and measurement is less good for various reasons, but also because the description of the diffuse double layer theory fails at short distances from the surface (Sposito, 1984). Apparently non-DLVO forces such as solvation, hydration and capillary effects may become operative at separations below 5 nm. Early validations of the theory were based on coagulation studies of monodisperse suspensions.

Critical Coagulation Concentration, ccc in Coagulation. Under simplifying conditions it is possible to derive the relation that ccc of mono-, di- and trivalent ions vary – as we have seen – in the ratio (1/z) or 100 : 1.6 : 0.13 or at low potentials in the ratio $(1/z)^2$ or 100 : 25 : 11.

Liang and Morgan (1990), for example, found the following critical coagulation concentrations, ccc, for hematite (diameter = 70 nm; pH_{pznpc} = 8.5) at pH = 6.5 : Cl^- : 80 mM; SO_4^{2-} : 0.45 mM; HPO_4^{2-} : 0.016 mM. Such results are not in accord with the Schulze-Hardy rule. (The experimental ccc's are in the ratio 100 : 0.56 (SO_4^{2-}): 0.02 (HPO_4^{2-}).) The reason is the chemical interaction of the coagulant ions with the colloids, i.e., SO_4^{2-} forms moderately stable und the HPO_4^{2-} very stable surface complexes with the hematite. As we have seen, such chemical interaction causes a stoichiometric relationship between ccc and the (incipient) colloid surface area concentration. The addition of an excess of coagulant can cause a charge reversal and a restabilization of the colloid (Stumm and O'Melia, 1968).

The presence of polymers or polyelectrolytes have important effects on the Van der Waal interaction and on the electrostatic interaction. Bacterial adhesion, as discussed in Chapter 7.9 may be interpreted in terms of DLVO theory. Since the interaction in bacterial adhesion occurs at larger distances, this interaction may be looked at as occurring in the secondary minimum of the net interaction energy (Fig. 7.4). *Particle Size.* The DLVO theory predicts an increase of the total interaction energy with an increase in particle size. This effect cannot be verified in coagulation studies.

In summary, the DLVO theory seems to break down at very close separation where interfacial phenomena such as particle–particle interaction (coagulation) and particle-surface interaction (deposition) are important.

7.6 Filtration compared with Coagulation

Filtration is analogous to coagulation in many respects. This is illustrated by juxtaposing the basic kinetic equations on particle removal:

$$\text{orthokinetic coagulation:} \quad -\frac{dN}{dt} = \frac{4}{\pi} \alpha G \phi N \qquad (7.12)$$

$$\text{packed-bed filtration:} \quad -\frac{dN}{dL} = \frac{3}{2} \frac{(1-f)}{d} \alpha \eta N \qquad (7.13)$$

where t the time, $1 - f$, the volume of filter media per unit volume of filter bed where f is porosity, η, a "single-collector efficiency" that reflects the rate at which particle contacts occur between suspended particles and the filter bed by mass transport, L, the bed depth, d, the diameter of the filter grain. The effectiveness of particle aggregation in coagulation and particle removal in filtration (attainable on integration of Eqs. (7.2) and (7.3) depends on a dimensionless product of the variables shown in Table 7.4, which have comparable meanings for both processes. Particle transport,

Table 7.4   Comparison between Coagulation and Filtration [a]

Coagulation	α	ϕ	G	t
	Collision efficiency	volumetric concentration of suspended particles	Velocity gradient	time
Filtration	α	$1 - f$	η	$\dfrac{L}{d}$
		volumetric concentration of filter medium	single-collector efficiency (v,d)	number of collectors
			contact opportunities	
Design and operational variables	Chemicals	Coagulation aids, media size, sludge recirculation	Energy input, mass transport	Residence time filter length and media diameter

[a] From W. Stumm (1977). The table is based at least partially on ideas expressed by C.R. O'Melia. The effectiveness of coagulation and filtration depends on a related dimensionless product.

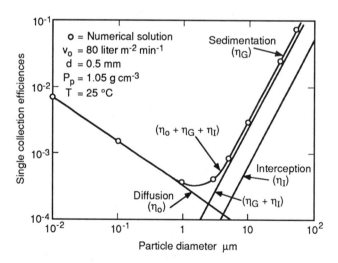

Figure 7.13

Theoretical dependence of filter efficiency of a single collector (proportional to the rate at which particle contacts occur between particles and the filter grain by mass transport) on particle diameter. For particles of small diameters transport by diffusion increases with decreasing size. Contact opportunities of the larger particles with the filter grain are due to interception and sedimentation; they increase with increasing size.

(From Yao, Habibian and O'Melia, 1971)

that is, by forces of fluid-mechanical origin, is required in both processes either to move the particles toward each other or to transport them to the surface of the filter grain or to the surface of a previously deposited particle (O'Melia, 1974). The contact opportunities of the particles for collisions with one another or with filter grains depend on Gt or η(L/d). The detention time, t, is somewhat related to L/d, the ratio of bed depth to medium diameter (O'Melia, 1974).

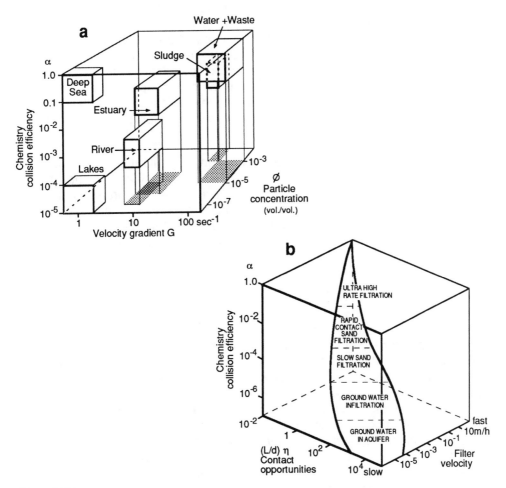

Figure 7.14

Variables that typically determine the efficiency of coagulation and filtration in natural waters and in water and waste treatment systems
a) How the variables determine the coagulation efficiency
b) Marked increase in filtration rate can be achieved by counterbalancing a reduction in contact opportunities by chemically improving the contact efficiency, with similar efficiency in particle removal. (Compare with Table 7.4.)

(From Stumm, 1977)

Fig. 7.14a illustrates how coagulation in natural systems and in water and waste treatment systems depends on the variables in Table 7.4. In natural waters long detention times may provide sufficient contact opportunities despite very small collision frequencies (small G and small ϕ). In fresh water the collision efficiency is usually also low ($\alpha \sim 10^{-3}$ to 10^{-6}; that is, only 1 out of 10^3 to 10^6 collisions leads to a successful agglomeration). In seawater, colloids are less stable ($\alpha \cong 0.1$ to 1) because of the high salinity; κ^{-1} for seawater is about 0.36 nm, hence seawater double layers are nondiffuse. Estuaries with their salinity gradients and tidal movements represent gigantic natural coagulation reactors where much of the dispersed colloidal matter of rivers settles. In water and waste treatment systems we can reduce detention time (volume of the tank) by adding coagulants at a proper dosage ($\alpha \rightarrow 1$) and by adjusting the power input (G). If the concentration ϕ is too small, it can be increased by adding additional colloids as so-called coagulation aids. O´Melia has shown that similar trade-offs in operation and design exist for optimizing particle removal in filtration.

Fig. 7.14b shows how, in the development from very slow filtration in groundwater percolation to ultrahigh-rate contact filtration, a relatively constant efficiency in particle removal (constant product $\alpha\eta(L/d)$) is maintained despite a dramatic increase in filtration rate. This is achieved by counterbalancing decreased contact opportunities (decreasing η and L and increasing d) by improving the effectiveness of particle attachment (through natural release or addition of suitable chemical destabilizing agents that increase α to 1).

Soils

Similar considerations as used for colloids in natural water systems can be used in soil systems. The stability of soil colloids is affected by the effect of the electrolyte concentration on the diffuse double layer. But chemical factors that have to be considered in addition to those of the electric double layer are of utmost importance. Acid-base reactions influence the pH_{pzc} and the pH-dependent surface charge. Specific adsorption of small ions (surface complexation) change the surface charge. Adsorption of polymer ions (incl. humic acids) change the surface charge; furthermore, coagulation by the polymer bridging mechanism may take place at extremely low electrolyte concentrations. Alternatively the colloids coated with polymer ions may be stabilized because of steric effects and/or resulting repulsive electrostatic and solvation forces.

A special case of coagulation is the "quasi crystal" formation by unit layers of montmorillonite bearing exchangeable Ca^{2+} cations (cf. Fig. 3.10). As Sposito (1989) points out, "one can imagine that the competition between the repulsive electrostatic forces and the attractive van der Waals force will, along with random thermal motions, largely determine the behavior of two siloxane surfaces approaching each other to a distance of separation > 10 nm. However, at a separation distance of

0.95 nm, which is characteristic of the outersphere surface complex in Fig. 3.10, it can be expected that the force required to bring the particle surfaces together must have a component that reflects the effort necessary to desolvate the exchangeable cations. Indeed between 10 and 0.95 nm, the force bringing the siloxane surfaces into close proximity must displace all the water molecules from the second solvation shell of the Ca^{2+} cation. When the two surfaces coalesce into the quasi crystal configuration, these outer most solvating water molecules which have been ejected from the interlayer region, and the force required to accomplish this task, must to some extent, depend on the structure of the cation solvation complex at the molecular level."

Colloid Transport in Aquifers and Packed Bed Filtration

Substantial concentrations of colloids have been found in aquifers. To what extent are such colloids transported in packed bed filtration and in aquifers and to what extent are such colloids generated within the filtration media. Do colloids, capable of adsorbing substantial amounts of contaminants, contribute to the migration of organic pollutants, heavy metals and radionuclides in groundwater? As shown in Fig. 7.12, for colloids, the major transport mechanism is convective diffusion and the capture efficiency should increase with decreasing diameter, whereas for larger than ~ 1 μm particles interception and sedimentation are dominant. The mechanisms by which attached particles can detach from filter grains are not very well understood. In very porous filter grains or in systems with cracks colloids may become chromatographically separated from ions by *size exclusion*. (While ions diffuse readily into the pores of the filter grains, colloids cannot enter into the porous structure and thus, colloid retardation (with regard to water) may under certain circumstances be smaller than that of solutes). For a discussion on colloid transport in aquifers, see O'Melia (1990), Gschwend and Reynolds (1987) and Ryan and Gschwend (1990).

7.7 Coagulation in Lakes [1]

With hydraulic residence times ranging from months to years, lakes are efficient settling basins for particles. Lacustrine sediments are sinks for nutrients and for pollutants such as heavy metals and synthetic organic compounds that associate with settling particles. Natural aggregation (coagulation) increases particle sizes and thus particle settling velocities (Eq. 7.1) and accelerates particle removal to the bottom sediments and decreases particle concentrations in the water column.

---

[1] The writing of this section has been stimulated by the work of O'Melia and Weilenmann (Weilenmann et al., 1989, and O'Melia, 1990).

In field studies on coagulation and sedimentation in lakes (Weilenmann et al., 1989), the particle size distribution at concentrations at several depths in the water column of Lake Zurich were measured (Fig. 7.15).

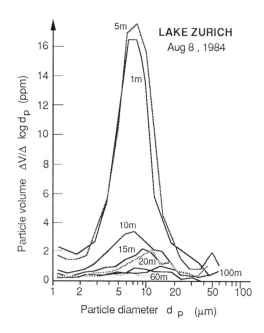

Figure 7.15

Observed particle volume concentration distributions in the water column of Lake Zurich. Samples obtained on August 8, 1984 at the depths indicated.
(From Weilenmann et al., 1989)

In this figure particle concentrations are expressed as the incremental change in particle volume concentration (ΔV) that is observed in an incremental logarithmic change in particle diameter ($\Delta \log d_p$). Plots such as in Fig. 7.15 have the useful characteristics that the area under a curve between any two particle sizes is related to the total volume concentration of particles between these two sizes (Lerman, 1979). The particle volume distributions in this lake have a single size peak; the particle diameter increases somewhat with depth. Particle volume concentration decreases substantially with depth. The total particle volume concentration in the epilimnion (as indicated by the area under the curves for 1 or 5 m) is about 10 cm³ m⁻³ (10 ppm), while the distributions in the bottom waters indicate total volume concentration of about 1 cm³ m⁻³ (1 ppm) in the hypolimnion. How do these results compare with present theories for the kinetics of coagulation and settling?

Experimental evidence obtained from Swiss lakes were compared with model simulations so as to evaluate effects that coagulation can have in lakes. In the course of this study special attention was directed towards the chemical factors that influence colloidal stability in natural waters.

The experimental aspects of this study were focussed on two hard-water lakes in Switzerland, namely, the northern basin of Lake Zurich and Lake Sempach. The hydraulic residence time of Lake Zurich is 1.2 years. Most of the particles in the lake are produced directly or indirectly by biological processes within the lake itself (e.g., photosynthesis, $CaCO_3$ precipitation). Phosphorus removal has been implemented in recent years at all wastewater treatment plants discharging into the lake; at present Lake Zurich can be described as between meso- and eutrophic. Lake Sempach has an average hydraulic residence time of 15.8 years; as in Lake Zurich, particles in the lake waters are primarily autochthonous. Phosphorus concentrations have increased substantially and the lake is eutrophic.

Experimental measurements in each lake included particle concentration and size measurements in the water column, sedimentation fluxes in sediment traps, and chemical and size characteristics of materials recovered from sediment traps. The colloidal stability of the particles in the lake waters was determined with laboratory coagulation tests. *Colloidal stability* was described by the stability ratio (α). (For a perfectly stable suspension, $\alpha = 0$; for a complete unstable one, $\alpha = 1$.)

Box models were used in a model to simulate particle transport in lacustrine systems that involve fluid flow, coagulation, and gravity.

Particle stabilities in Lake Zurich ($\alpha \approx 0.1$) and Lake Sempach ($\alpha \approx 0.01$) differed by about an order of magnitude, with the particles in Lake Sempach being more stable. Dissolved natural organic substances similar to humic acids, have been reported as stabilizing agents and divalent metal ions as destabilizing agents in aquatic systems. Both lakes contain appreciable and similar concentrations of calcium, about 1.2×10^{-3} M. The difference in colloidal stability between the two lakes is consistent with the difference in the concentrations of dissolved organic carbon (DOC) in these lakes. DOC is low in Lake Zurich, ~ 1 mg liter^{-1}; in Lake Sempach it is about 4 mg liter^{-1}. These results suggest that solution chemistry retards coagulation in Lake Sempach, reducing coagulation rate coefficients by an order of magnitude compared to Lake Zurich. The low particle stability (high attachment probability) in Lake Zurich favors coagulation, particle growth, and enhanced sedimentation.

Results of model simulations of effects of coagulation ($\alpha = 0.1$) and sedimentation at steady state in Lake Zurich during summer are presented in Fig. 7.16. Particle volume concentrations in the epilimnion and hypolimnion are plotted as functions of the particle production flux in the epilimnion. Biological degradation and chemical dissolution of particles are neglected in these calculations. Predicted particle concentrations are sensitive to the physical characteristics of the particles produced in the epilimnion, to their aggregation, and to their removal by sedimentation. As shown in Fig. 7.15 observed particle volume concentrations are $5 - 10$ cm^3 m^{-3} in the epilimnion and ~ 1 cm^3 m^{-3} in the hypolimnion. Model simulations indicate that these concentrations can result from a particle production flux of $7 - 10$ cm^3 m^{-2} d^{-1} and a particle or aggregate density of 1.5 g cm^{-3}, in agreement with estimates from the sediment trap data. Aggregates with a density of 1.5 g cm^{-3} are considered to result from coagulation of calcite particles with phytoplankton. A particle production flux of 5 cm^3 m^{-2} d^{-1} in the epilimnion would produce, in the absence of particle removal by sedimentation assisted by coagulation, a particle volume concentration in the epilimnion of 43 cm^3 m^{-3} (Fig. 7.16), considerably in excess of observations. The model for coagulation and sedimentation used in this research provides a physical connection among particle production rates, deposition fluxes, and water column concentrations observed in Lake Zurich. In Lake Sempach, coagulation occurs and is important in establishing the decrease in particle concentration with depth observed from the epilimnion to the hypolimnion, but it is sufficiently slow that biological degradation and chemical dissolution of particulate materials exert substantial effects in reducing suspended concentrations and settling fluxes. The origin of this slow coagulation rate in Lake Sempach lies in solution chemistry and

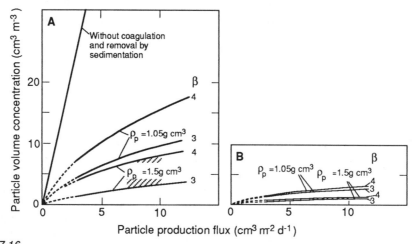

Figure 7.16

Model simulations of particle volume concentrations in the summer as functions of the particle production flux in the epilimnion of Lake Zurich, adapted from Weilenmann, O'Melia and Stumm (1989).

Predictions are made for the epilimnion (A) and the hypolimnion (B). Simulations are made for input particle size distributions ranging from 0.3 to 30 μm described by a power law with an exponent of β. For β = 3, the particle size distribution of inputs peaks at the largest size, i.e., 30 μm. For β = 4, an equal mass or volume input of particles is in every logaritmic size interval. Two particle or aggregate densities (ρ_p) are considered, and a colloidal stability factor (α) of 0.1 us used. The broken line in (A) denotes predicted particle concentrations in the epilimnion when particles are removed from the lake only in the river outflow. Shaded areas show input fluxes based on the collections of total suspendet solids in sediment traps and the composition of the collected solids.

specifically in the high DOC (organic substances similar to humates) concentration in the lake water.

The enhanced stability of natural particles in natural waters containing natural organic matter is a consistent observation without a clear cause. Some speculation is presented here. Humic substances comprise the principal fraction of dissolved organic matter in most natural waters. These molecules can be considered as flexible polyelectrolytes with anionic functional groups; they also have hydrophobic components. In fresh waters such molecules assume extended shapes due to intra-molecular electrostatic repulsive interactions. When adsorbed at interfaces at low ionic strenght, they assume flat configurations. Adsorption on inorganic surfaces such as metal oxides e.g., iron oxides, could result from the ligand exchange of functional groups on the humic substances (carboxylic, phenolic) with surface hydroxyl groups on the metal oxides, supplemented by a hydrophobic interaction involving nonpolar components of the humic molecules (cf. Example 7.3 and Fig. 7.9). The result would be an accumulation of negative charge on the surface of the oxide due to the adsorbed organic substances.

Humic substances in lakes result from autochthonous biological processes within the lake and from allochthonous inputs from terrestrial sources. The macromole-cular biological debris produced by biological wastewater treatment plants can have chemical and physical characteristics similar to natural organic substances and might provide a source of stabilizing organic matter when discharged into lakes.

An interesting corollary to these arguments is that extensive sewage treatment not only removes organic C and phosphate (which in turn causes a decrease in DOC) but may affect the entire lake metabolism by indirectly enhancing the coagulation and in turn increasing sediment fluxes and suspended particle concentrations significantly. In a lake of low humic acids (DOC) and high $[Ca^{2+}]$, photosynthetically produced biomass and other particles are transported much faster into the hypo-limnion.

Colloids in the Ocean

Non-living submicrometer particles (0.4 − 1 μm) have been found in the North-Pacific, they are present in concentrations of 10^6 to 10^7 particles m ℓ^{-1} (Koike et al., 1990). More recently, Wells and Goldberg (1991) report that very small marine colloids (d < 120 nm) are, by at least three order of magnitudes, more abundant. A vertical stratification of these particles was found (very high concentrations in the thermodine and season-dependent in the bottom-waters). This stratification indi-cates that these very small colloidal particles are reactive. The apparent close association of metals with these colloids suggests that they may play an important part in the transport and fate of trace elements in seawater (Wells and Goldberg, 1991).

7.8 Some Water-Technological Considerations in Coagulation and Flotation

Coagulation is an unit process in water supply treatment. Coagulants are used to remove color (humic and fulvic acids) and turbidity from the raw water. Coagulation techniques have also been used in waste treatment and sludge conditioning. Often hydrolyzing metal ions, Al(III) or Fe(III), are used as coagulants (Fig. 7.7c and Fig. 7.17). The Al(III) salt (or Fe(III) salt) added undergoes hydrolysis and forms polymeric or oligomeric charged hydroxo complexes of various structures such as, for example,

$$\left(\begin{array}{c} HO \\ \diagdown \\ H_2O \end{array} \diagup \begin{array}{c} \diagup O \diagdown \\ Al \\ \diagdown OH \diagup \end{array} Al - OH - Al - \right)_n^{m+}$$

Such species adsorb specifically and modify the surface charge of the colloids. As shown in Fig. 7.7 these hydrolysis products have a different effect than Al^{3+} (at low

Precipitation, coagulation, flocculation

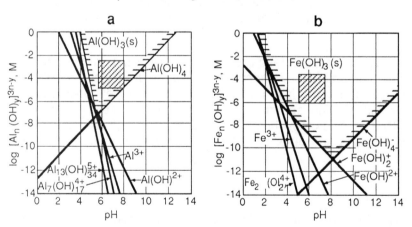

Figure 7.17

Equilibrium composition of solutions in contact with freshly precipitated, $Al(OH)_3$ and $Fe(OH)_3$, calculated, using representative values for the equilibrium constants for solubility and hydrolysis equilibria. Shaded areas are approximate operating regions in water treatment practice; coagulation in these systems occurs under conditions of oversaturation with respect to the metal hydroxide.

Oversaturation is induced by adjusting the pH of incipiently acidic Fe(III) and Al(III) solutions. After initiation of the oversaturation, the hydrolysis of Al(III) and Fe(III) progresses and charged multimeric hydroxo Al(III) and Fe(III) species, the actual coagulants, are formed as intermediates; ultimately these intermediates polymerize to solid Al(III) and Fe(III)(hydr)oxides, respectively.

(From Stumm and O'Melia, 1968)

pH); they are able to reverse the surface charge. While non hydrolyzed Al^{3+} (at low pH) compresses the double layer, hydrolyzed Al(III) coagulants interact with the colloids through specific adsorption; the dose necessary for coagulation depends on the concentration of the colloids (acually their surface area concentration) (Fig. 7.18). Stoichiometry prevails also in the removal of color. The Al added reacts largely by ligand exchange with the hydroxo groups of the coagulant hydroxo Al-species.

Al(III) and Fe(III) salts are acids. The addition of these chemicals to water is similar to an acidimetric titration of the water. As a result, the pH of the system after the addition of these coagulants will depend upon the coagulant dosage and the alkalinity of the water or waste water. Coagulation by Fe(III) and Al(III) polymers would be expected to exhibit restabilization (overdosing) and stoichiometry, if adsorption is important.

Schematic curves of residual turbidity as a function of coagulant dosage at constant pH for natural waters treated with Al(III) or Fe(III) salts are presented in Fig. 7.18. Four curves are presented, each for a water containing a different concentration of colloidal material. The colloid concentration is represented by the concentration of colloidal surface/unit volume of suspension, $\bar{S}$ (e.g., as m^2/liter). Each of these schematic coagulation curves is subdivided into four zones. In zone 1, corresponding to low dosage, insufficient coagulant has been added to bring about destabilization. Increasing the coagulant dosage produces destabilization and permits rapid aggregation (zone 2). A further increase in dosage can restabilize the dispersions at some pH level (zone 3).

In zone 4, a sufficient degree of oversaturation occurs to produce a rapid precipitation of a large quantity of aluminum or ferric hydroxide, enmeshing the colloidal particles in what has been termed a "sweep floc". Fig. 7.18b is a schematic representation of the interrelationships between coagulant dosage and colloid concentration (m^2/liter) at constant pH. At low colloid concentrations ($\bar{S}_1$ in Fig. 7.18b), coagulation requires the production of a large excess of amorphous hydroxide precipitate. For systems containing a low concentration of colloid it has been proposed that insufficient contact opportunities exist to produce aggregates of even completely destabilized particles in a reasonable detention time. Such conditions may prevail in many water treatment plants when the turbidity of the raw water is low.

At higher concentrations of colloid ($\bar{S}_2$ and $\bar{S}_3$ in Fig. 7.18b), a smaller coagulant dosage is required than for coagulation which involves the precipitation of the metal hydroxide. In this region, increasing colloid concentration requires an increasing coagulant dosage; that is, a stoichiometry in coagulation is observed. The destabilization zone (zone 2) is observed to widen with increasing colloid concentration (curves for $\bar{S}_2$ and $\bar{S}_3$ in Fig. 7.18a). At very high colloid concentrations of the order encountered in sludge conditioning in wastewater treatment plants, a high coagulant dosage is required to destabilize the colloid ($\bar{S}_4$). Under these circumstances the coagulant (metal polymer) dosage required for destabilization may equal or exceed that required for the precipitative coagulation of dilute suspension.

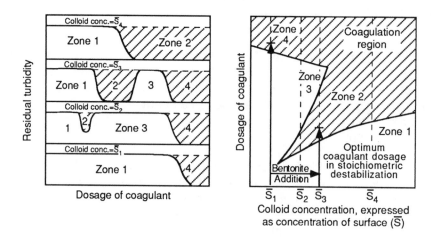

Figure 7.18

Schematic representation of coagulation observed in jar tests using aluminum(III) or iron(III) salts at constant pH.

(From Stumm and O'Melia, 1968 and O'Melia, 1972)

As pointed out by O'Melia, an understanding of these properties of hydrolyzing metal salts can be useful in coagulation practice. Consider first that Al(III) salts can be effective as coagulants in present practice in two ways, by adsorption to produce charge neutralization, and by enmeshment in a precipitate. The chemical dosage which is required depends upon how destabilization is achieved. High dosages of coagulant to produce a gelatinous metal hydroxide precipitate can be effective in these and other situations where low but objectionable concentrations of colloidal materials are present. Al(III) coagulants are also used to remove "color" from the water. The color is usually due to the presence of humic or fulvic acids. The removal of color is due to the formation of mixed Al(III) hydroxo-, humato-aggregates. At a given pH, the Al(III) needed is proportional to the color constituents of the waters. In waters that are both turbid and colored, the required total alum dosage is usually dictated by the concentration of color constituents.

An additional method exists for the effective coagulation of low turbidity waters; that is, the addition of more particles in the form of a coagulant aid. Bentonite (a finely divided clay) and some forms of activated silica serve this function.

Flotation

Flotation is a solid-liquid separation process, that transfers solids to the liquid surface through attachment of gas bubbles to solid particles. Flotation processes are used in the processing of crushed ores, whereby a desired mineral is separated from the gangue or non-mineral containing material. Various applications in solid separation processes are also in use in waste treatment.

In flotation, all three states of matter – solid, liquid, gas – are involved and each of these involves surface chemistry. Among the various process steps that are involved are:
1) The generation of gas (air) bubbles and the selective attachment of these bubbles to the particles to be removed;
2) the addition of chemicals (additives) that adsorb (selectively) on the particles to render the attachment of air bubbles possible;
3) the actual flotation process, the production of froth, the continuous operation in a flow through reactor.

The Attachment of Air Bubbles. Particles are carried upward and are held in the froth if air bubbles can be attached to them. The adhesion of bubbles is only possible if the particle surface is sufficiently hydrophobic. The contact angle (see Appendix Chapter 4) is an important variable related to the adhesion of bubbles to solid surfaces. As shown in Fig. A.4.1, Chapter 4, and by Youngs Equation (Eq. A.4.3, Chapter 4)

$$\gamma_{LV} \cos \theta = \gamma_{SV} - \gamma_{SL} \tag{7.14}$$

As mentioned in the Appendix of Chapter 4, the contact angle θ increases ($\cos \theta$ decreases) with increasing hydrophobic character of the solid surface ($\gamma_{SV} < \gamma_{SL}$), i.e., extensive adsorption at the air-solid surface and minimum adsorption at the solid water interface is needed.

Collectors are additives which adsorb on the particle (mineral) surface and prepare the surface for attachment to an air bubble so that it will float to the surface. Collectors must adsorb *selectively* to render a fractionation of different solids possible.

Suitable collectors can render hydrophilic minerals such as silicas or hydroxides hydrophobic. An ideal collector is a substance that attaches with the help of a functional group to the solid (mineral) surface often by ligand exchange or electrostatic interaction, and exposes hydrophobic groups toward the water. Thus, amphipatic substances (see Chapter 4.5), such as alkyl compounds with C_8 to C_{18} chains are widely used with carboxylates, or amine polar heads. Surfactants that form hemicelles on the surface are also suitable. For sulfide minerals mercaptanes, monothiocarbonates and dithiophosphates are used as collectors. Xanthates or their oxidation products, dixanthogen $(R - O - C - S -)_2$ are used as collectors for
$$\overset{\parallel}{S}$$

many ores. The S-group can specifically sorb (ligand exchange) for example to lead and copper ores in reactions such as

$$\equiv Pb - OH + S(CSOR)_2 \longrightarrow \equiv Pb - S(CSOR)_2 + OH^- \tag{7.15}$$

Frothing Agents are intended to stabilize the particle (mineral)-air mixture (foam) at the surface of the flotation tank. Alkyl or amyl alcohols in the C_5 to C_{12} range are typical frothers. They lower γ_{SV} which is beneficial to the stability of the foam; to some extent collectors and frothers may counteract in their effects so that compromise conditions must be selected.

7.9 Bacterial Adhesion; Hydrophobic and Electrostatic Parameters

In many natural and technical systems metabolically active bacteria are found to be associated with interfaces. Bacterial adhesion is of importance in the formation of biofilms. Interaction of bacteria in soils is of great importance. *Bioflocculation*, the aggregation of microorganisms in biological waste treatment, is essential for the functioning of biological waste treatment. It has been proposed by many investigators, e.g., Busch and Stumm (1968), that polymers which are either excreted by microorganisms or exposed at the surface of cells are responsible for the aggregation of microorganisms. Such polymers were shown to accumulate under conditions of declining bacterial growth. These polymers were also shown to destabilize hydrous oxide colloids.

Van Loosdrecht et al. (1990) have investigated systematically the adhesion of microorganisms to solid surfaces in aquatic environments and describe the initial adhesion process in terms of a colloid-chemical theory. Obviously, bacteria are not inert colloidal particles. Their cell surfaces and their characteristics can change with alterations in environmental conditions.

Following the arguments by Van Loosdrecht et al. (1990) schematically the interaction between the (negatively charged) cell surface and the solid surface – a sulphated polystyrene surface (hydrophobic, negatively charged) – in presence of polymers, that have adsorbed to both surfaces can be shown to be caused by various interaction energies between the two partially or totally polymer-coated surfaces (Fig. 7.19). The effects of polymers have been interpreted (Van Loosdrecht et al., 1990) as follows:
  i)   The presence of the adsorbed polymer layer may quantitatively change the van der Waals interaction (G_a) and the electrostatic interaction (G_{el}).
  ii) When polymers adsorb and coat both bacteria and solid surface completely, an extra repulsive interaction (G_s) has to be added due to steric hindrance. Such an effect is schematically shown in Fig. 7.19a.

a   **b**

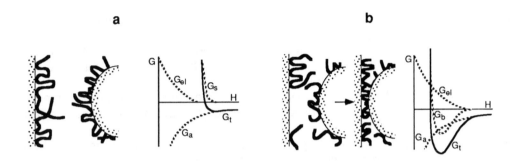

Figure 7.19

a) Schematic representation of interaction between likewise-charged completely polymer coated-surfaces.
   G_a: van der Waals interaction
   G_{el}: electrostatic interaction
   G_s: steric interaction
   G_t: total interaction
   H: distance between the two surfaces

b) Schematic representation of interaction between likewise-charged partly polymer coated-surfaces.
   G_a: van der Waals interaction
   G_{el}: electrostatic interaction
   G_b: bridging interaction
   G_t: total interaction
   H: distance between the two surfaces

(From Van Loosdrecht, Norde, Lyklema and Zehnder, 1990)

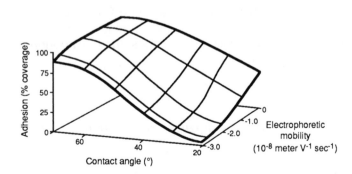

Figure 7.20

Relation between bacterial adhesion to sulphated polystyrene (A) and cell surface characteristics as determined by electrophoretic mobility and contact angle measurement. Results were obtained by interpolating the data points for the adhesion of 17 different strains of bacteria.

(From Van Loosdrecht et al., 1990)

iii) If only one of the surfaces is covered with polymers, or if both surfaces are partly covered with polymers, one and the same polymer molecule may attach to both surfaces, thereby forming a "bridge" between the two surfaces. This bridging represents a strong binding and therefore lowers the Gibbs energy (Fig. 7.19b).

iv) When the adsorbed polymers bear charges, i.e., when they are polyelectrolytes, as with proteins, the effects of these charges have also to be considered.

For ii) and iii) loosely structured layers are required, and the chains must protrude into the solution over a distance exceeding the thickness of the electrical double layer so that on approach of the surfaces, the adsorbed layers interfere before the electrical double layers overlap.

Hydrophobicity of the bacterial surface affects the van der Waals interaction by changing (increasing) the Hamaker Constant A_{bsw} (valid for the interaction of a bacterium (b) with a surface (s) in water (w) (Eq. (3) and footnote c in Table 7.3). As we have seen in the Appendix of Chapter 4, the extent of hydrophobicity is related to the contact angle which is formed by a drop of water on a layer of a bacterial cell. (The larger the contact angle the more hydrophobic is the surface.)

Electrophoretic Mobility is a measure of the electrostatic interaction between the surfaces.

The relation between cell surface characteristics and bacterial adhesion to polystyrene is shown in Fig. 7.20. Obviously, the adhesion process can be satisfactorily described. Interestingly different bacteria are characterized by remarkably different contact angles; furthermore, the contact angle, i.e., the hydrophobility may change depending on growth conditions. As can be seen from Fig. 7.20, surface hydrophobility is the dominant characteristic. At high contact angle complete adhesion is found. However, at more hydrophilic cell surfaces, the electrophoretic mobility (Zeta Potential) becomes more influental.

7.10 Colloids; The Use of (Membrane) Filtration to separate "Particulate" from "Dissolved" Matter

In natural waters and soil and sediment systems one needs to distinguish analytically between dissolved and particulate material. Fig. 7.1 classifies various types of particulate and dissolved materials. Obviously, operational distinguishing (e.g., based on filtration or centrifugation) between "dissolved" and "particulate" matter merely by filtration is often not able to discriminate between particles and solutes, because size distribution of aquatic components vary in a continuous matter from Ångstrøms to microns.

The use of filters and membranes of different pore size to accomplish a sequential size fractionation is in principle, and under certain circumstances, possible; it was proposed (for literature see Buffle, 1988 and 1991) to estimate the size of the various colloids and macromolecules; and to determine to which extent trace elements (particularly metals) are associated with various size categories of colloids and macromolecules. Such sequential size fractionation techniques need to be applied with extreme caution; we list some of the reasons why these techniques may yield errorous results (for details consult Buffle, 1991):

1) Although most particles, larger than a given pore size, are normally retained, many smaller particles (sometimes 10 − 1000 times smaller than the pore size) may also be retained. If above and within a pore depth filtration occurs (see Chapter 7.6 and Fig. 7.12); colloidal particles smaller than the pore size become attached to the larger particles. Furthermore the pore size distribution of membrane filters is often non-narrow.

2) Coagulation occurs in the bulk sample and in the filter. Because of the long times involved in the filtration through < 1 μm pores, coagulation in the bulk sample (and in the suspension above the filter) occurs. For the particle concentrations typically encountered in many natural waters ($10^5 − 10^9$ particles cm^{-3}) coagulation of half of the particles occur over a period of hours or days, depending on chemical conditions (Eqs. 7.9 − 7.11); i.e., filtration must be done as quickly as possible after sampling. Furthermore, coagulation on and in the filter occurs; its extent increases with the flow-rate used for filtration (Eq. 7.12 and Fig. 7.2).

3) The interaction of solutes (especially adsorption) with the filter material and the retained particles. The problem is especially serious with trace elements (heavy metals) that adsorb especially at pH values of 7 and above on filter walls and on the filter materials (glass, acrylic copolymers, cellulose esters, polycarbonates etc.). One also needs to consider that

   i) trace metal concentrations are often below 10^{-8} M (Figs. 7.21 a, b, c, d illustrate some of the problems mentioned for iron(III)hydroxiphosphate particles); and

   ii) that the retained particles and the filter material is charged, that the concentration of solutes and colloids change during filtration, and that adsorption occurs as a consequence of double layer properties and of chemical interaction with functional groups.

There is no perfect way to distinguish between what is conceptually dissolved and what is non-dissolved (particulate). Ion-selective electrodes respond selectively to solute ions but are often not sufficiently selective. One other possibility is to use voltammetric techniques on Hg (or other) electrodes in presence of colloids without prior centrifugation or filtration. Gonçalves et al. (1985, 1987), Müller and Sigg (1990) (cf. Chapter 11.3).

The arbitrary limit of a pore size of 0.2 − 0.5 μm in many sampling procedures has − besides all disadvantages − also some advantages: particles not retained by such filters do not settle down in natural waters within days and "move" with the solutes; most bacteria and other organisms (except viruses) are retained by these filters and thus the filtered sample is often nearly sterile and less subject to microbially mediated changes. On the other hand, one needs to realize that the thermodynamic basis of all solution and heterogeneous equlibria (including adsorption and solubility equlibria) refers to the conceptually defined solutes.

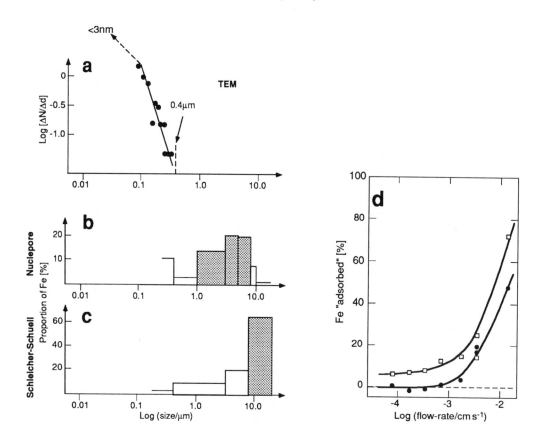

Figure 7.21

Operational problems in size fractionation by membrane filters

a,b,c) Size distributions of iron oxihydroxiphosphate particles obtained by transmission electron microscopy (true distribution), syringe filtration on Nuclepore polycarbonate filters and Schleicher and Schuell cellulose ester depth filters. Iron particles formed at the oxic/anoxic interface of an eutrophic Lake Bret (Switzerland). (From Buffle, Perret and Newman, 1992)

d)   Fraction of iron particles retained on 3.0 μm membranes, as a function of flow-rate. ● Nuclepore polycarbonate; □ Schleicher and Schuell cellulose nitrate (from Perret, 1989). In the absence of coagulation or adsorption, no particle should be retained. (From Buffle, Perret and Newman, 1992)

Interstitial Water. The differentiation between solutes and particles is of great importance in the sampling of interstitial water. Most conveniently so-called peepers are used. These consist usually of plexiglass plates in which small compartments (0.5 cm deep and 0.5 – 1 cm high) are separated from the sediments by a dialysis membrane. The compartments are initially filled with degassed distilled water. After 1 – 2 weeks for equilibration subsequent to the retrieval of the peeper, the "dissolved" components are measured in each component. For this type of application the pore size does not seem to be very critical; colloids do not seem to accumulate in the compartments (low diffusion coefficients).

Davison et al. (1991) have proposed to use devices with probing compartments filled with gels. Very high resolutions on vertical concentration gradients can be obtained this way.

Appendix

Distribution Coefficients, Possible Artefacts

Distribution coefficients describe in a summarizing way the distribution of an element between the dissolved and solid phases. They are conditional constants valid for a given pH, temperature and other conditions; they are independent of the concentrations of solids in water. They are usually defined (Chapter 4.8) as

$$K_D = \frac{C_S}{C_W} \; [m^3 \, kg^{-1}] \qquad\qquad (A.7.1)^{1)}$$

where C_S is the concentration in the solid particles [mol kg^{-1}] and C_W is the (dissolved) concentration in water [mol m^{-3}]. Distribution coefficients for metal ions (and ligands) can be derived from surface complex formation equilibira (see Table 11.1).

Particle Concentration Effect. Experimental data often show a "particle concentration effect", i.e., the K_D decreases with the concentration of particles. Such observations are in contradiction with thermodynamics. Various "artefacts" could account for this observation (Morel and Gschwend, 1987):
1) equilibrium was not attained in these measurements;
2) the porous solid contains, within the separated solid phase, water and this water contains solutes; or
3) the analytically determined dissolved concentration (i.e., within the water phase separated by filtration or centrifugation) contains colloids.

We will analyze the latter case and follow the argumentation given by Morel and Gschwend (1987). Distinguishing between particles (that are retained in filters or that are separated in centrifugation) and colloids (that are in the filtrate or supernatant) we can characterize an "observed" distribution coefficient.

$$K_D^{observed} \left[\frac{m^3}{kg}\right] = \frac{\dfrac{\text{mol adsorbed to particles}}{\text{mass of particles}}}{\dfrac{\text{mol dissolved}}{\text{volume water}} + \dfrac{\text{mol adsorbed to colloids}}{\text{mass of colloids}} \times \dfrac{\text{mass of colloids}}{\text{volume water}}} \qquad (A.7.2)$$

If we define

$$K_P = \frac{\dfrac{\text{mol adsorbed to particles}}{\text{mass of particles}}}{\dfrac{\text{mol dissolved}}{\text{volume of water}}} \qquad [m^3 \, kg^{-1}] \qquad\qquad (A.7.3)$$

---

1) The unit $\ell \, kg^{-1}$ is often used.

and

$$K_C = \frac{\frac{\text{mol adsorbed to colloids}}{\text{mass of colloids}}}{\frac{\text{mol dissolved}}{\text{volume of water}}} \quad [\text{m}^3 \text{ kg}^{-1}] \qquad (A.7.4)$$

and we use the symbol M_C for $\frac{\text{mass of colloids}}{\text{volume of water}}$ [kg m^{-3}]

we obtain $\left(\text{divide in Eq. (A.7.2) numerator and denominator by } \frac{\text{mol dissolved}}{\text{volume water}}\right)$

$$K_D^{\text{observed}} = \frac{K_P}{1 + K_C M_C} \qquad (A.7.5)$$

If, as might be expected, the unfilterable solid concentration covaries with the total solid concentration, then at high solid concentrations ($K_C M_C > 1$) a decrease in the apparent partition coefficient with increasing M_C will result (see Morel and Gschwend, 1987).

The "Solid Concentration" Effect due to Polydispersed Solutes. An adsorption constant or a distribution coefficient usually characterizes the adsorption equilibrium of a single solute. In many natural circumstances a large number of solutes compete with each other for the available surfaces sites. This is especially so for organic solutes such as fulvic or humic acids or naturally present surfactants. The concentrations of such polydisperse substances are characterized by collective parameters, such as humic acids, fulvic acids, dissolved organic carbon. Langmuir, Frumkin or Freundlich adsorption equations are often used to represent adsorption equilibria. Although mathematically convenient, adsorption data may often show a good fit to these isotherms, a rigorous molecular interpretation is rendered difficult. Especially the values of *apparent* adsorption constants, e.g., evaluated from Langmuir plots (or distribution coefficients), decrease with increasing particle concentration while the maximum adsorption density, Γ_{max}, is nearly independent of particle concentration. This is readily explained (Zutić andTomaić, 1988) with a preferential adsorption of higher molecular weight fractions (adsorption constants increase with increasing molecular weight). The fractionation effect increases with increasing particle concentration. Thus, it is inappropriate to predict the adsorption behavior of humic acids or of other polydispersed solutes by mere extrapolation of laboratory experiments performed with high particle concentrations.

Reading Suggestions

Buffle, J. (1991), Ed., *Chemical and Biological Regulation of Aquatic Processes,* in press.

Gschwend, Ph. M., and M. D. Reynolds (1987), "Monodisperse Ferrous Phosphate Colloids in an Anoxic Groundwater Plume", *J. Contam. Hydrol.* **1/3**, 309-327.

Kim, J. I. (1991), "Actinide Colloid Generation in Groundwater", *Radiochim. Acta* **52/53**, 71-81.

Liang, L., and J. J. Morgan (1990), "Chemical Aspects of Iron Oxide Coagulation in Water: Laboratory Studies and Implications for Natural Systems", *Aquatic Sciences* **52/1**, 32-55.

Morel, F. M. M., and P. M. Gschwend (1987), "The Role of Colloids in the Partitioning of Solutes in Natural Waters", in W. Stumm, Ed., *Aquatic Surface Chemistry*, Wiley-Interscience, New York, pp. 405-422.

O'Melia, C. R. (1987), "Particle-Particle Interactions", in W. Stumm, Ed., *Aquatic Surface Chemistry*, Wiley-Interscience, New York, pp. 385-403.

O'Melia, Ch. R. (1990), "Kinetics of Colloid Chemical Processes in Aquatic Systems", in W. Stumm, Ed., *Aquatic Chemical Kinetics, Reaction Rates of Processes in Natural Waters*, Wiley-Interscience, New York, pp. 447-474.

Schnoor, J. L. (1990), "Kinetics of Chemical Weathering: A Comparison of Laboratory and Field Weathering Rates", in W. Stumm, Ed., *Aquatic Chemical Kinetics*, Wiley-Interscience, New York, pp. 475-504.

Stumm, W., and Ch. R. O'Melia (1968), "Stoichiometry of Coagulation," *J. AWWA* **60**, 514-539.

Weilenmann, U., Ch. R. O'Melia, and W. Stumm (1989), "Particle Transport in Lakes: Models and Measurements," *Limnology & Oceanography* **34**, 1-18.

Chapter 8

Carbonates and their Reactivities

8.1 Introduction

Carbonate minerals are among the most reactive minerals found at the earth's surface. Dissolution of calcite and dolomite, the two most preponderant carbonates exposed to weathering, represents ca 50 % of the chemical denudation of the continents. $CaCO_3$ and other carbonates are also important minerals in aquifers. Ca^{2+}, Mg^{2+} and HCO_3^- are the most abundant ions present in natural waters (Wollast and Mackenzie, 1983). The carbonate minerals are an essential part of the natural carbon cycle and its alteration by human activities. The formation of carbonate minerals, their nucleation and crystal growth, and their dissolution and the diagenesis of carbonate sediments are important processes in the global cycling of elements. An understanding of these processes depends on an appreciation of the surface chemistry of carbonate minerals.

To highlight one example of a surface process of global significance, we repeat here (Chapter 5.1) a brief account on carbonate weathering and $CaCO_3$ precipitation:

The rate of chemical weathering is related to the rate of CO_2 consumption. Consider first the dissolution of carbonates

$$CaCO_3(s) + CO_2 + H_2O \rightleftharpoons Ca^{2+} + 2\,HCO_3^- \tag{8.1}$$

but the HCO_3^- and Ca^{2+} ions produced by carbonate weathering are precipitated again in the ocean (mostly through incorporation into marine organisms) as $CaCO_3$ (reversal of reaction (8.1)); thus, the CO_2 consumed in the dissolution is released again upon formation of $CaCO_3$ in the ocean. Although the weathering of carbonate minerals is much faster than that of silicate weathering, it has been shown (Berner and Lasaga, 1989) that silicate weathering is more important than carbonate weathering. The weathering of calcium silicate (written here in a simplified way)

$$CaSiO_3(s) + 2\,CO_2 + 3\,H_2O = Ca^{2+} + 2\,HCO_3^- + H_4SiO_4 \tag{8.2}$$

also produces Ca^{2+} and HCO_3^- which form $CaCO_3$ in the sea. But in (8.2) we need stœchiometrically 2 CO_2 per calcium bicarbonate produced; thus, only half of the CO_2 consumed in the silicate weathering is released and returned to the atmosphere upon $CaCO_3$ formation. Thus, silicate weathering results in a net loss of atmospheric CO_2. Of course, ultimately, the cycle is completed by metamorphic and magmatic breakdown deep in the earth, of $CaCO_3$ with the help of SiO_2, a reaction that may be represented in a simplified way as $CaCO_3 + SiO_2 = CaSiO_3 + CO_2$. Knowledge of the dissolution and precipitation of minerals is also necessary for the quantitative evaluation of geochemical processes in the oceanic system.

We already touched on some aspects of carbonate surface chemistry e.g., in Chapter 3.4. We have already illustrated some of the factors that affect surface charge and the point of zero charge, pH_{pzc}, in Chapter 3.5, and have discussed certain elementary aspects of $CaCO_3$ nucleation in Chapter 6.5 and of coprecipitation (and solid solution formation) in Chapter 6.7.

8.2 Dissolution and Crystal Growth of Carbonates

Dissolution of carbonates can only occur if the solution is thermodynamically undersaturated, pH is an important variable affecting the saturation ratio (Appendix 8.1 gives a brief review of the $CaCO_3$ solubility characteristics in open and closed systems).

At very low pH, the rate of dissolution is so fast that the rate is limited by the transport of the reacting species between the bulk of the solution and the surface of the mineral (Berner and Morse, 1974). The rate can then be described in terms of transport (molecular or turbulent diffusion) of the reactants and products through a stagnant boundary layer, δ. The thickness of this layer depends on the stirring and the local turbulence. (See Chapter 5 for a discussion of transport vs surface controlled processes.)

But within the pH range of natural waters, the dissolution (and precipitation) of carbonate minerals is surface controlled; i.e., the rate of dissolution is rate determined by a chemical reaction at the water-mineral interface. Fig. 8.1 gives the data on the dissolution rates of various carbonate minerals in aqueous solutions obtained in careful studies by Chou and Wollast (1989).

The dissolution rate for calcite and aragonite have been described in terms of the following rate law (Plummer et al., 1978; Busenberg and Plummer, 1986; Chou and Wollast, 1989.[1]

$$\text{rate [mol cm}^{-2}\text{ s}^{-1}] = k_1[H^+] + k_2[H_2CO_3] + k_3 - k_{-3}[Ca^{2+}][CO_3^{2-}] \tag{8.3}$$

The last term in Eq. (8.3) is only important close to the carbonate saturation equilib-

---

[1] These authors used activities instead of concentrations but at constant ionic strength (constant ionic medium scale), the rate law can be written in terms of concentrations.

rium. Fig. 8.2 illustrates the pH–p_{CO_2} domains where the [H⁺] dependence (first term on the right hand side of Eq. (8.3)), the p_{CO_2}-dependence (second term) and the third term (H_2O) are primarily operative in the $CaCO_3$ dissolution.

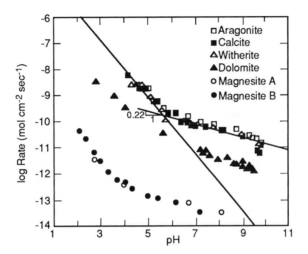

Figure 8.1

Dissolution rate of carbonates as a function of pH. These experiments were carried out with a continuous flow reactor in open systems with controlled p_{CO_2}.

(After Chou and Wollast, 1989)

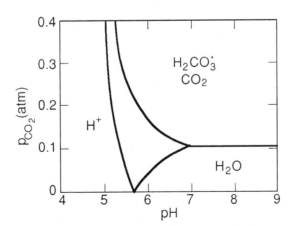

Figure 8.2

Predominance areas where the first, second or third term of Eq. (8.3) is important. In the triangle region all three terms have to be considered.

The rate law of Eq. (8.3) has been interpreted in terms of the following dissolution reactions which occur as parallel reactions.

$$CaCO_3(s) + H^+ \quad \underset{k_{-1}}{\overset{k_1}{\rightleftharpoons}} \quad Ca^{2+} + HCO_3^- \qquad (8.4)$$

$$CaCO_3(s) + H_2CO_3^* \quad \underset{k_{-2}}{\overset{k_2}{\rightleftharpoons}} \quad Ca^{2+} + 2\,HCO_3^- \qquad (8.5)$$

$$CaCO_3(s) \quad \underset{k_{-3}}{\overset{k_3}{\rightleftharpoons}} \quad Ca^{2+} + CO_3^{2-} \qquad (8.6)$$

In the rate law (8.3), the back reactions of (8.4) and (8.5) have been neglected because usually the system is sufficiently far from equilibrium.

Surface Reactions. As we have seen from the dissolution of oxides the surface-controlled dissolution mechanism would have to be interpreted in terms of surface reactions; in other words, the reactants become attached at or interact with surface sites; the critical crystal bonds at the surface of the mineral have to be weakened, so that a detachment of Ca^{2+} and CO_3^{2-} ions of the surface into the solution (the decomposition of an activated surface complex) can occur.

Eq. (8.3) and its relationship to the reactions (8.4) – (8.6) may give us valuable clues on the formation of the precursors of the surface activated complex. As we have seen with oxides and silicates, charge (or potential) determining ions are of great influence on the dissolution kinetics because – as their definition implies – they interact chemically with the surface. Charge determining species for $CaCO_3(s)$ (cf. Chapter 3.5) are, H^+, Ca^{2+}, HCO_3^- and H_2CO_3.

Recent studies with the Atomic Force Microscope (Gratz et al., 1991) revealed interesting insight into calcite dissolution reactions at the near-atomic scale. Generally, surface dissolution was observed only at steps (both at pits and to a lesser extent at monomolecular ledges) and not in terraces. Dissolution occurred by removal of molecules from pre-existing steps. These authors also observed simple growth sequence wherein $CaCO_3$ monolayers were deposited on the calcite surface by uniformely advancing ledges.

In making experiments with $CaCO_3$, be it on dissolution or on crystal growth, it is essential that we know the variables or keep as many of them constant. Since these rates are dependent on pH and on surface charge, it is essential that we realize that the $CaCO_3$ in equlibrium with the solution at a given pH, has different concentrations of H^+, Ca^{2+}, HCO_3^-, CO_3^{2-} and $CO_2 \times$ aq and different surface charge characteristics depending on how this pH has been adjusted. Much of the discrepancy in the literature can be accounted for by not considering whether the pH has been adjusted i) in a closed system, ii) in an open system and constant p_{CO_2} by adding acid or base, or iii) by simply changing in a $CaCO_3$, H_2O, CO_2 system the nearly zero (see Appendix 8.1), while in case (i) a decrease in $[H^+]$ will result in a concommittant increase in $[CO_3^{2-}]$ and decrease in $[Ca^{2+}]$. Thus, in the latter case there are some compensating effects, in the sense that the two oppositely charged bivalent charge determining ions change with pH according to

$$\frac{d\,[Ca^{2+}]}{d\,pH} = -\frac{d\,[CO_3^{2-}]}{d\,pH} = 1 \; ; \; \text{while} \; \frac{d\,[HCO_3^-]}{d\,pH} = 0 \tag{8.7}$$

In case of dolomite and magnesite dissolution a dependence on a fractional order of $[H^+]$ has been observed, indicating that adsorption of charge determining ions is involved.

Walter and Morse (1984) were able to document the relative importance of *microstructure* for the dissolution of biogenic carbonates. Biogenic magnesian calcites are structurally disordered and chemically heterogeneous. Both these factors play a role in the reactivity of these minerals in natural systems.

8.3 The Crystal Growth of CaCO₃ (Calcite)

The crystal growth of calcite has been studied by Plummer and coworkers (1978), by Kunz and Stumm (1984) and by Chou and Wollast (1989) to correspond to the backward rate of dissolution (Eq. 8.3).

$$R_{tot} = \frac{d\,[CaCO_3]}{d\,dt} = k_3\,[Ca^{2+}]\,[CO_3^{2-}] + k_{-1}\,[Ca^{2+}]\,[HCO_3^-] \tag{8.8}$$

Since most natural waters are near saturation this reversal of the rate law would appear to be in line with the concept of microscopic reversibility of the forward and backward reaction steps. Some data are given in Fig. 8.3.

As we have seen in Chapter 6, the crystal growth has to be preceded by nucleation (Fig. 6.7). In fresh waters this nucleation occurs heterogeneously on particle surfaces. In seawater nucleation occurs primarily upon the templates of calcareous organisms.

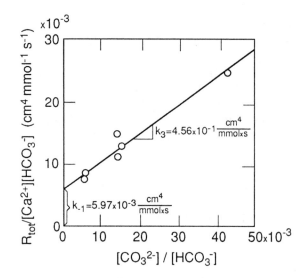

Figure 8.3

Data on the $CaCO_3$ crystal growth (20° C) carried out under conditions of constant p_{CO_2}. The data are plotted in line with Eq. (8.8) as

$$\frac{R_{tot}}{[Ca^{2+}][HCO_3^-]} = k_{-1} + k_3 \frac{[CO_3^{2-}]}{[HCO_3^-]}$$

where k_{-1} and k_3 are obtained from intercept and slope.

(From Kunz and Stumm, 1984)

8.4 Saturation State of Lake Water and Seawater with Respect to $CaCO_3$

Photosynthesis occurring in the upper layers of the oceans and of lakes removes CO_2 from the water and raises the saturation ratio of $CaCO_3$. In a simplified way:

$$CO_2 + H_2O \underset{R}{\overset{P}{\rightleftharpoons}} CH_2O + O_2$$

$$Ca^{2+} + 2 HCO_3^- \rightleftharpoons CaCO_3 + CO_2 + H_2O$$

$$\overline{Ca^{2+} + 2 HCO_3^- \underset{R}{\overset{P}{\rightleftharpoons}} CaCO_3 + CH_2O + O_2} \qquad (8.9)$$

a

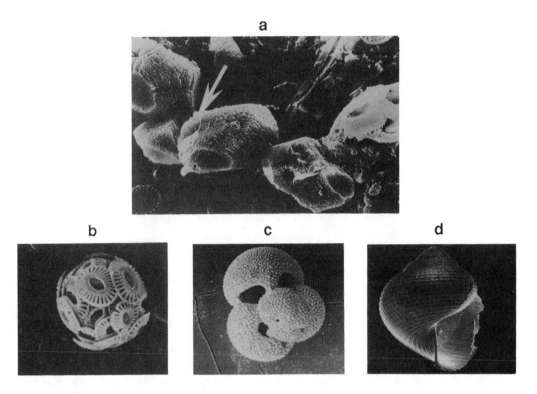

b   c   d

Figure 8.4

Forms of CaCO$_3$ in lake water and in ocean
a) calcites from lake water interconnected into aggregates. Diatoms (arrow) are overgrown with calcite
b) calcareous pelagic plants (cocolithophores) (diameter ~ 10 μm)
c) animals (foraminifera 100 μm diameter)
d) pelagic pteropod (1000 μm diameter)
(Figs. b), c), d) from Morse and Mackenzie, 1990)

As a consequence of photosynthesis, P, CaCO$_3$ is precipitated in the top layers. In the deeper layers respiration, R, of the phytoplankton and its debris in the deeper water layers causes the dissolution of CaCO$_3$ (cf. 8.1). Fig. 8.4 illustrates the type of carbonates formed and Fig. 8.5 gives for Lake Constance saturation ratios ($\Omega = \frac{IAP}{K_{s0}}$; cf. Chapter 6)[1] and the corresponding sedimentation fluxes of CaCO$_3$ into the sediments. Obviously the settling of CaCO$_3$ represents a significant sedimentation load ("conveyor belt")

---

[1] In Chapter 6 the saturation ratio was denoted as S.

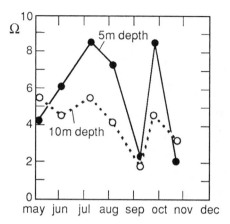 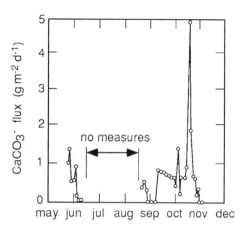

Figure 8.5

Saturation ratios of CaCO₃ in Lake Constance and the corresponding sedimentation load.
(From Sigg, Sturm, Davis and Stumm, 1982).

The saturation ratio, Ω, as a function of depth is given for the Atlantic, Indian and Pacific Oceans in Fig. 8.6. The equilibrium saturation is also given in this figure for aragonite. Because of the greater solubility of aragonite, the water column becomes undersaturated with regard to aragonite at smaller depths than for calcite. The Mediterranean Sea is supersaturated everywhere with respect to calcite as well as to aragonite.

Planktonic foraminifera and cocolithophores are composed of low magnesian calcite (< 1 mol % $MgCO_3$). Benthic foraminifera are formed of either aragonite or high magnesian calcite. Pteropods are the most abundant aragonite organisms.

Growth of Concretions. In sedimentary rocks we commonly find concretions, that is, material formed by deposition of a precipitate, such as calcite or siderite, around a nucleus of some particular mineral grain or fossil. The origin of most concretions is not known but their often-spherical shape with concentric internal structure suggests diffusion as an important factor affecting growth. The rate of growth, if diffusion-controlled is readily amenable to mathematical treatment. Berner (1968) has provided some idea of the time scale involved in the growth of postdepositional concretions. His calculations illustrate, for example, that for a typical slowly flowing groundwater, with a supersaturation in $CaCO_3$ of 10^{-4} M (assumed to be constant), the time of growth of calcite concretions ranges from 2500 years for concretions of 1-cm radius to 212,000 years for those of 5-cm radius. Hence concretion growth, if diffusion and convection is the rate-controlling step, is relatively rapid when considered on the scale of geological time.

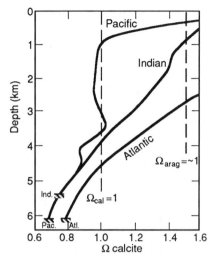

Figure 8.6

Saturation profiles for the northern Atlantic and Pacific oceans, and the central Indian Ocean (GEOSECS stations 31, 221, 450).

(From Morse and Mackenzie, 1990)

8.5 Some Factors in the Diagenesis of Carbonates

In carbonate diagenesis[1] we deal usually with a combination of low supersaturation and absence of mechanical agitation. Homogeneous nucleation will certainly not occur. The important factors to be investigated are heterogeneous nucleation and rates of growth and dissolution of crystals.

Growth Inhibitors. Because the rate-determining step is frequently controlled at the interface, small amounts of soluble foreign constituents may alter markedly the growth rate of crystals and their morphology. The retarding effect of substances that become adsorbed may be explained as being due primarily to the obstruction by adsorbed molecules to the deposition of lattice ions. In some cases it has been shown that the rate constant for crystal growth is reduced by an amount reflecting the extent of adsorption.

The effects of trace concentrations of dissolved organic matter and of orthophosphates (Berner and Morse, 1974) and polyphosphates as "crystal poisons" (e.g., inhibiting the spread of monomolecular steps on the crystal surface by becoming

---

[1] Diagenesis refers to the sum total of processes that bring about charges in sediment or sedimentary rock subsequent to deposition in water. The processes may be physical or chemical or biological in nature and may occur at any time (Berner, 1986).

adsorbed on active growth sites such as kinks) on the nucleation and growth of calcite have been investigated in some details.

Another category of inhibition may be exerted by cations or anions that become adsorbed on active growth sites. Well-hydrated Mg^{2+} interferes with the formation of calcite, apatite, and many other minerals.

Finally, ions that become adsorbed on the surface of nuclei or crystallites may become incorporated into the growing crystals, that is, solid solutions are formed. These solid solutions may be more soluble than the pure solid phases; the inhibition may then be due to an reduction in the degree of supersaturation.

It is interesting to note that many crystal poisons not only interfer with nucleation and the growth of crystals but may also retard their dissolution. As we have seen (Chapter 6), precipitation and dissolution of solids proceed by the attachment or detachment of ions most favorably at kink sites of the crystalline surface. Solutes such as organic substances, or phosphates may upon adsorption immobilize kinks and thus retard dissolution.

8.6 Coprecipitation Reactions and Solid Solutions

Carbonate minerals in natural systems precipitate in the presence of various other solutes: This trace amounts of all components present in the solution may get incorporated into the solid carbonate minerals ("coprecipitation").

The uptake of a cation into a carbonate is thought to proceed via adsorption, eventually leading to surface precipitation and formation of a solid solution. Kinetics of cation adsorption occurs usually in subsequent steps; the specific adsorption, i.e., the transfer at a carbonate surface from the solution phase into the adsorbed state must be assumed for most cations to be very fast. Most likely its rate is related to the rate of water exchange (cf. Chapter 4.4). Two examples of the rates of uptake or carbonate surfaces are given in Fig. 8.7.

Davis et al. (1987) proposed a model for the kinetics of Cd(II) sorption on $CaCO_3$. In an initial fast step Cd(II) becomes adsorbed on a hydrated layer of the calcite surface (see Chapter 3.4 and Eqs. 3.11a and 3.16). Subsequently, in a slow step surface precipitation occurs. In separate experiments on the Ca^{2+} exchange rate (accomplished with ^{45}Ca isotopes) these authors showed that there is a slow long term recrystallization taking place. In the presence of Cd^{2+} the new crystalline material may grow as a solid solution rather than as pure calcite. If it is assumed that the rate of crystal growth remains unchanged in the presence of traces of Cd^{2+}, then the ratio of mol fractions of $CdCO_3$ and $CaCO_3$ in the new crystalline material may be equated with the ratio of the *rates* of Cd sorption and Ca isotopic exchange

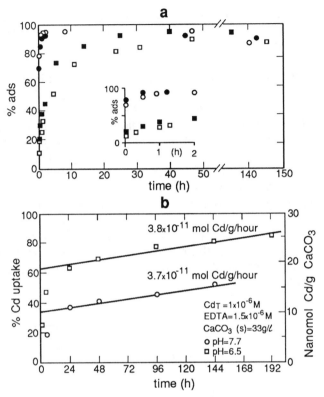

Figure 8.7

Rates of metal ion adsorption on carbonate surfaces:
a) Mn(II) on siderite;
b) Cd(II) on calcite.

In both cases the solid carbonate was in saturation equilibrium (constant p_{CO_2}). The data show that the removal of Mn(II) or Cd(II) from solution occurs in at least two stages.

■ log TOT Mn = –3.61,   log TOT Fe = –1.12
◻ log TOT Mn = –3.38,   log TOT Fe = –1.02
○ log TOT Mn = –4.59,   log TOT Fe = –1.19
● log TOT Mn = –4.34,   log TOT Fe = –1.24

Fig. a) from Wersin, Charlet, Karthein and Stumm (1989)
Fig. b) from Davis, Fuller and Cook (1987)

in this second stage of adsorption. The distribution coefficient X_{CdCO_3} / X_{CaCO_3} on the $CaCO_3$ surface was found for a given situation (Davis et al., 1987) to be ca. 1500 ± 300.

The equilibrium constant for the solid solution reaction

$$CaCO_3(s) + Cd^{2+} = CdCO_3(s) + Ca^{2+} \; ; \; K_x = \frac{10^{-8.3}}{10^{-11.3}} \cong 1000 \qquad (8.10)$$

The equilibrium value in (8.10) and the experimentally determined value is of the same order of magnitude. Plausibly, the activity coefficient of $CdCO_3(s)$ could be somewhat smaller than 1.

The processes described and their kinetics is of importance in the accumulation of trace metals by calcite in sediments and lakes (Delaney and Boyle, 1987) but also of relevance in the transport and retention of trace metals in calcareous aquifers. Fuller and Davis (1987) investigated the sorption by calcareous aquifer sand; they found that after 24 hours the rate of Cd^{2+} sorption was constant and controlled by the rate of surface precipitation. Clean grains of primary minerals, e.g., quartz and alumino silicates, sorbed less Cd^{2+} than grains which had surface patches of secondary minerals, e.g., carbonates, iron and manganese oxides. Fig. 6.11 gives data (time sequence) on electron spin resonance spectra of Mn^{2+} on $FeCO_3(s)$.

Recent work by Stipp and Hochella (1991) provide evidence for the processes of reconfiguration and hydration at the calcite surface. These results may provide a basis for future spectroscopic studies of trace metal adsorption and subsequent solid-solution formation.

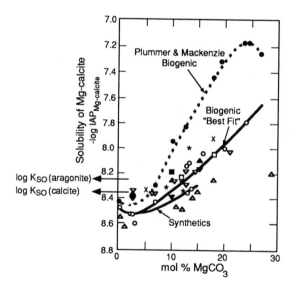

Figure 8.8

Solubilities of the Mg-calcite as a function of $MgCO_3$ constant. The solubility is expressed in line with Eq. (8.11) as $IAP_{Mg\text{-}calcite} = (Ca^{2+})^{(1-x)} (Mg^{2+})^x (CO_3^{2-})$. The solid curves represent the general trend of results on dissolution of biogenic and synthetic Mg-calcites. The curve fitting the data of Plummer and Mackenzie (1974) is dashed. The various points refer to the results of different researches. (For the origin of the data see Morse and Mackenzie, 1990.) (IAP = ion activity product.)
(Modified from Morse and Mackenzie, 1990)

Non-Ideality of Surface Precipitates or of Solid Solutions

The phenomena of surface precipitation and isomorphic substitutions described above and in Chapters 3.5, 6.5 and 6.6 are hampered because equilibrium is seldom established. The initial surface reaction, e.g., the surface complex formation on the surface of an oxide or carbonate fulfills many criteria of a reversible equilibrium. If we form on the outer layer of the solid phase a coprecipitate (isomorphic substitutions) we may still ideally have a metastable equilibrium. The extent of incipient adsorption, e.g., of HPO_4^{2-} on $FeOOH(s)$ or of Cd^{2+} on calcite is certainly dependent on the surface charge of the sorbing solid, and thus on pH of the solution etc.; even the kinetics of the reaction will be influenced by the surface charge; but the final solid solution, if it were in equilibrium, would not depend on the surface charge and the solution variables which influence the adsorption process; i.e., the extent of isomorphic substitution for the ideal solid solution is given by the equilibrium that describes the formation of the solid solution (and not by the rates by which these compositions are formed). Many surface phenomena that are encountered in laboratory studies and in field observations are characterized by partial, or metastable equilibrium or by non-equilibrium relations. Reversibility of the apparent equilibrium or congruence in dissolution or precipitation[1] can often not be assumed.

8.7 Magnesian Calcite

It is doubtful that formation and dissolution of any mineral in low temperature aqueous solutions has been more fully investigated than the magnesian calcite. This mineral is a preponderant carbonate phase, mostly of biogenic origin, in seawater. Fig. 8.8 gives some data on the solubilities of Mg-calcites as a function of $MgCO_3$ content.

Another source of divergence is the use of different models for the aqueous carbonate systems. Precipitation and dissolution experiments can be carried out in closed or open systems and various ways of pH-adjustments (see 8.2).

Analyses of natural calcites, formed at low temperatures, show $MgCO_3$ contents of up to 30 mol %.

Let's consider first a formal (equilibrium) approach to the solubility of Mg-calcite and compare its solubility with that of $CaCO_3$ (calcite or aragonite)

$$Ca_{(1-x)} Mg_x CO_3(s) = (1-x)Ca^{2+} + xMg^{2+} + CO_3^{2-} \quad ; K_{eq(x)} \tag{8.11}$$

---

[1] The term incongruent is generally used, if a mineral upon dissolution reacts to form a new solid or if the reversal of a dissolution process leads to a different composition. In natural environments incongruent solubility is probably more prevalent, e.g., in weathering of many clays, than congruent dissolution.

$$Ca^{2+} + CO_3^{2-} = CaCO_3(s) \qquad\qquad ; K_{s0(CaCO_3)}^{-1} \qquad (8.12)$$

$$Ca_{(1-x)} Mg_x CO_3(s) + xCa^{2+} = CaCO_3(s) + xMg^{2+} ; K_{eq(x)} K_{s0(CaCO_3)}^{-1} \qquad (8.13)$$

$$\left(\frac{[Mg^{2+}]}{[Ca^{2+}]}\right)^x = K_{eq(x)} K_{s0(CaCO_3)}^{-1} \qquad (8.14)$$

If a Mg-calcite is in contact with a solution whose $([Mg^{2+}] / [Ca^{2+}])^x$ ratio is smaller than $K_{eq(x)}K_{s0(CaCO_3)}^{-1}$, the Mg-calcite is less stable than $CaCO_3(s)$. For example (at 25° C) a 10 mol % Mg-calcite ($-\log K_{eq(10)} \approx 8.0$) is less stable than calcite ($-\log K_{s0} = 8.35$) or aragonite ($-\log K_{s0} = 8.22$) if the concentration or activity ratio is smaller than 10^2 or $10^{3.5}$, respectively. Thus, high Mg-calcite should be converted in marine sediments into calcite or aragonite. The conversion to aragonite has been observed. As the Mg-calcite is dissolved, Mg^{2+} becomes enriched in the solution (incongruent dissolution) and a purer $CaCO_3(s)$ is precipitated. As Fig. 8.8 suggests low Mg-calcites ($x = 3-4$ mol %) are probably stable in comparison to calcite. Higher Mg-calcites – although thermodynamically unstable – may persist for considerable time periods.

Some of the differences in solubilities are also related to different disordering of the crystal surface. As shown by Bishop et al. (1987) with Raman investigations, the biogenic phases are characterized by greater positional disorder than synthetic minerals of the same composition.

The fact that the carbonates of foraminifera burried in marine sediments can be used as a "memory storage" for Cd^{2+} present in the sea when the foraminifera were formed (Delaney and Boyle, 1987) is evidence for the non-reversibility or extremely slow reversibility of biogenic mineral carbonates.

The experimental controversy on surface and solubility characteristics goes on. It is beyond the scope of this chapter to review the various theories. Some of the discrepancies can be accounted for that surface processes attain relatively fast certain degree of metastability while the attainment of an equilibrium even within long-time periods cannot be accomplished.

Morse and Mackenzie (1990) point out the two fundamental problems:
1) In most experiments to calculate solubilities the magnesian calcites have been treated as solids of fixed compositions of one component, whereas they are actually a series of at least two-component compounds forming a partial solid solution series.
2) Magnesian calcite phases dissolve incongruently, leading to a formation of a phase different in composition from the original reactant solid.

Dolomite is one of the most abundant sedimentary carbonate minerals but its mode of formation and its surface properties are less well known than for most other carbonate minerals. As we have mentioned, the nucleation of dolomites and its structural ordering is extremely hindered. There is a general trend for the "ideality" of dolomite to increase with the age of dolomite over geological time (Morse and Mackenzie, 1990). Most dolomites that are currently forming in surficial sediments and that have been synthesized in the laboratory are calcium-rich and far from perfectly ordered. Such dolomites are commonly referred to as "protodolomites". Morse and Mackenzie (1990) have reviewed extensively the geochemistry (including the surface chemistry of dolomites and Mg-calcites.

Appendix

Carbonate Solubility Equilibria

Solubility Equilibrium in an Open System CaCO₃(s) (calcite)

The following species are in equilibrium: Ca^{2+}, H^+, HCO_3^-, CO_3^{2-}, OH^- and $H_2CO_3^*$. We need to consider simultaneously the following eqs. (K values at 25° C):

$$[H_2CO_3^*] = K_H p_{CO_2}; \qquad K_H = 3 \times 10^{-2} \text{ atm}^{-1} \text{ M} \qquad (A.8.1)$$

$$[H^+][HCO_3^-] / [H_2CO_3^*] = K_1; \quad K_1 = 5 \times 10^{-7} \qquad (A.8.2)$$

$$[H^+][CO_3^{2-}] / [HCO_3^-] = K_2; \quad K_2 = 5 \times 10^{-11} \qquad (A.8.3)$$

$$[H^+][OH^-] = K_W; \qquad K_W = 10^{-14} \qquad (A.8.4)$$

$$[Ca^{2+}][CO_3^{2-}] = K_{s0}; \qquad K_{s0} = 5 \times 10^{-9} \qquad (A.8.5)$$

plus the charge balance (or proton balance):

$$[H^+] + 2[Ca^{2+}] = [HCO_3^-] + 2[CO_3^{2-}] + [OH^-] \qquad (A.8.6)$$

All these equations (A.8.1 to A.8.6) are plotted in Fig. A.8.1. If in this system p_{CO_2} = const, the pH is adjusted by adding acid or base the following concentration relations prevail:

$$\frac{d \log [H_2CO_3^*]}{d \text{ pH}} = 0; \quad \frac{d \log [HCO_3^-]}{d \text{ pH}} = 1; \quad \frac{d \log [CO_3^{2-}]}{d \text{ pH}} = 2; \quad \frac{d \log [Ca^{2+}]}{d \text{ pH}} = -2 \qquad (A.8.7)$$

If the pH adjustment is made by changing p_{CO_2} alone (no addition of base or acid) then the following relations prevail:

$$\frac{d \log p_{CO_2}}{d \text{ pH}} = -0.5; \quad \frac{d \log p_{CO_2}}{d \log [Ca^{2+}]} = 0.5; \quad \frac{d \log p_{CO_2}}{d \log [HCO_3^-]} = 1 \qquad (A.8.8)$$

In this situation the CaCO₃ surface is characterized by zero surface charge (pH = pH_{pzc})

304

Table A.8.1 Equilibrium Constants for Carbonate and $CaCO_3$ (calcite) Equilibria

		$-\log K$					
		5° C	10° C	15° C	20° C	25° C	40° C
$CaCO_3(s)$	$= Ca^{2+} + CO_3^{2-}$	8.35	8.36	8.37	8.39	8.42	8.53
$CaCO_3(s) + H^+$	$= HCO_3^- + Ca^{2+}$	−2.22	−2.13	−2.06	−1.99	−1.99	−1.69
$H_2CO_3^*$	$= H^+ + HCO_3^-$	6.52	6.46	6.42	6.38	6.35	6.30
$CO_2(g) + H_2O$	$= H_2CO_3^*$	1.20	1.27	1.34	1.41	1.47	1.64
HCO_3^-	$= H^+ + CO_3^{2-}$	10.56	10.49	10.43	10.38	10.33	10.22

Table A.8.2 Solubility Products of Carbonates (25° C)

		$-\log K_{s0}$	
Calcite	$CaCO_3$	8.42	(8.3)
Aragonite	$CaCO_3$	8.22	
Vaterite	$CaCO_3$	7.73	
Magnesite	$MgCO_3$	7.46	(8.2)
Nesquehonite	$MgCO_3 \times 3\,H_2O$	4.67	(5.2)
Hydromagnesite	$Mg_4(CO_3)_3(OH)_2 \times 3\,H_2O$	36.47	
Dolomite	$CaMg(CO_3)_2$	17.09	
Huntite	$CaMg_3(CO_3)_4$	30.46	
Witherite	$BaCO_3$	7.63	
Rhodochrosite	$MnCO_3$	9.2	(10.54)
Siderite	$FeCO_3$	10.68	(10.50)

Values in ( ) are data from the compilation of Robie et al. (1979).

Solubility Equilibrium in a closed System

In this system

$$TOTC = C_T = [H_2CO_3^*] + [HCO_3^-] + [CO_3^{2-}] = constant \qquad (A.8.9)$$

plus the same equations as Eqs. (A.8.2) to (A.8.6) are valid. The equations are plotted in Fig. A.8.2 for TOTC = 10^{-3} M.

If pH is changed (TOTC = constant) upon addition of acid or base the following con-centration relations prevail in the range $pK_1 > pH < pK_2$:

$$\frac{d \log [HCO_3^-]}{d \ pH} = 0 \ ; \frac{d \log [CO_3^{2-}]}{d \ pH} = 1 \ ; \frac{d \log [Ca^{2+}]}{d \ pH} = -1 \qquad (A.8.10)$$

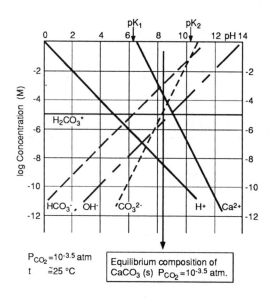

Figure A.8.1

Equilibrium composition of a solution in presence of $CaCO_3$(s) (calcite) at constant particle pressure of CO_2 (p_{CO_2} = $10^{-3.5}$ atm) at 25° C. If no acid or base is added, the equilibrium composition is indicated by the arrow. The calcite at this composition is characterized by a net surface charge of zero.

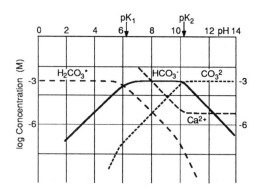

Figure A.8.2

$CaCO_3$(s) equilibrium in a closed system for TOTC = 10^{-3} M (25° C).

Reading Suggestions

Berner, R. A., and A. C. Lasaga (1989), "Modeling the Geochemical Carbon Cycle", *Scientific Amer.* **260**, 74-81.

Busenberg, E., and L. N. Plummer (1986b), "A Comparative Study of the Dissolution and Crystal Growth Kinetics of Calcite and Aragonite", in F. A. Mumpton, Ed., *Studies in Diagenesis*, U.S. Geol. Surv. Bull. 1578, pp. 139-168.

Morse, J. W., and F. J. Mackenzie (1990), *Geochemistry of Sedimentary Carbonates*, Elsevier, Amsterdam. (Comprehensive treatment of carbonate geochemistry, covering the range from electrolyte chemistry of carbon-containing waters to the global cycles of carbon.)

Stipp, S. L., and M. F. Hochella, (1991), "Structure and Bonding Environments at the Calcite Surface as Observed with X-Ray Photoelectron Spectroscopy (XPS) and Low Energy Electron Diffraction (LEED)", *Geochim. Cosmochim. Acta* **55/6**, 1723-36.

Wollast, R. (1990), "Rate and Mechanism of Dissolution of Carbonates in the System $CaCO_3$-$MgCO_3$", in W. Stumm, Ed., *Aquatic Chemical Kinetics, Reaction Rates of Processes in Natural Waters*, Wiley-Interscience, New York, pp. 431-445.

Chapter 9

Redox Processes Mediated by Surfaces

9.1 Specific Adsorption of Oxidants and Reductants

As was shown in Chapter 2.3 surfaces can interact chemically with H^+, OH^-, cations and anions (ligands). In Fig. 9.1 some examples of surface complexes with oxidants and reductants are given. The functional OH^- groups – in Fig. 9.1 we use as an example a hydrous ferric oxide – facilitate the binding of reductant cations (I); or they exchange surface OH-groups with oxidizing or reducing ligands to form surface complexes with reductants or oxidants (II,III). Some reductants can become bound with the help of bridging ligands (V). Surface complexes with transition elements may facilitate the (outer-spheric) binding of O_2. Similar considerations apply to non-oxidic surfaces. As Luther (1987) has postulated, pyrite can form surface complexes both with oxidants and reductants (VI,VII in Fig. 9.1).

The binding of a reductant or oxidant species to form an inner-spheric surface complex changes its electronic structure and thus influences its reductive and oxidative reactivity. As a consequence the following differences in thermodynamic and kinetic properties between dissolved and adsorbed species may be observed (Wehrli et al., 1989).
1) Surface OH-groups act as σ-donor ligands which increase the electron density at the adsorbed transition metal. E.g., Fe(II) bound to a hydrous oxide becomes a better reductant than Fe^{2+} in solution (or than Fe(II) outer-spherically bound). Higher oxidation states will generally form more stable surface complexes than the lower oxidation state. In other words, the specific adsorption of Cu(II) should lower its redox potential with regard to Cu(I)aq. The situation, once again, is comparable with the effect of complex formation in solution chemistry, where usually a ligand with an oxygen donor atom stabilizes the higher oxidation state, e.g., the ferric complex thus lowering the redox potential of the complex bound couple (see Fig. 9.2) and making the Fe(II)-complex a better reductant than Fe^{2+}(aq).
2) Coordinated donor ligands accelerate the ligand exchange kinetics, especially in the trans-position (cf. Chapter 5.3). For example, the water molecules that are in a trans-position to the sigma bond are expected to exchange faster. The water exchange rate of Fe(III) hydroxo complexes increases in the order Fe^{3+} > $Fe(OH)^{2+}$ > $Fe(OH)_2^+$ with rate constants of 10^2, 10^5 and > 10^7 s^{-1}, respectively (Schneider and Schwyn, 1987). In Chapter 4.4 it was shown that hydroxo complexes (because of the larger water exchange rate) adsorb faster than the corresponding free metal ions.

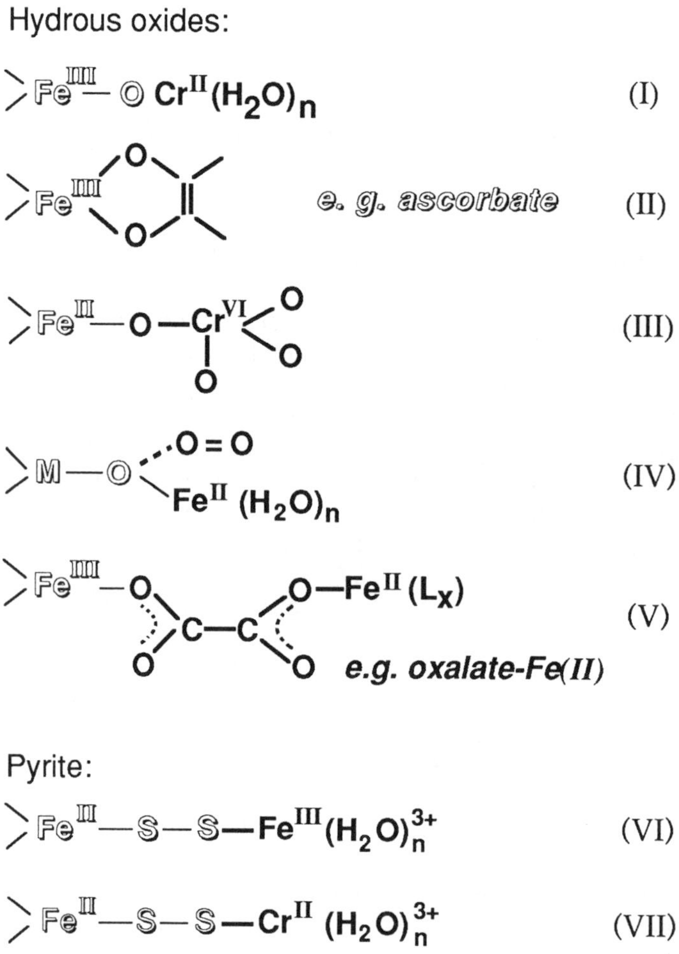

Hydrous oxides:

$$>Fe^{III}-O\ Cr^{II}(H_2O)_n \tag{I}$$

$$>Fe^{III}\underset{O}{\overset{O}{<}}|| \qquad \textit{e. g. ascorbate} \tag{II}$$

$$>Fe^{III}-O-Cr^{VI}\underset{O}{\overset{O}{<}} \tag{III}$$

$$>M-O\overset{\cdot\cdot O=O}{<}_{Fe^{II}(H_2O)_n} \tag{IV}$$

$$>Fe^{III}-O\underset{O}{\overset{}{:}}C-C\overset{O-Fe^{II}(L_x)}{<}_{O} \tag{V}$$

e.g. oxalate-Fe(II)

Pyrite:

$$>Fe^{II}-S-S-Fe^{III}(H_2O)_n^{3+} \tag{VI}$$

$$>Fe^{II}-S-S-Cr^{II}(H_2O)_n^{3+} \tag{VII}$$

Figure 9.1

Schematic illustration on the specific adsorption on Fe-minerals of oxidants and reductants directly or through ligand bridges. The formation of these surface complexes (which is usually fast) facilitates the subsequent electron transfer.

Inert M(III) cations will exchange their ligands faster once they are adsorbed to an oxide surface.

3) *Coordination of a Ligand to a Metal Center* of the surface decreases the electron density of the ligand donor atom and will affect its chemical and redox property. The *trans-effect* of the ligand brings labilizing effects on the oxobonds of the central metal ions in the surface of the lattice. Furthermore, the attachment of the ligand enhances surface protonation (Examples 2.4 and 5.1).
   The electron transfer from the ligand to the central metal ion (e.g., Mn(IV,III),

Fe(III)) may be facilitated, i.e., the activation energy of the redox reaction lowered.

Bridged Surface Ligands may mediate electron transfer. The oxidant and reductant have to encounter each other in a suitable structural arrangement to make an electron exchange possible, often a bridging ligand can link the redox partners and assist the electron transfer between them. In pure aqueous solutions OH^- is an effective bridging ligand. That is the reason why electron transfer between transition metal ions usually increases with pH. Similar to dissolved OH^-, the hydroxyl group at the mineral-water interface may act as a bridging ligand between the redox couples. Some organic ligands (such as V in Fig. 9.1) can mediate the electron transfer.

9.2 Some Thermodynamic Considerations

We first compare the consequences of solute and surface complexation from a thermodynamic point of view. We use for exemplification Fe(II) and Fe(III) because
1) more data are available with this redox pair than with others, and
2) the transformations of iron are especially important in the redox cycling of electrons in natural environments.
As Fig. 9.2 shows, the Fe(III)/Fe(II) redox couple can adjust with appropriate ligands to any redox potential within the stability of water. The principles exemplified here are of course also applicable to other redox systems.

As illustrated in Fig. 9.2 the redox potential at pH = 7, E_H^o (pH = 7), decreases in presence of most complex formers, especially chelates with oxygen donor atoms, such as a citrate, EDTA and salicylate because these ligands form stronger complexes with Fe(III) than with Fe(II). Phenanthroline which stabilzes Fe(II) more than Fe(III) is an exception, explainable in terms of the electronic configuration of the aromatic N-Fe(II) bond (Luther et al., 1991). But Fe(II) complexes are usually stronger reductants than Fe^{2+}. This stabilization of the Fe(III) oxidation state is also observed with hydroxo complexes and by the binding to O^{-II} in solid phases. Thus, Fe(II) minerals are, thermodynamically speaking, strong reductants. For example, the couple Fe_2SiO_4 (fayalite) $-Fe_3O_4$ (magnetite) has an E_H similar to that for the reduction of $H_2O(\ell)$ to $H_2(g)$ (Baur, 1978), Fig. 9.2. A surface complex of Fe^{2+} adsorbed inner-spherically onto a hydrous oxide surface is more reducing than $Fe^{2+}(aq)$.

Redox reactions are of importance in the dissolution of Fe-bearing minerals (Table 9.1). Reductive dissolution of Fe(III)(hydr)oxides can be accomplished with many reductants, especially organic and inorganic reductants, such as ascorbate, phenols, dithionite, HS^-, etc. Fe(II) in presence of complex formers can readily dissolve

Fe(III)(hydr)oxides. The Fe(II) bound in magnetite and silicate and adsorbed to oxides can reduce Fe(III) (reactions 4, 6, 8) and O_2 (reactions 7 and 9).

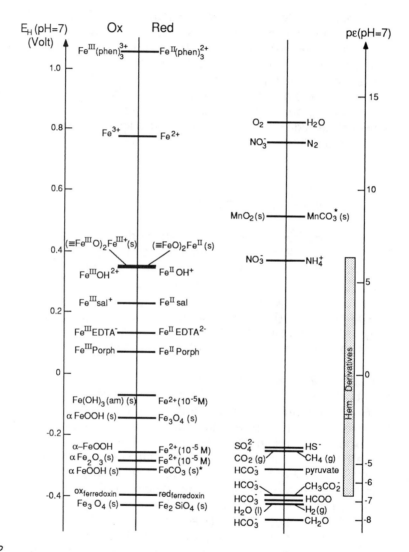

Figure 9.2

Representative Fe(II)/Fe(III) redox couples at pH = 7. $p\varepsilon$ (= $-\log\{e^-\}$) = $\frac{F}{2.3\,RT}\,E_H$. (phen = phenanthroline; sal = salicylate; porph = porphyrin; * = valid for $[HCO_3^-]$ = 10^{-3} M)

Complex formation with Fe(II) and Fe(III) both on solid and solute phases has a dramatic effect on the redox potentials; thus, electron transfer by the Fe(II), Fe(III) system can occur at pH = 7 over the entire range of the stability of water; E_H (−0.5 V to 1.1 V).

The range of redox potentials for hem derivatives given on the right illustrates the possibilities involved in bioinorganic systems.

Table 9.1   Redox Reactions with solid Fe Phases as Oxidants or Reductants

Reductant		Oxidant		
1 Ascorbate [1)	+	$\equiv Fe^{III}-OH$	=	$Fe^{2+}(aq) + Asc^{\bullet} + \equiv$ [2)
2 $Fe^{II}X$ [3)	+	$\equiv Fe^{III}-OH$	=	$Fe^{III}X(aq) + Fe^{2+}(aq) + \equiv$
3 $Fe^{II}Ox_n$	+	$\equiv Fe^{III}-OH$	=	$Fe^{III}Ox_n(aq) + Fe^{2+}(aq) + \equiv$
4 $3\,(Fe^{II}Fe_2^{III})O_4(s)$ [4)	+	$(Fe^{II}Fe_2^{III})O_4(s) + 8\,H^+$	=	$4\,\gamma Fe_2O_3 + 4\,Fe^{2+}(aq) + 4\,H_2O$
5 $2*Fe^{2+}$	+	$(Fe^{II}Fe_2^{III})O_4(s) + 8\,H^+$	=	$3\,Fe^{2+}(aq) + 2*Fe^{3+}(aq) + 4\,H_2O$
6 $3\,(Fe^{II}Fe_2^{III})O_4(s)$	+	$2*Fe^{3+}$	=	$4\,\gamma Fe_2O_3(s) + 2*Fe^{2+}(aq) + Fe^{2+}$
7 $\equiv Fe^{III}-OFe^{II}$	+	$\frac{1}{4}\,O_2 + H^+$	=	$\equiv Fe^{III}-OFe^{III} + \frac{1}{2}\,H_2O$
8 $Fe^{II}Ca_{0.5}$ silicate (s)	+	$*Fe^{3+}$	=	Fe^{III} silicate (s) $+ \frac{1}{2}\,Ca^{2+} + *Fe^{2+}$
9 $Fe^{II}M_{0.5}^{II}$ silicate (s)	+	$\frac{1}{4}\,O_2 + H^+$	=	Fe^{III} silicate (s) $+ \frac{1}{2}\,M^{2+} + \frac{1}{2}\,H_2O$

[1)   other typical reductants include phenols, dithionite, S–(II), CNS⁻ etc.
[2) $\equiv$ is the reconstituted surface of the hydrous oxide after detachment of Fe(II) into solution;
      Asc• is the (oxidized) ascorbate radical
[3)   X = ligand, preferably chelate with O-donor atom, oxalate, NTA, citrate, EDTA, etc.
[4)   magnetite

9.3 Catalysis of Redox Reactions by Surfaces

In *aqueous solution*, electron transfer between two solute redox reactants A^+ and B may occur either by an outer-spher (o.s.) or an inner-sphere (i.s.) redox mechanism (ET stands for electron transfer)

Outer-sphere:

$$A^+ + B \underset{\text{diffusion}}{\rightleftharpoons} A^+ \cdots B \underset{\text{ET}}{\rightleftharpoons} A \cdots B^+ \underset{\text{diffusion}}{\rightleftharpoons} A + B^+ \qquad (9.1)$$

Inner-sphere:

$$AX^+ + B \underset{\substack{\text{complex} \\ \text{formation}}}{\rightleftharpoons} AX^+B \underset{\text{ET}}{\rightleftharpoons} AXB^+ \underset{\text{dissociation}}{\rightleftharpoons} A + XB^+ \qquad (9.2)$$

In the o.s. reaction, the ion pair $A^+ \cdots B$ is formed in a first step. The corresponding equilibrium constant can usually be obtained from simple electrostatic models. In this "ideal" case specific chemical interactions can be neglected and the rate constant of the E.T. step follows the theory of R.A. Marcus (see for example Marcus, 1975, or Cannon, 1980). In the i.s. reaction each of the three steps in reaction (9.2) may determine the reaction rates. The lability of the coordinated ligands at the reactant B and the thermodynamic stability of the complex AX^+ are the main factors in the formation kinetics of the bridged precursor complex AX^+B. In some cases the slow decomposition of the successor AXB^+ may slow down the overall rate.

In heterogeneous redox reactions similar reaction sequences are observed; usually an encounter (outer-sphere or inner-sphere) surface complex is formed to facilitate the subsequent electron transfer.

A classical case of heterogeneous inner-spheric electron transfer has been demonstrated by Gordon and Taube (1962) on the oxidation of U(IV) by PbO_2; by using ^{18}O in PbO_2, they could show that both O^{-II} ions in the product UO_2^{2+} are derived from the oxide lattice:

$$
\begin{aligned}
&\equiv Pb^{IV}\text{-OH} \\
&\equiv Pb^{IV}\text{-OH}
\end{aligned} \;+\; U^{IV} \;\rightleftharpoons\;
\begin{aligned}
&\equiv Pb^{IV}\text{-O} \\
&\equiv Pb^{IV}\text{-O}
\end{aligned}\!\!\Big\rangle U^{IV} \;+\; 2\,H^+
$$

$$(9.3)$$

$$
\begin{aligned}
&\equiv Pb^{IV}\text{-O} \\
&\equiv Pb^{IV}\text{-O}
\end{aligned}\!\!\Big\rangle U^{IV} \;\xrightarrow{\;ET\;}\;
\begin{aligned}
&\equiv Pb^{II}\text{-O} \\
&\equiv Pb^{IV}\text{-O}
\end{aligned}\!\!\Big\rangle U^{IV} \;\xrightarrow{\;H^+.H_2O\;}\; UO_2^+ \;+\; \equiv Pb^{IV} - OH + Pb^{2+}(aq)
$$

A typical redox reaction sequence between an (usually organic) reductant R and iron(III) (hydr)oxide can be schematically expressed as

Surface sites		Reactants	fast	Surface species	
$\equiv Fe^{III}\text{-OH}$	$+$	Reductant R	$\longrightarrow$	$\equiv Fe^{III}R$	(9.4a)

Surface species	electron transfer	Reduced surface species + oxidized reactant	
$\equiv Fe^{III}R$	$\rightleftharpoons$	$\equiv Fe(II) + {}^\bullet O$	(9.4b)[1]

Reduced surface species	slow detachment of Fe(II)	Fe(II)(aq) + surface site	
$\equiv Fe^{II}$	$\xrightarrow{}$		(9.4c)

The scheme of Eqs. (9.4) suggests that either the electron transfer (9.4b) or the detachment (9.4c) are rate determining. The steps (b) and (c) may be coupled.

Reductive Dissolution of Higher Valent Oxides

We exemplify the reductive dissolution of minerals by illustrating the reductive dissolution of hydrous ferric oxide. With this oxide (Sulzberger et al.; 1989, Suter et al.,

---

[1] if the reductant is an organic ligand, its oxidation product is usually a radical.

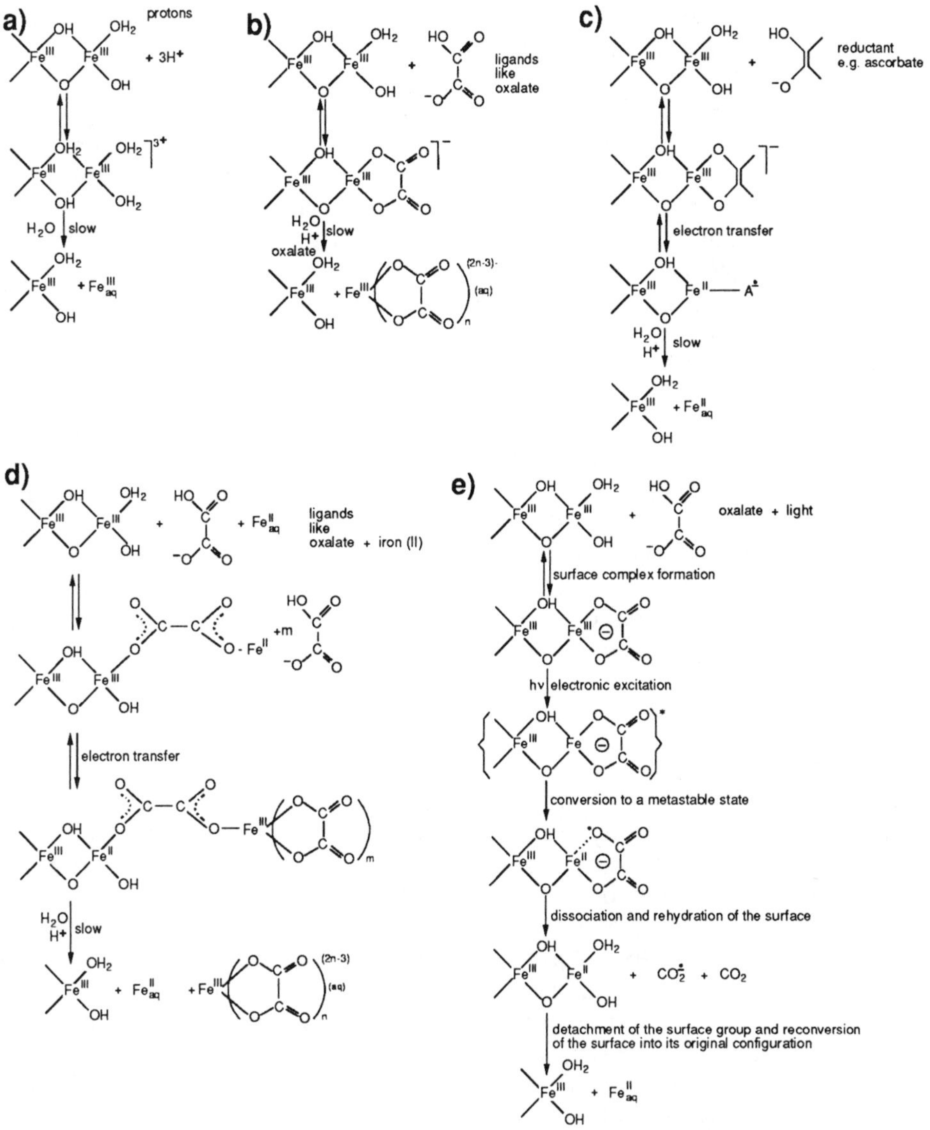

Figure 9.3

Schematic representation of the various reaction modes for the dissolution of Fe(III)(hydr)oxides:

a) by protons; b) by bidentate complex formers that form surface chelates. The resulting solute Fe(III) complexes may subsequently become reduced, e.g., by HS⁻; c) by reductants (ligands with oxygen donor atoms) such as ascorbate that can form surface complexes and transfer electrons inner-spherically; d) catalytic dissolution of Fe(III)(hydr)oxides by Fe(II) in the presence of a complex former; e) light-induced dissolution of Fe(III)(hydr)oxides in the presence of an electron donor such as oxalate. In all of the above examples, surface coordination controls the dissolution process. (Adapted from Sulzberger et al., 1989, and from Hering and Stumm, 1990.)

1991; Blesa et al., 1987) and with Mn(III,IV)(hydr)oxides much experimental data (Stone, 1987; Stone and Morgan, 1987) are available.

Several possible pathways for Fe(III)(hydr)oxide dissolution are outlined in Fig. 9.3. The structures in this figure are highly schematic. They are not intended to give the details of structural or coordinative arrangements but simply to illustrate the presence of surface hydroxyl groups and of oxo- and hydroxo-bridges and to indicate correct relative charges in a convenient way. This schematic presentation illustrates the diversity of reaction pathways. Steady-state dissolution kinetics are expected if the original surface structure is restored after detachment of the surface iron atom. In this case, the dissolution rate will be a function of the concentration of the appropriate surface species.

The first two pathways (a) and (b) show, respectively, the influence of H^+ and of surface complex forming ligands on the non-reductive dissolution. These pathways were discussed in Chapter 5. Reductive dissolution mechanisms are illustrated in pathways (c) – (e) (Fig. 9.3). Reductants adsorbed to the hydrous oxide surface can readily exchange electrons with an Fe(III) surface center. Those reductants, such as ascorbate, that form inner-sphere surface complexes are especially efficient. The electron transfer leads to an oxidized reactant (often a radical) and a surface Fe(II) atom. The Fe(II)-O bond in the surface of the crystalline lattice is more labile than the Fe(III)-O bond and thus, the reduced metal center is more easily detached from the surface than the original oxidized metal center (see Eqs. 9.4a – 9.4c).

Pathway (d) in Fig. 9.3 provides a possible explanation for the efficiency of a combination of a reductant and a complex former in promoting fast dissolution of Fe(III) (hydr)oxydes. In this pathway, Fe(II) is the reductant. In the absence of a complex former, however, Fe^{2+} does not transfer electrons to the surface Fe(III) of a Fe(III) (hydr)oxide to any measurable apparent extent. The electron transfer occurs only in the presence of a suitable bridging ligand (e.g., oxalate). As illustrated in Fig. 9.3d, a ternary surface complex is formed and an electron transfer, presumably inner-sphere, occurs between the adsorbed Fe(II) and the surface Fe(III). This is followed by the rate-limiting detachment of the reduced surface iron. In this pathway, the concentration of $Fe(II)_{aq}$ remains constant while the concentration of dissolved Fe(III) increases; thus, $Fe(II)_{aq}$ acts as a catalyst to produce Fe(II)(aq) from the dissolution of Fe(III)(hydr)oxides.

Although thermodynamically favorable, reductive dissolution of Fe(III)(hydr)oxides by some metastable ligands (even those, such as oxalate, that can form surface complexes) does not occur in the absence of light. The photochemical pathway is depicted in Fig. 9.3e. In the presence of light, surface complex formation is followed by electron transfer via an excited state (indicated by *) either of the iron oxide bulk phase or of the surface complex. (Light-induced reactions will be discussed in Chapter 10.)

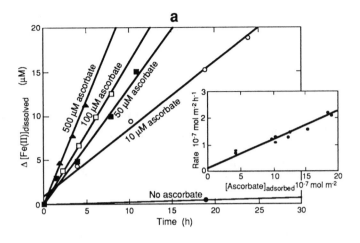

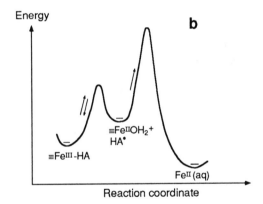

Figure 9.4

a) Reductive dissolution of hematite (0,5 g/ℓ) with ascorbate at pH = 3. See pathway c in Fig. 9.3 for the mechanism proposed. As the insert shows the dissolution rate is proportional to the ascorbate adsorbed, i.e., ⟨≡FeHA⟩ (compare Eq. 9.11).

b) Activated complex diagram for the dissolution of Fe(III)(hydr)oxides by ascorbate, HA, in accordance with the mechanisms given in Eqs. (9.6) – (9.11). The fast formation of a surface complex, ≡FeHA, is followed by a reversible (assumed) electron transfer leading to a Fe(II) in the surface lattice which then in a slow (rate determining) reaction step becomes detached into solution. Since the electron transfer process is treated as a preequilibrium to the detachment step, under steady state assumption the overall reaction rate is proportional to the surface complex ⟨≡FeHA⟩ and to the redox equilibrium.

Several pathways may contribute to the overall dissolution reaction. Over the course of the dissolution, the relative importance of the contributing pathway may change, possibly due to accumulation of some reactant (Sulzberger et al., 1989; Siffert and Sulzberger, 1991). Many general reviews on reductive and oxidative dissolution are available (e.g., Valverde and Wagner, 1978; Segal and Sellers, 1984; Stone and Morgan, 1987; Hering and Stumm, 1990; White, 1990).

Rate Laws

The following reaction sequence in reductive dissolution is plausible and is exemplified (Suter et al., 1991) here for the reaction of Fe(III)(hydr)oxide with ascorbate. It follows the general scheme given in Eqs. (9.4a) – (9.4c).

1) The fast adsorption (surface complex formation) of ascorbate, HA⁻, to the surface of the Fe(III)(hydr)oxide

$$\equiv Fe^{III}\text{-}OH_2^+ + HA^- \underset{k_{-1}}{\overset{k_1}{\rightleftharpoons}} \equiv Fe^{III}\text{-}HA + H_2O \tag{9.5}$$

Since this step is fast in comparison to the subsequent ones, this reaction can be considered as a pre-equilibrium. k_1/k_{-1} is the surface complex formation (equivalent to the Langmuir) constant.

2) Electron transfer accompanied by release of the oxidized ascorbate, the ascorbate radical, HA•

$$\equiv Fe^{III}\text{-}HA + H_2O \underset{k_{-2}}{\overset{k_2}{\rightleftharpoons}} \equiv Fe^{II}\text{-}OH_2^+ + HA^{\bullet} \tag{9.6}$$

3) The detachment of the Fe(II) into solution. This detachment is facilitated by the increase in surface protonation that is accompanied by the ligand binding. The fact that iron(II) is more readily detached than an iron(III) site from the mineral surface is due to the lower Madelung energy of the Fe(II)-oxygen bond than the Fe(III)-oxygen bond

$$\equiv Fe^{II}\text{-}OH_2^+ \underset{k_3}{\overset{H^+ \text{ slow}}{\rightleftharpoons}} \text{new surface site} + Fe^{2+}(aq) \tag{9.7}$$

4) The scavenging of the radical HA• (in the case of ascorbate). HA• reduces relatively fast (in a non-rate determining step) another Fe(III); the product is dehydro-ascorbate. The rate expression for the production of dissolved iron(II) at a given pH is written as follows:

$$\frac{d[Fe^{2+}]}{dt} = k_3 \langle \equiv Fe^{II}OH_2 \rangle \tag{9.8}$$

The surface concentration of iron(II) is in turn dependent on the surface concentration of the ascorbato-iron(III) complex. Thus, we write the rate expression for the production of surface iron(II).

$$\frac{d\langle\equiv Fe^{II}OH_2\rangle}{dt} = k_2\langle\equiv Fe^{III}HA\rangle - (k_2[HA^\bullet] + k_3)\,\langle\equiv FeOH_2^+\rangle \tag{9.9}$$

Using the steady-state approximation by setting the time differential equal to zero allows the resulting algebraic equation to be solved for the surface concentration of iron(II).

$$\langle\equiv Fe^{II}OH_2^+\rangle = \frac{k_2}{k_{-2}[HA^\bullet] + k_3}\,\langle\equiv Fe^{III}HA\rangle \tag{9.10}$$

This expression can be substituted into the rate expression for the production of dissolved iron(II) yielding a pseudo first-order rate expression written in terms of the surface concentration of ascorbate.

$$\frac{d[Fe^{2+}]}{dt} = \frac{k_2 k_3}{k_{-2}[HA^\bullet] + k_3}\,\langle\equiv Fe^{III}HA\rangle \tag{9.11}$$

Since the dissolution rate is constant (Fig. 9.4a) over the time interval observed, it is assumed there is also a steady-state for the intermediate product $HA^\bullet$ [1]. As illustrated in Fig. 9.4a, at a given pH, the rate is proportional to $\langle\equiv Fe^{III}HA\rangle$.

There is a pH dependence of the dissolution rate; the rate decreases with increasing pH. Various explanations can be given:
1) H^+ facilitates the detachment of Fe(II). The reduction in charge from Fe(III) to Fe(II) in the surface of the solid phase causes a charge deficiency in the solid phase which is balanced by H^+ uptake;
2) at higher pH values Fe(II) becomes adsorbed and may block surface reactive sites;
3) thermodynamically, the adsorption of the reductant is pH-dependent; furthermore, the free energy of the reaction of the electron transfer decreases with increasing pH.

Although the experiments (Suter et al., 1991) were carried out at relatively low pH, there are experimental results at pH = 7 and pH = 8 which clearly demonstrate reductive dissolution in the neutral pH-range.

If $k_{-2}[HA^\bullet] > k_3$, Eq. (9.13) – compatible with a scheme on the activation energies involved (Fig. 9.4b) – reduces to

---

[1] The presence of a radical scavenger such as mesitylene accelerates the reduction (Ruf, 1992).

$$\frac{d[Fe^{2+}]}{dt} \cong \frac{k_3}{[HA^\bullet]}\frac{k_2}{k_{-2}}\langle\equiv Fe^{III}HA\rangle \tag{9.12}$$

Since k_2/k_{-2} corresponds to the equilibrium constant of the redox reaction (redox potential), Eq. (9.12) suggests that the dissolution reaction may depend both on the tendency to bind the reductant to the Fe(III)(hydr)oxide surface and (even if the electron transfer is not overall rate determining), on the redox equilibrium (see Fig. 9.4b).

Reduction with a Fe(II) Complex

A similar rate mechanism as that given above can be derived (Suter et al., 1991) if the Fe(II)-complex is used (instead of ascorbate) as the reductant. In case of Fe(II) oxalate we propose a mechanism where Fe^{II} binds to the goethite surface through oxalate as a ligand bridge. We assume that the electron is transferred within the ternary surface complex through the oxalate bridge to the iron(III). Possibly a ternary surface complex such as I or II is involved.

$$\tag{9.13}$$

The reaction can be described by the sequence of reaction steps given in Fig 9.3d. One interesting feature of this reaction sequence is that it is essentially a catalytic reaction. Fe(II) added to the system remains constant while the dissolution reaction produces Fe(III) in solution (compare Fig. 9.5).

The Rate of reductive Dissolution of Hematite by H_2S as observed between pH 4 and 7 is given in Fig. 9.6 (dos Santos Afonso and Stumm, in preparation). The HS^- is oxidized to SO_4^{2-}. The experiments were carried out at different pH values (pH-stat) and using constant P_{H_2S}. 1.8 – 2.0 H^+ ions are consumed per Fe(II) released into solution, as long as the solubility product of FeS is not exceeded, the product of the reaction is Fe^{2+}. The reaction proceeds through the formation of inner-sphere $\equiv Fe-S$. The dissolution rate, R, is given by

$$R = k_e \langle\equiv FeS\rangle + k_e' \langle\equiv FeHS\rangle \tag{9.14}$$

The data set given in Fig. 9.6 shows a rate dependence on $[HS^-]$; this part of the figure reflects the initial linear "Langmuir" type on the adsorption equilibrium:

$$\equiv Fe\text{-}OH + HS^- \rightleftharpoons \equiv Fe\text{-}S + H_2O$$

It is remarkable that this reductive dissolution, a heterogeneous multi-electron transfer, is so fast.

It is interesting to juxtapose the reduction of a solid Fe(III)(hydr)oxide with a Fe(II) complex, e.g., with an oxalato complex of Fe(II) ($Fe^{II}Ox$) on one hand with the oxidation of a Fe(II)bearing solid phase, e.g., a Fe(II) silicate with Fe(III) on the other hand (compare reactions (3) and (8), Table 9.1). In both cases the electron transfer (ET) occurs heterogeneously between Fe(III) and a Fe(II) complex. In one case the oxidant is the solid phase, in the other case the reductant is the solid phase. In simplified schematic notation:

$$\equiv Fe^{III} - OH + Fe^{II}Ox(aq) \rightleftharpoons \equiv Fe^{III} - OxFe^{II} \overset{ET}{\rightleftharpoons} \equiv Fe^{II} - OxFe^{III}$$
$$\longrightarrow Fe^{II}(aq) + Fe^{III}Ox(aq) + \equiv \tag{9.15}$$

$$Fe(III)(aq) + \equiv Si - Fe^{II} \rightleftharpoons \equiv Si - Fe^{II} Fe^{III} \overset{ET}{\rightleftharpoons} \equiv SiFe^{III} Fe^{II}$$
$$\longrightarrow Fe^{III}(aq) + Fe^{II}(aq) + \equiv Si \tag{9.16}$$

(where $\equiv$ is the reconstituted surface of the hydrous oxide after detachment of Fe(II) and Fe(III).)

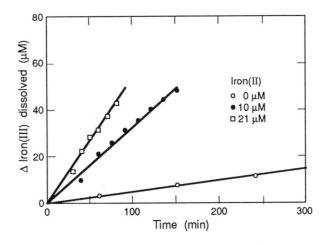

Figure 9.5

Dissolution of goethite by oxalate in the presence of different concentrations of ferrous iron. The reaction mechanism proposed is that of Fig. 9.3d. The change in the concentration of Fe(III) is given (preconditioning of the surface introduces some incipient Fe(III)). pH = 3.0, goethite: 0.46 g/ℓ, oxalate: 0.001 M.

(From Suter et al., 1991)

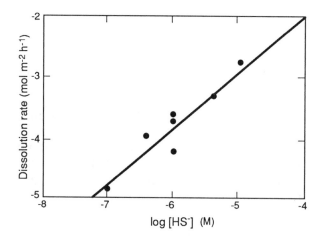

Figure 9.6

Reductive dissolution of hematite by H_2S. Individual rates were determined at constant pH and constant p_{H_2S} (pH = 4 – 6 and p_{H_2S} = 10^{-2} – 10^{-4} atm).
(From dos Santos Afonso and Stumm, in preparation)

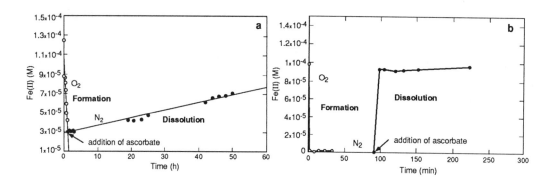

Figure 9.7

Reactivity of Fe(III)(hydr)oxide as measured by the reductive dissolution with ascorbate. "Fe(OH)$_3$" is prepared from Fe(II) (10^{-4} M) and HCO$_3^-$ (3 × 10^{-4} M) by oxygenation (p_{O_2} = 0.2 atm) in presence of a buffer imidazol pH = 6.7 (Fig. a) and in presence of TRIS and imidazol pH $\cong$ 7.7 (Fig. b). After the formation of Fe(III)(hydr)oxide the solution is deaerated by N$_2$, and ascorbate (4.8 × 10^{-2} M) is added. The reactivity of "Fe(OH)$_3$" differs markedly depending on its preparation. In presence of imidazole (Fig. a) the hydrous oxide has properties similar to lepidocrocite (i.e., upon filtration of the suspension the solid phase is identified as lepidocrocite). In presence of TRIS, outer-sphere surface complexes with the native mononuclear Fe(OH)$_3$ are probably formed which retard the polymerization to polynuclear "Fe(OH)$_3$" (von Gunten and Schneider, 1991).
(From Deng, Ruf and Stumm, in preparation)

In both reactions, electron transfer induces the dissolution of the solid phase; i.e., reductive and oxidative dissolution, respectively. Although no kinetic implications follow directly from the thermodynamic considerations, there are cases where the redox rate is related to the redox equilibrium (see e.g., Eq. 9.12).

Different Reactivities of different Fe(III)(Hydr)oxide Surfaces

Different modifications of hydrous oxides, even if present in solution with the same surface area concentrations, are characterized by significantly different reactivities (e.g., dissolution rate). This depends above all on the different coordination geometry of the surface groups. For a given pH (on surface protonation) the reactivity of a Fe^{III}-center is likely to increase with the number of terminal ligands (Wehrli et al., 1990), i.e., groups such as $-Fe-OH$ are less acid and react faster than

$$\equiv Fe \diagdown \diagup OH \qquad \equiv Fe \diagdown \equiv Fe - OH \diagup \equiv Fe \qquad \text{groups respectively}$$

Figs. 9.7a, b show the difference in reactivity of two different freshly formed (by oxygenation of Fe(II)) "Fe(OH)$_3$") (Deng and Stumm, in preparation). In one case a structure similar to lepidocrocite may be formed; in the other case an oligomeric Fe(OH)$_3$ is formed where (according to a recommendation by von Gunten and Schneider, 1991) extensive polymerization of incipiently formed mononuclear Fe(OH)$_3$ has been prevented by the presence of TRIS (= $H_2NC(CH_2OH)_3$) which most likely forms outer-sphere complexes on the surface of the colloidal Fe(OH)$_3$. As has been shown by von Gunten and Schneider (1991) the "Fe(OH)$_3$" formed under these conditions is colloidally stable, and is characterized by a high number of terminal OH$^-$ ligands per Fe^{III}-center.

Enhanced Reactivity of Fe(III) formed at Surfaces. Another way to keep the Fe(III) hydroxide formed from oxygenation of Fe(II) from extensive polymerization, is to oxidize (O$_2$) the adsorbed Fe(II). Apparently the Fe(III) formed on the surface (or part of it), plausibly because of a different coordinative arrangement of the adsorbed ions, does not readily polymerize fully to a "cross-linked" three-dimensional structure and is thus more reactive than freshly formed lepidocrocite.

Hydrous Manganese(III,IV) Oxide

Hydrous Manganese oxides, widely distributed in natural systems, are stronger oxidants than iron(III)(hydr)oxide. These oxides readily oxidize many natural and xenobiotic organic compounds. Various substituted phenols, naturally present in surface water and in soils as degradation products of lignin and other plant materials are capable of reducing manganese oxides. Specific interaction between reductant molecules and oxide surface sites is necessary for the redox reaction.

The following reaction scheme (Stone and Morgan, 1987; Stone, 1987) may be considered:

Precursor complex formation:

$$\equiv Mn^{III} + ArOH \underset{k_{-1}}{\overset{k_1}{\rightleftharpoons}} (\equiv Mn^{III}, ArOH) \qquad (9.17a)$$

Electron Transfer:

$$(\equiv Mn^{III}, ArOH) \underset{k_{-2}}{\overset{k_2}{\rightleftharpoons}} (\equiv Mn^{II}, ArO^{\bullet}) + H^+ \qquad (9.17b)$$

Release of phenoxy radical:

$$(\equiv Mn^{II}, ArO^{\bullet}) \underset{k_{-3}}{\overset{k_3}{\rightleftharpoons}} \equiv Mn^{II} + ArO^{\bullet} \qquad (9.17c)$$

Detachment of released Mn(II):

$$\equiv Mn^{II} \underset{k_{-4}}{\overset{k_4}{\rightleftharpoons}} Mn^{2+} (+ \equiv Mn^{III}) \qquad (9.17d)$$

Stone (1987) proposes – in agreement with his experimental data – the following rate law for the dissolution rate:

$$\frac{d[Mn^{2+}]}{dt} = k_2 \langle \equiv Mn^{III}, ArOH \rangle \qquad (9.18)$$

and he shows that Eq. (9.18) can be written (assuming steady state phenol surface coverage) in terms of S_T (= total concentration of surface sites) and the phenol concentration

$$\frac{d[Mn^{2+}]}{dt} = \frac{k_1 k_2 S_T [ArOH]}{k_{-1} + k_2 + k_1 [ArOH]} \qquad (9.19)$$

This equation, containing the Langmuir expression for the adsorption (surface complex formation) of phenol on the Mn(III)(hydr)oxide, is similar to the principles discussed for reductive dissolution of Fe(III)(hydr)oxide.

Fig. 9.8 illustrates that the reduction rates are linearly related to the half-wave potentials of the substituted phenols. Half-wave potentials are the electrode potentials on mercury, Hg, or graphite electrodes, where the substances are oxidized (specifically where the oxidation current density is half of the current density observed at maximum oxidation rate). The half-wave potentials (for phenols the data are from Suatoni et al., 1961) measure the tendency of an anode to oxidize the phenols; in a first approximation the half-wave potentials are related to the redox potential of the phenol and its oxidation product. Thus, Fig. 9.8 implies that the larger the tendency of the electrode at the selected electrode potential is to oxidize a subsituted phenol, the larger is the tendency of the manganese oxide to oxidize this phenol. Various explanations for this redox potential dependence, including a reaction scheme as that given in Fig. 9.4b, are possible.

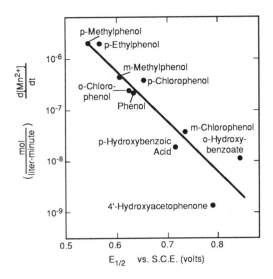

Figure 9.8

Rates of reductive dissolution of amorphous manganese (III,IV) oxide particles decrease as the electrode half-wave potentials of the substituted phenols (as reported by Suatoni et al., 1961) increase (4.8×10^{-5} M total manganese, pH 4.4).

(From Stone, 1987)

9.4 Oxidation of Transition Ions; Hydrolysis and Surface Binding Enhance Oxidation

The oxidation of Fe(II), VO^{2+}, Mn^{2+}, Cu^+ by oxygen is favored thermodynamically and kinetically by hydrolysis and by specific adsorption to hydrous oxide surfaces. As suggested in formula (IV) of Fig. 9.1 Fe(II) and the other transition elements Mn(II), VO(II), Cu(I) may more readily associate (probably outer-spherically) with

O_2 if they are present as complexes with OH^- (i.e., hydrolysis species) or as complexes with hydroxo surface groups of hydrous oxides (Wehrli and Stumm, 1989; Wehrli, 1990; Luther, 1990). As explained by Luther (1990), the OH^- ligands donates electron density to the reduced metal ion through both the σ and π systems, which results in metal basicity and increases reducing power, e.g., the Fe(III) oxidation state is stabilized by the OH^- ligands (Fig. 9.2). Fe(II) bound to silicates are also more readily oxidized by O_2 than Fe^{2+}. This has been demonstrated for example for hornblende (White and Yee, 1985).

For the oxygenation of Fe(II) *in solution* (at pH > 5), the empirical rate law shows a second-order dependence on OH^- concentration (Stumm and Lee, 1961), that is,

$$-\frac{d[Fe(II)]}{dt} = k\,[Fe(II)]\,[OH^-]^2\,[O_2] \tag{9.17}$$

This dependence is consistent with the predominant reaction of O_2 with the hydrolyzed, solute species, $Fe(OH)_2$, within the pH range of interest, or

$$-\frac{d[Fe(II)]}{dt} = k'\,[Fe(OH)_2]\,[O_2] \tag{9.18}$$

Similarly, the rate of oxygenation of Fe(II) bound to the surface hydroxyl groups of a hydrous oxide can be expressed in terms of the surface species. Thus,

$$-\frac{d[Fe(II)]}{dt} = k''\,\langle Fe(OM\equiv)_2\rangle\,[O_2] \tag{9.19}$$

The surface hydroxyl group facilitates – similar to $OH^-(aq)$ – the electron transfer to O_2.

Eq. (9.17) for the oxygenation of Fe(II) in solution, valid at pH > 5, can be extended (Millero, 1985) to the entire pH range (Fig. 9.9a).

$$-\frac{d[Fe(II)]}{dt} = \left(k_0[Fe^{2+}] + k_1\,[Fe(OH)^+] + k_2\,[Fe(OH)_2]\right)(O_2) \tag{9.20}$$

i.e., the pH dependence changes from $R \propto [H^+]^{-2}$ to $R \propto [H^+]^{-1}$ below pH 5 and the rate is pH independent below pH = 3. Parallel occurring rate determining steps for the various iron species and their experimentally determined reaction rates k [M^{-1} s^{-1}] and their equilibrium constants can be compared (Fig. 9.9b) (data from Singer and Stumm, 1979; Millero et al., 1987; Tamura et al., 1976; Wehrli, 1990).

$$Fe^{2+} + O_2 \longrightarrow Fe^3 + O_2^- \qquad ;\ \log k_0 = -5.1;\ \log K_0 = -15.7 \tag{9.21}$$

$$Fe(OH)^+ + O_2 \longrightarrow Fe(OH)^2 + O_2^- \qquad ;\ \log k_1 = 1.4;\ \log K_1 = -8.45 \tag{9.22}$$

$$Fe(OH)_2 + O_2 \longrightarrow Fe(OH)_2^+ + O_2^- \qquad ; \; \log k_2 = 6.9; \quad \log K_2 = -3.04 \quad (9.23)$$

$$(\equiv Fe^{III}O)_2Fe^{II} + O_2 \longrightarrow (\equiv Fe^{III}O)_2 Fe^{III+} + O_2^- \; ; \; \log k_3 = 0.7; \quad \log K^s = -9.0 \quad (9.24)$$

As Wehrli (1990) has shown, the value of log K^s can be obtained from extrapolation (Fig. 9.9b).

The couple $(\equiv FeO)_2Fe^{III+} + e^- = (\equiv FeO)_2Fe^{II}$; $E_H^o = 0.36$ V is of similar magnitude as that of $Fe^{III}OH^{2+} + e^- = Fe^{II}OH^+$; $E_H^o = 0.34$ V. Fig. 9.10 shows the effect of adsorption on the rate of transition metal oxygenation. Fig. 9.11 compares the catalytic effects of hydrolysis and adsorption for VO(II), Fe(II) and Mn(II).

This idea can be extended to Fe(II) centers within or near the surface lattice of Fe(II) silicates. White and Yee (1985) have shown that the equilibrium

$$\equiv Si\text{-}Fe^{III+} + e^- = \equiv Si\text{-}Fe^{II} \quad E_H^o = 0.33 - 0.52 \text{ V} \qquad (9.25)$$

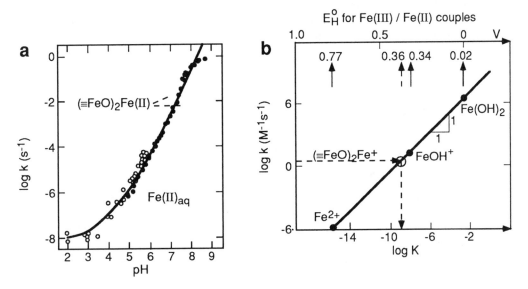

Figure 9.9

a) Oxidation of Fe(II) by O_2. Open circles, data are from Singer and Stumm (1970); dots are data from Millero et al. (1987).

b) Log of the reaction rate constants (Eqs. 23 – 25) are plotted vs log of equilibrium constants on the redox potential, E_H^o, for the Fe^{III}/Fe^{II} couple. From the observed rate of the oxygenation of Fe(II) inner-sperically adsorbed on a goethite surface (data of Tamura et al., 1976) an equilibrium constant on redox potential can be estimated. Data for Fe^{2+} and $Fe(OH)^+$ from Singer and Stumm (1979), for $Fe(OH)_2$ from Millero et al. (1987), for $(\equiv FeO)_2Fe^+$ goethite surface from Tamura et al. (1976). Fig. b) is modified from Wehrli (1990).

is characterized by a similar E_H^o as that of Fe(II) bound to iron oxide surfaces. The equilibrium for the first step in the reaction with O_2 is

$$\equiv Si\text{-}Fe^{II+} + O_2 \longrightarrow \equiv Si\text{-}Fe^{III} + O_2^- \; ; \; \log K \approx -8.3 \text{ to } -11.4 \quad (9.26)$$

As reported by White and Yee (1985) structural Fe(II) at the surface of hornblende or augite is oxidized faster by O_2 than Fe(II) in solution at pH < 5; i.e., Fe(II) bound to silicates is kinetically a better reducing agent than Fe^{2+} in solution. The reader is referred to White (1990) for an excellent treatment of heterogeneous reactions with Fe(II)-minerals.

Pyrite Oxidation. The oxidation of Fe(II) minerals by Fe^{3+} is also of importance in the oxidation of pyrite by O_2. This process is mediated by the Fe(II)–Fe(III) system. Pyrite is oxidized by Fe^{3+} (which forms a surface complex with the pyrite (cf. formula VI in Fig. 9.1) (Luther, 1990). The rate determining step at the relatively low pH values encountered under conditions of pyrite dissolution is the oxygenation of Fe(II) to Fe(III) usually catalyzed by autotrophic bacteria (Singer and Stumm, 1970; Stumm-Zollinger, 1972). Thus, the overall rate of pyrite dissolution is insensitive to the mineral surface area concentration. Microbially catalyzed oxidation of Fe(II) to Fe(III) by oxygen could also be of some significance for oxidative silicate dissolution in certain acid environments.

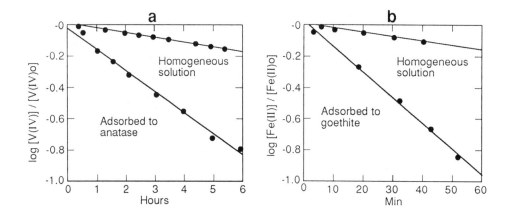

Figure 9.10

First-order plots demonstrating the catalysis of transition-metal oxygenation by oxide surfaces.
a) Oxygenation of vanadyl, VO^{2+}, at pH 4 and p_{O_2} = 1 atm (from Wehrli and Stumm, 1988).
b) Oxygenation of ferrous iron, Fe^{2+}, at pH 6.4 and p_{O_2} = 0.7 atm (data from Tamura et al., 1980).
In both cases the presence of a solid phase accelerates the oxidation rate.
(Modified from Wehrli, Sulzberger and Stumm, 1989)

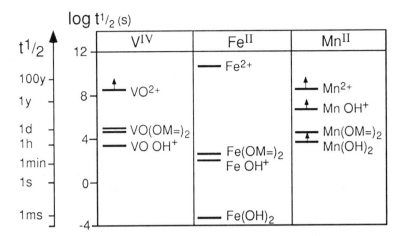

Figure 9.11

Effects of hydrolysis and adsorption on the oxygenation of transition-metal ions. *Arrows* indicate lower limit. (From Wehrli and Stumm, 1989)

References:

Fe^{2+} (George, 1954); $Fe(OH)^+$ and $Fe(OH)_2$ (Millero, 1985, based on Stumm and Lee, 1961); $Fe(OFe\langle)_2$ (amorphous $FeOOH$ and β-$FeOOH$) (Tamura et al., 1980).

Data represent order of magnitude because surface area concentrations were not determined; Mn $(OH)^+$ (Diem and Stumm, 1984, based on observations that $Mn(II)_{aq}$ at pH = 8.3 persists for a half-time of > 10 yr); $Mn(OFe\langle)_2$ γ-$FeOOH$ and Mn $(OAl\langle)_2$ δ-Al_2O_3 (Davies and Morgan, 1988).

9.5 Mediation of Redox Reactions of Organic Substances by the Hydrous Ferric Oxide Surface and Fe(II)-Surfaces

In sediments, soils and anoxic waters natural biogenically produced reductants resulting essentially from the biodegradation of biomass are abundant. These reductants include in addition to Fe(II), H_2S, sulfides, organic thio compounds and organic substances. Although they have shown to interact directly with reducible organic substances including organic pollutants, such reactions are generally slow because of the difficulties involved in establishing a direct precursor complex between the reductant and the reducible organic substance. Often redox reactions among such partners in natural systems are fast either because of microbial mediation or in abiotic systems because the electron transfer is mediated by one or more electron carriers.

Fe(II)/Fe(III) as a Mediator in Electron Transfer. The Fe(III)–Fe(II) system often acts as an electron carrier. A possible schematic example is given by

$$2 \langle Fe^{II}Ca_{0.5}\text{-silicate}\rangle + Cl_3C\text{-}CCl_3 \longrightarrow$$

Electron transfer mediator

As we have seen, redox reactions in the Fe(II)/Fe(III) system – e.g., $Fe(OH)_3$ – Fe^{2+}; $Fe(OH)_3$ – $FeCO_3$; Fe(III) – FeS_x; Fe(III) – Fe(II) porphyrins – are relatively fast. For example, Kriegmann and Reinhard (1989) have shown that chlorinated hydrocarbons ($Cl_3C\text{-}CCl_3$ and CCl_4) can be reduced in the presence of biotite and vermiculite at pH 7 – 8 and 50° C:

$$2 \langle Fe^{II}Ca_{0.5}\text{-silicate}\rangle + Cl_3C\text{-}CCl_3 \longrightarrow$$
$$2 \langle Fe^{III}\text{-silicate}\rangle + Cl_2C = CCl_2 + 2 Cl^- + Ca^{2+} \tag{9.28}$$

They propose that structural oxidation is accompanied by regeneration of the Fe(II) sites by HS^-.

$$2 \langle Fe^{III}\text{-silicate}\rangle + 2 HS^- \longrightarrow 2 \langle Fe^{II}H^+\text{-silicate}\rangle + \text{oxidized sulfur} \tag{9.29}$$

Schwarzenbach et al. (1990) have shown that reductive transformations of a series of monosubstituted nitrobenzenes and nitrophenols in aqueous solutions containing reduced sulfur species occur readily in presence of small concentrations of an iron prophyrin as an electron transfer catalyst.

Organic Carbon Oxidation Microbially Coupled to Reduction of Fe(III)(hydr)oxide. More and more evidence is accumulating that bacteria can grow anaerobically by coupling organic carbon oxidation to the dissimilatory reduction of iron(III) oxides (Nealson, 1982; Arnold et al., 1986; Lovely and Philips, 1988; Nealson and Myers, 1990).

Pure cultures of a Fe(III) reducing bacterium have been shown to obtain energy for growth by oxidizing benzoate, toluene, phenol or p-cresol with Fe(III) as the sole electron acceptor (Lovely et al., 1989, 1991). Such redox reactions are important because, at the onset of anaerobic conditions, e.g., in sediments and subsurface environments, Fe(III) oxides are the most abundant oxidants.

9.6 Redox Reactions of Iron and Manganese at the Oxic Anoxic Boundary in Waters and Soils

Fig. 9.12 gives representative redox intensity ranges of importance in soils, sediments and in surface and ground waters. The heterogeneous redox couples involving the solid phases Mn(III,IV) oxides, Fe(II,III) and FeS and FeS_2 provide significant redox buffering in natural systems.

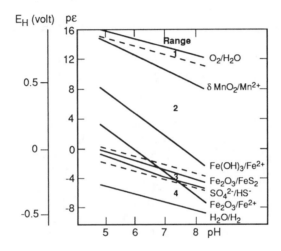

Figure 9.12

Representative ranges of redox intensity in soil and water

- Range 1 is for oxygen bearing waters.
- The pε-range 2 is representative of many ground and soil waters where O_2 has been consumed (by degradation of organic matter), but SO_4^{2-} is not yet reduced. In this range soluble Fe(II) and Mn(II) are present; their concentration is redox-buffered because of the presence of solid Fe(III) and Mn(III,IV) oxides.
- The pε-range 3 is characterized by SO_4^{2-}/HS or SO_4^{2-}/FeS, SO_4^{2-}/FeS$_2$ redox equilibria.
- Range 4 occurs in anoxic sediments and sludges.

(Modified from Drever, 1988)

Protons and electrons are coupled. An increase in pε [= –log {e⁻} = (F/2.3 RT) E_H, where F/2.3 RT = 1/= 0.059 V at 25° C] is accompanied by a decrease in pH.

In soils the organic matter (range 2 in Fig. 9.12) is a significant pH and pε buffer because it represents a reservoir of bound H⁺ and e⁻. When organic matter is mineralized; alkalinity and [NO_3^-], [SO_4^{2-}] increase and Fe(II) and Mn(II) become mobilized. Phosphate, incipiently bound to Fe(III)(hydr)oxides, is released as a consequence of the partial reductive dissolution of the Fe(III) solid phases. At lower pε values (range 3 in Fig. 9.12) the concentration of Fe(II) and Mn(II) further increases. SO_4^{2-} reduction is accompanied by precipitation of FeS and MnS and by the formation of pyrite.

At the oxic-anoxic boundaries a rapid turnover of iron takes place. This oxic-anoxic boundary may occur in deeper layers of the water column of fresh and marine waters, at the sediment-water interface or within the sediments.

Fig. 9.13 (from Davison, 1985) shows the redox transformations schematically for the water column. The important items are:
1) The principal reductant is the biodegradable biogenic material that settles in the deeper portions of the water column;
2) Electron transfer becomes more readily feasible if, as a consequence of fermentation processes – typically occurring around redox potentials of .22 to −.22 V – molecules with reactive functional groups such as hydroxy and carboxyl groups are formed;
3) Within the depth-dependent redox gradient, concentration peaks of solid Fe(III) and of dissolved Fe(II) develop, the peak of Fe(III) overlying the peak of Fe(II);
4) As illustrated in Fig. 9.13 the Fe(II), forming complexes with these hydroxy and carboxyl ligands, encounter in their upward diffusion the settling of Fe(III)(hydr)-oxides and interact with these according to the catalytic mechanism, thereby dissolving rapidly the Fe(III)(hydr)oxides. The sequence of diffusional transport of Fe(II), oxidation to insoluble Fe(OH)$_3$ and subsequent settling and reduction to dissolved Fe(II) typically occurs within a relatively narrow redox-cline.

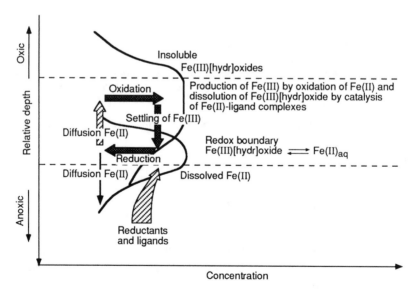

Figure 9.13

Transformation of Fe(II,III) at an oxic-anoxic boundary in the water or sediment column (Modified from Davison, 1983)

Some of the processes, mentioned above occur also in soils. Microorganisms and plants produce a larger number of biogenic acids. The downward vertical displacement of Al and Fe observed in the podsolidation of soils can be accounted for by considering the effect of pH and of complex formers on both the solubility and the dissolution rates.

The reductive dissolution of Fe(III)(hydr)oxides is also of importance in the iron uptake by higher plants. According to Brown and Ambler (1964), iron defiency causes a release of reducing exudates from the roots. These substances cause the reductive dissolution of particulate Fe(III) in the proximity of the roots. This reduction is followed by uptake of Fe(II) into the root cells.

Similarly the redox transformations Mn(III,IV) oxides / Mn^{2+} causes rapid electron cycling at suitable redox intensities. Two differences of Mn and Fe in their redox chemistries are relevant:
1) the reduction of Mn(III,IV) to dissolved Mn(II) occurs at higher redox potentials than does the Fe(III)/Fe(II) reduction;
2) the oxygenation of Mn(II) to Mn(III,IV) oxides, even if catalyzed by surfaces and/or microorganisms, is usually slower than the oxygenation of Fe(II).

The redox cycling of these elements have pronounced effects on the adsorption of trace elements onto oxide surfaces and trace element fluxes under different redox conditions. The hydrous Mn(III,IV) oxides are important mediators in the oxidation of oxidizable trace elements; e.g., the oxidation of Cr(III), As(III) and Se(III) is too slow with O_2; however, these elements subsequent to their relatively rapid adsorption on the Mn(III,IV) oxide are rapidly oxidized by Mn(III,IV).

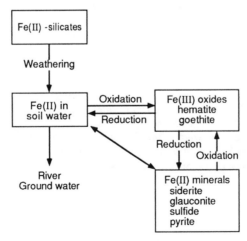

Figure 9.14

Important Fe-reservoirs in soils. Many of the reactions are influenced by microorganisms (Modified from Murad and Fisher, 1988)

NO₃⁻-Reducing. Fig. 9.15 shows data on groundwater below agricultural areas. The sharp decrease of O_2 and NO_3^- at the redox cline indicate that the kinetics of the reduction processes are fast compared to the downward water transport rate. Postma et al., 1991 suggest that pyrite, present in small amounts is the main electron donor for NO_3^- reduction (note the increase of SO_4^{2-} immediately below the oxic anoxic boundary). Since NO_3^- cannot kinetically interact sufficiently fast with pyrite a more involved mechanism must mediate the electron transfer. Based on the mechanism for pyrite oxidation discussed in Chapter 9.4 one could postulate a pyrite oxidation by Fe(III) that forms surface complexes with the disulfide of the pyrite (Fig. 9.1, formula VI) subsequent to the oxidation of the pyrite, the Fe(II) formed is oxidized direct or indirect (microbial mediation) by NO_3^-. For the role of Fe(II)/Fe(III) as a redox buffer in groundwater see Grenthe et al. (1992).

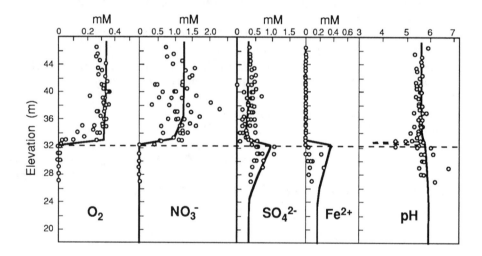

Figure 9.15

Redox components in groundwaters as a function of depth (unconfined sandy aquifer) below agricultural areas for 1988. NO_3^- contaminated groundwater emanate from the agricultural areas and spread through the aquifer. The redox boundary is very sharp which suggests that the redox process is fast compared to the rate of downward water transport. The investigators (Postma et al., 1991) suggest that reduction of O_2 and NO_3^- occur by pyrite. (The lines given are based on equilibrium models.) (Modified from Postma, Boesen, Kristiansen and Larsen, 1991)

Reading Suggestions

Behra, P., and L. Sigg (1990), "Evidence for Redox Cycling of Iron in Atmospheric Water", *Nature* **344**, 419-421.

Davison, W. (1985), "Conceptual Models for Transport at a Redox Boundary", in W. Stumm, Ed., *Chemical Processes in Lakes*, Wiley-Interscience, New York, pp. 31-53.

Gordon, G., and H. Taube (1962), "Oxygen Tracer Experiments on the Oxidation of Aqueous Uranium (IV) with Oxygen-Containing Oxidizing Agents", *Inorg. Chem.* **1**, 69-75.

Hering, J., and W. Stumm (1990), "Oxidative and Reductive Dissolution of Minerals," in M. F. Hochella Jr. and A. F. White, Eds., *Reviews in Mineralogy 23: Mineral-Water Interface Geochemistry*, pp. 427-465, Mineralogical Society of America.

Luther, G. W., (1990), "The Frontier-Molecular-Orbital Theory Approach in Geochemical Processes", in W. Stumm, Ed., *Aquatic Chemical Kinetics*, Wiley-Interscience, New York, pp. 173-198.

Postma, D., C. Boesen, H. Kristiansen, and F. Larsen (1991), "Nitrate Reduction in an Unconfined Sandy Aquifer: Water Chemistry, Reduction Processes, and Geochemical Modeling", *Water Resources Research* **27/8**, 2027-2045.

Sulzberger, B., D. Suter, C. Siffert, S. Banwart, and W. Stumm (1989), "Dissolution of Fe(III)(hydr)-oxides in Natural Waters; Laboratory Assessment on the Kinetics Controlled by Surface Coordination", *Marine Chemistry* **28**, 127-144.

Suter, D., S. Banwart, and W. Stumm (1991), "Dissolution of Hydrous Iron(III) Oxides by Reductive Mechanisms", *Langmuir* **7**, 809-813.

White, A. F. (1990), "Heterogeneous Electrochemical Reactions Associated with Oxidation of Ferrous Oxide and Silicate Surfaces", *Reviews in Mineralogy* **23**, 467-509.

Wehrli, B., B. Sulzberger, and W. Stumm (1989), "Redox Processes Catalyzed by Hydrous Oxide Surfaces," *Chemical Geology* **78**, 167-179.

Chapter 10

Heterogeneous Photochemistry

With this chapter we would like to convince the reader that heterogeneous photochemistry is an exciting field of surface science and that heterogeneous photoredox reactions are of importance in many naturally occurring processes.

First, we will illustrate the principle of photosynthesis with a simple molecular orbital picture and juxtapose biological photosynthesis with photosynthetic processes involving semiconducting minerals. In a second subchapter we will discuss some of the underlying principles of photoredox processes (either photosynthetic or photocatalytic) that occur at the semiconductor-electrolyte interface. The third subchapter will deal with the kinetics of heterogeneous photoredox reactions and with the role that surface complexation plays in these processes. In a fourth subchapter six case studies of heterogeneous photoredox processes will illustrate some principles that have been discussed in subchapters one to three. Finally, reference is made to the cycling of iron and its role in the geochemical cycling of elements.

10.1 Heterogeneous Photosythesis

In heterogeneous *photosynthesis* a thermodynamically unfavorable reaction ($\Delta G > 0$) is caused to occur by the presence of an illuminated solid, leading to conversion of radiant to chemical energy, whereas in heterogeneous *photocatalysis* the rate of a thermodynamically favorable reaction ($\Delta G < 0$) is increased by the presence of an illuminated solid. The principle of photosynthesis in most general terms is illustrated with a one-electron molecular orbital picture in Fig. 10.1. Absorption of light by the chromophore, C, leads to an electronically excited state, C*; i.e., an electron is transferred from the highest occupied molecular orbital (HOMO) (or from the valence band in case of a semiconductor) of C to the lowest unoccupied molecular orbital (LUMO) (or to the conduction band in case of a semiconductor) of C[1]. The photoelectron in the LUMO (or in the conduction band) is strongly reducing and the photohole that is left behind in the HOMO (or in the valence band) is strongly oxidizing; thus, the chromophore C in its electronically excited state is both a stronger reductant and a stronger oxidant than in its ground state; i.e., the ionization potential of C* decreases and the electron affinity increases approximately by the excita-

---

[1] The HOMO-LUMO principle is derived from molecular orbital theory (e.g., Purcell and Kotz, 1980); it emphasizes the importance of orbital symmetry (proper overlap of reductant and oxidant orbitals).

tion energy. The photoelectron and the photohole may recombine, either in a radia-tionless, thermal deactivation of the electronically excited state (internal conver-sion) or via radiative transition from the excited to the ground state. The probability that a photochemical reaction takes place, i.e., that the photoelectron is transferred to a thermodynamically suitable electron acceptor or that the photohole is filled with an electron from a thermodynamically suitable electron donor, depends on the rate constant of electron transfer as compared to the rate constants of radiative and/or radiationless transition. In order to increase the probability of the electron transfer reaction, the photoelectron/photohole pair that is formed through absorption of light must be locally separated. This can only be efficiently achieved in heterogeneous systems (see Figs. 10.6c and d) or in "supramolecular systems" (Balzani, 1987).

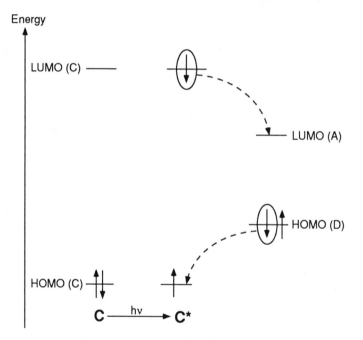

Figure 10.1

Photooxidation and photoreduction of an electron donor and an electron acceptor, respectively, as illustrated schematically with a one-electron molecular orbital scheme.

C = chromophore;
D = electron donor;   HOMO = highest occupied molecular orbital;
A = electron acceptor;   LUMO = lowest unoccupied molecular orbital.

We first juxtapose the most prominent photosynthetic process, biological photo-synthesis, with photosynthetic processes occurring on some inorganic surfaces. Energy storage by the photosynthesis of green plants and some other living sys-tems can be represented in simplified form by the following chemical reaction:

$$CO_2(g, 0.0032 \text{ atm}) + H_2O(\ell) \longrightarrow \tfrac{1}{6} C_6H_{12}O_6(s) + O_2(g, 0.21 \text{ atm}) \quad (10.1)$$

$$\Delta G^0_{298} = 496 \text{ kJ/mol}$$

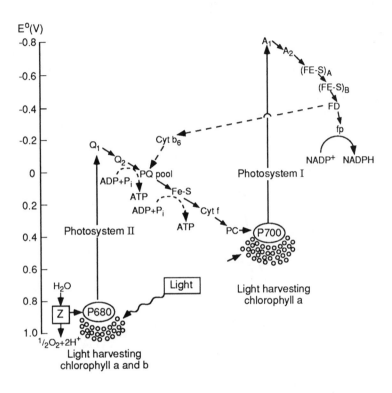

Figure 10.2

Z-scheme of photosynthesis (Archer and Bolton, in preparation). E^0 is the standard redox potential at pH 7. The essential features are the following: PhotosystemII which is associated with the oxidation of water and photosystem I which is associated with the reduction of $NADP^+$ are linked in series by an electron transport chain. Absorption of light by chlorophyll a and b of photosystem II mediates the transfer of an electron against the potential gradient from species Z (which is a manganese complex) to a speces Q_1, to give Z^+ and Q_1^-. Z^+ is a strong oxidant that is capable of oxidizing water to oxygen. From the reduced species Q_1^- an electron is transferred in spontaneous dark reactions to subsequent electron acceptors until it reaches photosystem I. Absorption of light by chlorophyll a in photosystem I enables the second uphill electron transfer to the highly reducing component A_1. A second series of dark electron-transfer reactions occurs via the protein ferredoxin and other electron traps and results finally in the reduction of $NADP^+$. NADPH acts as a reducing agent in dark reactions that reduce CO_2. These reactions occur in a series of catalyzed steps known as the Calvin cycle. The reactions of the Calvin cycle require ATP as well as NADPH. ATP is synthesized in reactions that proceed parallel to the Z-scheme and utilize some of the energy that would otherwise be wasted in the downhill electron transfer step. For the formation of one oxygen molecule from water 4 electrons are transferred which requires 8 photons, since 2 photons are needed to transfer an electron from Z to A_1.

For the formation of one O_2 molecule four electrons have to be transferred. This requires a "quantum storage device". In the photosynthetic system of green plants this is achieved with two photosystems that are linked through an electron transport chain, Fig. 10.2, and by means of the thylakoid-membrane that enables the separation of the photoproducts O_2 and the reduced form of nicotinamide adenine dinucleotide phosphate, NADPH.

The sunlight is absorbed by a number of pigments, of which chlorophyll a is the most important. The absorption spectrum of chlorophyll a is shown in Fig. 10.3. Apart from chlorophyll a there are accessory pigments which function as light-harvesting antennae and transfer the energy to the chlorophyll a. This is an elegant way to efficiently collect sunlight.

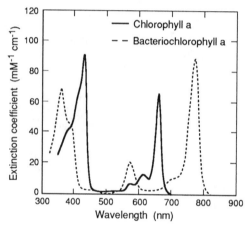

Figure 10.3

Absorption spectra of chlorophyll a and bacteriochlorophyll a.
(Modified from Archer and Bolton, in preparation)

There have been many attempts to mimic some features of photosynthesis with abiotic systems for purposes of artificial solar energy conversion. Ideally a fuel, e.g., H_2, is formed through a photosynthetic process. Photolysis of water is a highly endergonic process:

$$H_2O(\ell) \longrightarrow H_2(g) + \tfrac{1}{2} O_2(g) \tag{10.2}$$

$$\Delta G^o_{298} = 237.2 \text{ kJ/mol}$$

Thus, the energy required for this reaction to occur is very similar to the energy required fo form $\tfrac{1}{2} O_2$ and $\tfrac{1}{12} C_6H_{12}O_6$ from water and $\tfrac{1}{2} CO_2$, Eq. (10.1), in the photosynthesis of green plants. Since water absorbs light only in the UV and IR region of the solar spectrum, a light absorber other than H_2O is required that absorbs light in the visible region of the solar spectrum. Ideally the light absorber is a photocatalyst. A photocatalyst[1], PC, is a light-absorbing species that enables a photochemical reaction, e.g., a photoredox reaction, but remains unchanged after the overall process:

$$PC \xrightarrow{\text{hv}} PC*$$

$$PC* + D + A \longrightarrow PC + D^+ + A^- \tag{10.3}$$

This means that the photoelectron is transferred to an electron acceptor concomitantly with trapping of the photohole by an electron donor (Fig. 10.1). Semiconductor materials have been tested as photocatalysts for the photodissociation of water. Fig. 10.4 shows the energetics in terms of standard redox potential of some semiconductors as compared to the standard redox potential of H_2/H^+ and H_2O/O_2 at pH 0.

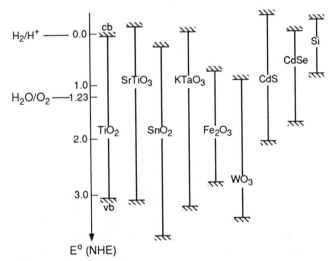

Figure 10.4
Standard redox potential at pH 0 of the valence band and of the conduction band of various semiconductors as compared to the standard redox potential of the redox couples H_2O/O_2 and H_2/H^+. (From Darwent, 1982)

According to the energetics shown in Fig. 10.4, Fe_2O_3 cannot be used as a photocatalyst for the photodissociation of water. However, it has been reported that

---

[1] Note, that the term "photocatalyst" is used independent of the sign of the free energy of the overall process.

α-Fe_2O_3 that was pretreated in such a way that the material contained some Fe_3O_4 could be used as a photocatalyst for the photodissociation of water with light of wavelengths shorter than 540 nm (Khader et al., 1987). In photosynthesis of the green plants two photocatalysts are involved: photosystem II and photosystem I. The linkage of two photosystems may also be advantageous for the photochemical splitting of water into oxygen and hydrogen involving artificial photocatalysts, where in one photosystem oxidation of water to form oxygen would take place and in the other the reduction of protons to form hydrogen (Beer et al., 1991).

In addition to photocleavage of water with photocatalysts, other photosynthetic processes such as photochemical CO_2 reduction, resulting in the formation of CO or methane, and the photochemical fixation of nitrogen are of great interest:

$$CO_2 \quad \xrightarrow[\text{photocatalyst}]{h\nu} \quad CO + \tfrac{1}{2}\,O_2 \qquad (10.4)$$

$$CO_2 + 2\,H_2O \quad \xrightarrow[\text{photocatalyst}]{h\nu} \quad CH_4 + 2\,O_2 \qquad (10.5)$$

$$N_2 + 3\,H_2O \quad \xrightarrow[\text{photocatalyst}]{h\nu} \quad 2\,NH_3 + \tfrac{3}{2}\,O_2 \qquad (10.6)$$

10.2 Semiconducting Minerals

The electronic properties of solids can be described by various theories which complement each other. For example band theory is suited for the analysis of the effect of a crystal lattice on the energy of the electrons. When the isolated atoms, which are characterized by filled or vacant orbitals, are assembled into a lattice containing ca. 5×10^{22} atoms cm^{-3}, new molecular orbitals form (Bard, 1980). These orbitals are so closely spaced that they form essentially continuous bands: the filled bonding orbitals form the *valence band* (vb) and the vacant antibonding orbitals form the *conduction band* (cb) (Fig. 10.5). These bands are separated by a forbidden region or band gap of energy E_g (eV).

An Intrinsic Semiconductor is characterized by an equal density of positive and negative charge carriers, produced by thermal excitation; i.e., the density of electrons in the conduction band, n_i, and of holes in the valence band, p_i, are equal

$$n_i = p_i \propto \exp\left(\frac{-E_g}{2\,kT}\right) \qquad (10.7)$$

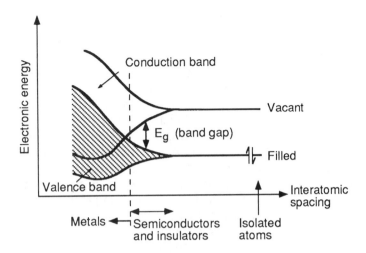

Figure 10.5

Formation of bands in solids by assembly of isolated atoms into a lattice (modified from Bard, 1980). When the band gap $E_g \ll kT$ or when the conduction and valence band overlap, the material is a good conductor of electricity (metals). Under these circumstances, there exist in the solid filled and vacant electronic energy levels at virtually the same energy, so that an electron can move from one level to another with only a small energy of activation. For larger values of E_g, thermal excitation or excitation by absorption of light may transfer an electron from the valence band to the conduction band. There the electron is capable of moving freely to vacant levels. The electron in the conduction band leaves behind a hole in the valence band.

Extrinsic Semiconductors are materials that contain donor or acceptor species (called doping substances) that provide electrons to the conduction band or holes to the valence band. If donor impurities (donating electrons) are present in minerals, the conduction is mainly by way of electrons, and the material is called an *n-type semiconductor*. If acceptors are the major impurities present, conduction is mainly by way of holes and the material is called a *p-type semiconductor*. For instance in a silicon semiconductor elements from a vertical row to the right of Si of the periodic Table, e.g. As, behave as electron donors (As $\longrightarrow$ As$^+$ + e$^-$) while elements from a vertical row to the left of Si, e.g. Ga, behave as hole donors (Ga + e$^-$ $\longrightarrow$ Ga$^-$); i.e., in the latter case, electrons are excited from the vb into the acceptor sites, leaving behind mobile holes in the vb with the formation of isolated negatively charged acceptor sites.

The energy of the *Fermi level*, E_F, is defined as that energy where the probability of a level being occupied by an electron is ½ (i.e., where it is equally probable that the level is occupied or vacant). For an intrinsic semiconductor E_F lies essentially midway between the cb and vb. For a n-type solid E_F lies slightly below the conduction band, while for a p-type solid E_F lies slightly above the valence band.

The Semiconductor-Electrolyte Interface

If a semiconductor is brought into contact with an electrolyte containing one or more redox couples, charge transfer between the two phases occurs until electrostatic equilibrium (equality of the free energies of the electron in both phases) is attained or equivalently, the Fermi levels in the solid and the solution become equal. As illustrated in Fig. 10.6, in the case where E_F lies above that in solution, electrons will flow from the semiconductor to the solution phase. Thereby the semiconductor becomes positively charged (the solution acquires an equivalent negative charge). The excess charge in the semiconductor does not reside on the surface, as it would in a metal, but instead is distributed in a space charge region, W (50 – 2000 Å wide). This charge distribution on the solid side is analogous to that encountered in the diffuse double layer on the solution side. The resulting electric field that forms in the space charge region is represented by a bending of the bands. The band bending provides efficient means of separating electron-hole pairs. In the electric field caused by the band bending, electrons move in the opposite direction from holes (compare Figs. 6c and d). In the dark, little reaction occurs because there are few holes in the valence band to accept electrons from the reduced form of the redox couple located at a potential within the gap.

Light-Induced Redox Processes at the Semiconductor-Electrolyte Interface

Upon absorption of light with energy equal to or higher than the band gap energy, a band to band transition occurs, i.e., an electron from the filled valence band is raised to the vacant conduction band leaving behind an electron vacancy, a hole, h^+, in the valence band. As a consequence of electronic excitation by light, a strongly reducing electron (photoelectron) and a strongly oxidizing hole (photohole) are formed. In the electrical field caused by the band bending, photoelectrons move in opposite direction from holes. The electrons and holes move in the semiconductor in a manner analogous to the movement of ions in solution, but their mobilities are orders of magnitude larger than those of ions in solutions. The charge carriers that reach the surface can undergo redox reactions with adsorbed species at the solid/liquid interface. The thermodynamic requirement for such a redox reaction to occur is that the redox potential of the electron donor is smaller than that of the valence band edge and that the redox potential of the electron acceptor exceeds that of the conduction band edge (Fig. 10.6). The redox reaction at the solid/liquid interface is in competition with the recombination of the photogenerated electron-hole pair; its efficiency depends usually on how rapidly the minority carriers – the photoholes in an n-type semiconductor and the photoelectrons in a p-type semiconductor – reach the surface of the solid and how rapidly they are captured through interfacial electron transfer from or to a thermodynamically suitable reductant or oxidant of the electrolyte. Plausibly, the redox reaction at the semiconductor-electrolyte interface is facilitated if the reductant and/or oxidant is adsorbed (by inner- or outer-spheric coordination) at the semiconductor surface.

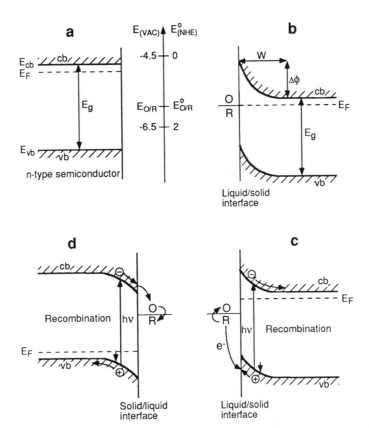

Figure 10.6

Formation of a space charge after electrostatic equilibration of a semiconductor with a solution containing a redox couple, O, R; and redox interaction with the electrolyte as a consequence of irradiation. Energy level diagrams are given

a) before contact with the electrolyte and in the dark, i.e., in the absence of any space charge;

b) of a (n-type) semiconductor-electrolyte interface in the dark; the space charge is due to a depletion of electrons in the surface region of the semiconductor; $\Delta\phi$ is the equilibrium band bending; and W the space charge region;

c) and d) of an irradiated ($h\nu \geq E_g$) n- and p-type semiconductor, respectively: photoholes in the valence band are able to oxidize R and photoelectrons in the conduction band are able to reduce O.

E_{cb}, E_{vb}, E_F, and E_g are, respectively, the energies of the conduction band, of the valence band, of the Fermi level, and of the band gap. R and O stand for the reduced and oxidized species, respectively, of a redox couple in the electrolyte. Note, that the redox system is characterized by its standard potential referred to the normal hydrogen electrode (NHE) as a reference point, $E^o_{(NHE)}$ (V) (right scale in Fig. 10.6a), while for solids the vacuum level is commonly used as a reference point, $E_{(VAC)}$ (eV) (left scale in Fig. 10.6a). Note, that the energy and the potential-scale differ by the Faraday constant, F, $E_{(VAC)} = F \times E^o_{(NHE)}$, where $F = 96'484.56$ C/mol $= 1.60219 \times 10^{-19}$ C per electron, which is by definition 1e. The values of the two scales differ by about 4.5 eV, i.e., $E_{(VAC)} = -eE^o_{(NHE)} -4.5$ eV, which corresponds to the energy required to bring an electron from the hydrogen electrode to the vacuum level.

Furthermore, one needs to consider that an inner-spherically adsorbed redox partner is characterized by a different redox potential than a solute redox partner (e.g., Fig. 9.2). Changing the concentration of specifically adsorbable (potential determining) ions changes the potential drop across the semiconductor/electrolyte interface and can modify the flat band positions. H^+ and OH^- are potential determining ions for oxide minerals. With semiconducting oxides, Nernstian behavior is often observed; thus the electrode potential of the semiconductor changes by 59 mV for every pH unit (25° C). Specifically, the flat band potentials decrease by 59 mV for $\Delta pH = +1$. Hence, electrons in the cb are better reductants in the basic pH range while holes in the vb are better oxidants in the low pH range.

The thermodynamic feasibility of redox reactions at the semiconductor-electrolyte interface can be assessed from thermodynamic considerations. Since typical redox potentials for many redox couples encountered in electrolytes of natural or technical systems often lie between the band potentials of typical semiconductors, many electron transfer reactions are (thermodynamically) feasible (Pichat and Fox, 1988). With the right choice of semiconductor material and pH the redox potential of the cb can be varied from 0.5 to 1.5 V and that of the vb from 1 to more than 3.5 V (see Fig. 10.4).

Semiconductor Colloids. The electric field at the semiconductor-electrolyte interface usually decreases with decreasing particle size. As shown by Grätzel (1989), the band bending becomes small or negligible in undoped colloids when the radius of the colloid becomes equal or smaller than the thickness of the space charge layer of the semiconductor. Under these circumstances the potential drop at the interface is mostly due to the potential drop on the solution side. Since the band bending is small in colloidal semiconductors, charge separation occurs via diffusion. The electron-hole pairs, generated by light absorption, may occur in random fashion along the optical path (Grätzel, 1989), and these charge carriers recombine or diffuse to the surface where they may undergo reactions with redox partners from the solution. When reductants and oxidants are adsorbed on a seminconductor particle, oxidative and reductive exchanges may occur at the same surface. Each particle can be pictured as a "short-circuited" photoelectrochemical cell where the semiconductor electrode and the counter electrode have been brought into contact (Bard, 1988).

Of major interest in geochemistry and in natural water systems are semiconducting minerals for which the absorption of light occurs in the near UV or visible spectral region and as a result of which redox processes at the mineral-water interface are induced or enhanced. Table 10.1 gives band gap energies of a variety of semiconductors.

Table 10.1 Band gap energies and corresponding wavelengths of light for a variety of semiconductors (quoted from Waite, 1986)

Semiconductor	Band Gap (eV)	Equivalent Wavelength (nm)	Semiconductor	Band Gap (eV)	Equivalent Wavelength (nm)
ZrO_2	5.0	248	CdS	2.4	516
Ta_2O_5	4.0	310	α-Fe_2O_3	2.34	530
SnO_2	3.5	354	ZnTe	2.3	539
$KTaO_3$	3.5	354	$PbFe_{12}O_{19}$	2.3	539
$SrTiO_3$	3.4	365	GaP	2.3	539
Nb_2O_5	3.4	365	$CdFe_2O_4$	2.3	539
ZnO	3.35	370	CdO	2.2	563
$BaTiO_3$	3.3	376	$Hg_2Nb_2O_7$	1.8	689
TiO_2	3.0 – 3.3	376 – 413	$Hg_2Ta_2O_7$	1.8	689
SiC	3.0	376	CuO	1.7	729
V_2O_5	2.8	443	PbO_2	1.7	729
Bi_2O_3	2.8	443	CdTe	1.4	885
$FeTi_3$	2.8	443	GaAs	1.4	885
PbO	2.76	449	InP	1.3	954
WO_3	2.7	459	Si	1.1	1127
$Wo_{3-x}Fe_x$	2.7	459	β-HgS	0.54	2296
$YFeO_3$	2.6	476	β-MnO_2	0.26	4768
$Pb_2Ti_{1.5}W_{0.5}O_{6.5}$	2.4	516			

10.3 Kinetics of Heterogeneous Photoredox Reactions and the Role of Surface Complexation

The quantum yield, Φ_λ, of a photochemical reaction is defined as the rate, R_λ, of formation of a primary photoproduct devided by the rate of light absorption, $I_{A\lambda}$, by the light-absorbing species (or chromophore) at a given wavelength, λ (Balzani and Carassiti, 1970). (Note that in a system more than one chromophore may be present, but a given photochemical reaction occurs mostly only from *one* electronically excited state of *one* chromophore). Thus, the rate of a photochemical reaction is then given by the following expression:

$$R_\lambda = I_{A\lambda} \times \Phi_\lambda \tag{10.8}$$

From Beer-Lambert's law follows:

$$I_{A\lambda} = 2.303 \times L \times \varepsilon_\lambda \times [C] \times I_{0\lambda} \tag{10.9}$$

where $I_{0\lambda}$ in moles of photons per liter per hour is the volume averaged light intensity that is available to the primary chromophore; ε_λ the molar extinction coefficient of the chromophore at a given wavelength in liter per mole per centimeter; L the light path length in centimeters, and [C] the concentration of the chromophore C. Equation (10.9) is only valid if $2.303 \times L \times \varepsilon_\lambda \times [C] \ll 1$. The rate of a photochemical reaction is then:

$$R_\lambda = 2.303 \times L \times \varepsilon_\lambda \times [C] \times I_{0\lambda} \times \Phi_\lambda \tag{10.10}$$

The rate expression given by Eq. (10.10) applies generally for homogeneous and for heterogeneous photochemical reactions, e.g. for heterogeneous photoredox reactions.

In heterogeneous photoredox reactions not only the solid phase, i.e. the semiconducting mineral, may act as the chromophore (as discussed in Chapter 10.2) but also a surface species: (i) a surface complex formed from a surface metal ion of a metal (hydr)oxide and a ligand that is specifically adsorbed at the surface of the solid phase, and (ii) a chromophore that is specifically adsorbed at the surface of a solid phase. In the following these three cases will briefly be discussed.

The (Bulk) Solid Phase Acting as Chromophore

This case is shown in Fig. 10.6c and d where through absorption of light a photohole in the vb and a photoelectron in the cb are formed. The probability that interfacial electron transfer takes place, i.e. that a thermodynamically suitable electron donor is oxidized by the photohole of the vb depends (i) on the rate constant of the interfacial electron transfer, k_{ET}, (ii) on the concentration of the adsorbed electron donor, $[R_{ads}]$, and (iii) on the rate constants of recombination of the electron-hole pair via radiative and radiationless transitions, $\sum k_i$. At steady-state of the electronically excited state, the quantum yield, Φ_λ, of interfacial electron-transfer can be expressed in terms of rate constants:

$$\Phi_\lambda = \frac{k_{ET}[R_{ads}]}{\sum k_i + k_{ET}[R_{ads}]} \tag{10.11}$$

Thus, via the quantum yield, the rate of a heterogeneous photoredox reaction, where the (bulk) solid phase acts as chromophore, depends on the surface concentration of the reductant (or oxidant).

The quantum yield of interfacial electron transfer does not only depend on the surface concentration of the reductant (or oxidant) but also on its adsorption property.

For instance, the more efficiently the photoholes are trapped from the valence band of an n-type semiconductor, the higher is the probability that the photoelectrons in the conduction band reach the surface and can reduce a thermodynamically suitable electron acceptor at the solid-liquid interface. This is illustrated with an example taken from a paper by Frei et al, 1990. In this example methylviologen, MV^{2+}, acts as the electron acceptor and TiO_2 as the photocatalyst. Upon absorption of light with energy equal or higher than the band-gap energy of TiO_2, a photoelectron is formed in the conduction band and a photohole in the valence band:

$$TiO_2 \underset{hv}{\rightleftharpoons} e_{cb}^- + h_{vb}^+ \qquad (10.12)$$

The photoelectron either recombines with the photohole or is transferred to the electron accetor, MV^{2+}. Thereby the reduced species $MV^{\cdot+}$ is formed. This $MV^{\cdot+}$ radical exhibits an absorption band with a maximum at 630 nm. The increase of the absorption at 630 nm of the $MV^{\cdot+}$ radical was measured as a function of time (in the microsecond range) after a laser pulse at 355 nm which creates high enough concentrations of photoelectrons and photoholes. In the experiment shown in Fig. 10.7a bare TiO_2 particles were examined where the photoholes are trapped by the surface OH^- groups:

$$e_{cb}^- + MV^{2+} \longrightarrow MV^{\cdot+} \qquad (10.13)$$

$$h_{vb}^+ + OH^-_{(surface)} \longrightarrow \dot{O}H \qquad (10.14)$$

In the experiment shown in Fig. 10.7b phenylfluorone acting as hole trapper was added:

$$e_{cb}^- + MV^{2+} \longrightarrow MV^{\cdot+} \qquad (10.15)$$

$$h_{vb}^+ + PF_{(adsorbed)} \longrightarrow PF^{\cdot+} \qquad (10.16)$$

In the absence of PF, the growth of the $MV^{\cdot+}$ absorption at 630 nm is slow, extending over at least 1 ms (Fig. 10.7a). Surface complexation with PF accelerates drastically the growth of the 630 nm absorption of $MV^{\cdot+}$. Two-thirds of the signal is formed already within 200 ns after the laser pulse (Fig. 10.7b). From FTIR spectroscopic investigations the authors suggest that phenylfluorone forms a bidentate surface complex at the surface of TiO_2. This example illustrates that a specifically adsorbed reductant that forms a bidentate surface complex as is the case for the adsorption of phenylfluorone at the TiO_2 surface seems to be specifically efficient for the trapping of the photoholes.

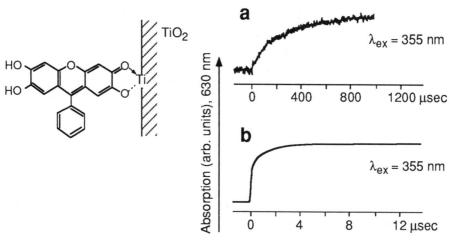

Figure 10.7

Effect of surface chelation on the kinetics of electron transfer from the conduction band of TiO_2 to methylviologen (MV^{2+}). Oscillograms showing the time-dependent growth of the $MV^{\overset{+}{\cdot}}$ absorption at 630 nm after laser excitation (at 355 nm) of aqueous solutions (pH 4.85) containing colloidal TiO_2 (1 g/ℓ) and 10^{-3} M MV^{2+}:
a)   bare TiO_2 particles;
b)   PF-chelated TiO_2 particles (PF concentration 3×10^{-5} M).
(Modified from Frei, Fitzmaurice, and Grätzel, 1990)

A Surface Complex Acting as the Chromophore

In heterogeneous photoredox systems also a surface complex may act as the chromophore. This means that in this case not a bimolecular but a unimolecular photoredox reaction takes place, since electron transfer occurs within the light-absorbing species, i.e. through a ligand-to-metal charge-transfer transition within the surface complex. This has been suggested for instance for the photochemical reductive dissolution of iron(III)(hydr)oxides (Waite and Morel, 1984; Siffert and Sulzberger, 1991). For continuous irradiation the quantum yield is then:

$$\Phi_\lambda = \frac{k_{PF}}{k_{IC} + k_{PF}} \tag{10.17}$$

where k_{PF} is the rate constant of *p*hoto*p*roduct *f*ormation, and k_{IC} the rate constant of the thermal deactivation of the charge-transfer state (*i*nternal *c*onversion). Combining Eqs. (10.10) and (10.17) yields for the rate of photoproduct formation, e.g. of dissolved iron(II) formation in the case of the photochemical reductive dissolution of iron(III)(hydr)oxides:

$$\frac{d[Fe(II)_{aq}]}{dt} = 2.303 \times L \times \varepsilon_\lambda \times [\equiv Fe^{III}R] \times I_{0_\lambda} \times \frac{k_{PF}}{k_{IC} + k_{PF}} \tag{10.18}$$

where $[\equiv Fe^{III}R]$ is the concentration of the surface complex formed from the specific adsorption of a reductant R at the surface of an iron(III)(hydr)oxide.

Hence, in both, in bimolecular and in unimolecular heterogeneous photoredox reactions the rate of photoproduct formation depends on the surface concentration of an electron donor or acceptor and on the adsorption properties of the adsorbing species.

An Adsorbed Compound Acting as Chromophore

Despite the relatively large band-gap energy, TiO_2 is often used as a photocatalyst because of its relative inertness with regard to dissolution. In order to render such materials more suitable as photocatalysts for the visible part of the solar spectrum, the surface of the solid phase is modified by the specific adsorption of a chromophore which has a lower energy gap between the HOMO and the LUMO as compared to the band-gap of the semiconductor. The principle is shown in Fig. 10.8 for the case of a photoelectrochemical cell.

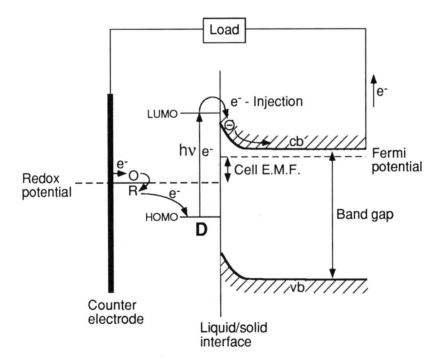

Figure 10.8

Mechanism of dye sensitization and charge transfer in a sensitized regenerative photoelectrochemical cell (cell E.M.F. = electromotoric force of the photoelectrochemical cell).

(Modified from Vlachopoulos, Liska, McEvoy, and Grätzel, 1987)

Upon absorption of light an electron from the HOMO of the adsorbed dye, D, is raised to the LUMO from where it is injected into the conduction band of the n-type semiconductor and transferred to a counter electrode where an oxidant, O, is reduced. From the reduced species, R, the electron is transferred to the HOMO of the adsorbed dye to fill the electron vacancy, so that after the overall photoelectrochemical process the dye is in its original oxidation state. Vlachopoulos et al. (1987) have reported on TiO_2 photoelectrodes that were sensitized to visible light with various dyes and that showed high quantum yields of interfacial electron transfer under visible irradiation.

10.4 Case Studies on Heterogeneous Photoredox Reactions

Photocatalytic Oxidation of Organic Pollutants

In heterogeneous photochemistry an increasing efford goes into the study of the photocatalytic degradation of organic pollutants with TiO_2 as a photocatalyst. One example is the photocatalytic oxidation of 4-chlorophenol (4-CP) (Al-Ekabi et al., 1989) according to the following overall stoichiometry:

$$C_4H_4ClO + 13\tfrac{1}{2} O_2 \xrightarrow[TiO_2]{h\nu} 6\,CO_2 + HCl + 2\,H_2O \qquad (10.19)$$

Fig. 10.9 shows the rate of 4-CP degradation as a function of the solution concentration of 4-CP.

As seen in Fig. 10.9 the rate reaches a plateau, consistent with the assumption that the rate is directly proportional to the surface concentration of 4-CP.

In these photocatalytic oxidation processes the organic compound is either directly oxidized by the valence band holes:

$$h_{vb}^+ + \text{organic compound} \longrightarrow \text{oxidized organic compound} \qquad (10.20)$$

or the organic compound is oxidized by $\dot{O}H$ radicals that are formed through reaction of the surface OH^- groups of TiO_2 with the photoholes:

$$h_{vb}^+ + OH^-_{(surface)} \longrightarrow \dot{O}H \qquad (10.21)$$

$$\dot{O}H + \text{organic compound} \longrightarrow \text{oxidized organic compound} \qquad (10.22)$$

The photoelectrons of the cb react with adsorbed molecular oxygen, eventually leading to formation of H_2O_2.

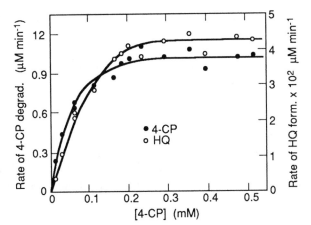

Figure 10.9

Effect of 4-chlorophenol (4-CP) concentration on the rate of degradation of 4-CP (●) and the rate of formation of hydroquinone (○), which is a major intermediate of the 4-CP degradation. Conditions: initial pH 5.8; flow rate, 250 ml min^{-1}; temperature, 30 ± 2° C.
(From Al-Ekabi, Serpone, Pelizzetti, Minero, Fox, and Draper, 1989)

Photocatalytic Production of Hydrogen Peroxide

H_2O_2 (and organic peroxides) can be formed by either the reduction of O_2 by e_{cb}^- or the oxidation of H_2O by h_{vb}^+ as follows (Hoffmann, 1990):

$$O_2 + 2\,e_{cb}^- + 2\,H^+ \longrightarrow H_2O_2 \tag{10.23}$$

$$2\,H_2O + 2\,h_{vb}^+ \longrightarrow H_2O_2 + 2\,H^+ \tag{10.24}$$

Isotopic labeling studies have shown that H_2O_2 formed in irradiated suspensions of ZnO contained oxygen atoms derived exclusively from O_2.

The overall rate of H_2O_2 formation on metal oxide (TiO$_2$, ZnO, etc.) surfaces was shown (Kormann et al., 1988; Hoffmann, 1990) to depend on p_{O_2}, on the concentration of organic electron donors and on the concentration of H_2O_2. The reason for the net rate of H_2O_2 formation depending also on the H_2O_2 concentration is that H_2O_2 may also be photodegraded by reactions such as

$$H_2O_2 + 2\,e_{cb}^- + 2\,H^+ \longrightarrow 2\,H_2O \tag{10.25}$$

$$H_2O_2 + 2\,h_{vb}^+ \longrightarrow O_2 + 2\,H^+ \tag{10.26}$$

The combination of the photocatalytic production and the photocatalytic degradation will result in a steady-state concentration of H_2O_2 during continuous irradiation. Steady-state concentrations of H_2O_2 in excess of 100 µM were obtained in illuminated suspensions of ZnO. TiO_2 is less effective for H_2O_2 production (Kormann et al., 1988; Hoffmann, 1990).

Electron donors, D, adsorbed on the particle surface react with valence-band holes

$$D + h_{vb}^+ \longrightarrow D^{\overset{\bullet}{+}} \tag{10.27}$$

Hole trapping by electron donors bound to the surfaces of the semiconductor particles competes with the $e_{cb}^- - h_{vb}^+$ recombination, allowing e_{cb}^- to react with molecular O_2 via Eq. (10.23). Fig. 10.10 shows that the quantum yield, Φ, of H_2O_2 and total peroxide formation increases with increasing concentration of the electron donor.

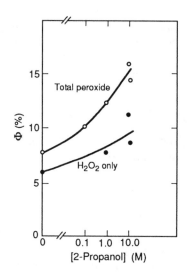

Figure 10.10

Quantum yield, Φ (in %), of H_2O_2 and organic peroxide formation with ZnO colloids as a function of the 2-propanol concentration.

(From Kormann, Bahnemann and Hoffmann, 1988).

Photocatalytic Oxidation of S(IV) (SO₂, HSO₃⁻ or SO₃²⁻) by O₂

Photocatalytic Oxidation of S(IV) (SO₂, HSO₃⁻ or SO₃²⁻) by O₂

The photocatalytic oxidation of organic and inorganic compounds and the photocatalytic production of H_2O_2 occurs also at the surface of iron(III)(hydr)oxides. It has been proposed (e.g., Hoffmann, 1990; Faust and Hoffmann, 1986) that the oxidation of S(IV) by O_2 in atmospheric water is catalyzed by iron(III)(hydr)oxide particles. It is assumed that the reductant (HSO₃⁻) is specifically adsorbed at the surface of an iron(III)(hydr)oxide, forming either a monodentate or a bidentate surface complex:

$$\equiv Fe\text{-}OH + HSO_3^- \;\rightleftharpoons\; \equiv FeSO_3^- + H_2O \tag{10.28}$$

or

$$\equiv Fe\begin{array}{c} OH_2^+ \\ \diagdown \\ \diagup \\ OH \end{array} + HSO_3^- \;\rightleftharpoons\; \equiv Fe\begin{array}{c} O \\ \diagdown \\ \diagup \\ O \end{array}S-O + 2H_2O \tag{10.29}$$

It is also assumed (Hoffmann 1990) that the adsorbed sulfite is oxidized by the valence band holes, h_{vb}^+, that are formed through absorption of light with photon energies exceeding the band-gap energy (ca. 2.2 eV) of an iron(III)(hydr)oxide, e.g., hematite (α-Fe_2O_3). This interfacial electron transfer reaction results in formation of the SO_3^- radical anion which reacts with another radical to form $S_2O_6^{2-}$, one of the end product, if the reaction is carried out under nitrogen.

Unlike TiO_2 some iron(III)(hydr)oxide phases are reductively dissolved upon photocatalytic oxidation of an adsorbed organic or inorganic compound. This means that reaction of the photoelectrons, trapped at surface iron centers, with oxygen is slower than detachment of surface iron(II) from the crystal lattice. Whether detachment of surface iron(II) or its reoxidation by oxygen predominates depends on the electrostatic energies between surface iron(II) and the neighboring oxygen ions.

Photocatalytic Reductive Dissolution of Hematite in the Presence of Oxalate

In the absence of oxygen the photocatalytic reductive dissolution of hematite in the presence of oxalate occurs according to the following overall stoichiometry (Siffert and Sulzberger, 1991):

$$\alpha\text{-}Fe_2O_3 + C_2O_4^{2-} + 6\,H^+ \xrightarrow{\;h\nu\;} 2\,Fe^{2+} + 2\,CO_2 + 3\,H_2O \tag{10.30}$$

In aerated suspensions no measurable (in the time-frame of typical experiments) reductive dissolution takes place and hematite acts as a photocatalyst for the oxidation of oxalate by O_2:

$$C_2O_4^{2-} + O_2 + 2\,H^+ \xrightarrow[\text{hematite}]{\;h\nu\;} 2\,CO_2 + H_2O_2 \tag{10.31}$$

The rate of the photochemical reductive dissolution of hematite in the presence of oxalate is strongly wavelength-dependent. In order to calculate quantum yields at various wavelengths according to Eq. (10.18), the rates of dissolved Fe(II) formation were experimentally determined at different wavelengths but at constant light intensity (Siffert and Sulzberger, 1991). As seen in Fig. 10.11 only light in the near-UV region ($\lambda < 400$ nm) leads to an enhancement of the dissolution brought about by the redox process at the surface. This is in line with what has been observed by

Faust and Hoffmann (1986) and Litter and Blesa (1988) who investigated the wavelength-dependence of the rate of photochemical reductive dissolution of iron(III)(hydr)oxides using hematite-bisulfite and maghemite-EDTA as model systems, respectively.

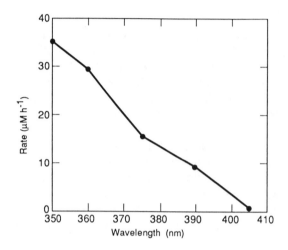

Figure 10.11

Rate of the photochemical reductive dissolution of hematite, $R_\lambda = d[Fe(II)]/dt$, in the presence of oxalate as a function of the wavelength at constant incident light intensity ($I_0 = 1000$ µeinsteins ℓ^{-1} h^{-1}). The hematite suspensions were deaerated; initial oxalate concentration = 3.3 mM; pH = 3. (In order to keep the rate of the thermal dissolution constant, a high enough concentration or iron(II), [Fe^{2+}] = 0.15 mM, was added to the suspensions from the beginning. Thus, the rates correspond to dissolution rates due to the surface photoredox process).

(From Siffert and Sulzberger, 1991)

From these rates, quantum yields were calculated assuming that the surface complex acts as the chromophore, Table 10.2. For the calculation of these quantum yields we used the extinction coefficients as estimated from the absorption spectrum of dissolved ferrioxalate. Thus, these quantum yields can only be considered as very rough estimate. Between 350 and 390 nm they are practically constant. This is what one would expect if the surface complex were the chromophore, namely a constant quantum yield in the wavelength-range of its ligand-to-metal charge-transfer transition. The upper limit value of the quantum yield is two, since the photochemically formed $CO_2^{\cdot-}$ radical is a strong reductant and can in a thermal reaction reduce a second surface iron(III), so that without loss processes, as the thermal deactivateion of the charge-transfer state, two iron(II) could be formed per absorbed photon.

The fact that these estimated quantum yields are between 350 and 390 nm not too different from the quantum yields of photolysis of the corresponding solution

Table 10.2 Quantum yield, Φ_λ, of dissolved iron(II) formation under the assumption that the iron(III) oxalato surface complex is the chromophore.

λ (nm)	$\Phi_\lambda = \dfrac{R_\lambda}{2.3 \times L \times I_{0\lambda} \times \varepsilon_\lambda \times [\equiv Fe^{III}C_2O_4^-]}$
350	1.4
360	1.5
375	1.4
390	1.5
405	0.2

$[\equiv Fe^{III}C_2O_4^-] = 20\ \mu M$; $L = 0.5$ cm; the ε_λ values are estimated from the absorption spectrum of dissolved ferrioxalate, and hence these quantum yields can only be considered as very rough estimate.
(Modified from Siffert and Sulzberger, 1991)

iron(III) oxalato complexes (Parker, 1968) is an indication that the surface complex may act as the chromophore but by no means a proof. As discussed for the S(IV) oxidation with α-Fe_2O_3 as a photocatalyst, hematite itself may also act as the chromophore. Then the photochemical reductive dissolution is induced by a band-to-band transition (presumingly between a band with oxygen-character and a band with iron-character), i.e., by a charge-transfer transition (Siffert and Sulzberger, 1991). The question of whether in the photochemical reductive dissolution of iron(III)(hydr)oxides the surface complex acts as the chromophore can only be answered with a suitable inner-sphere surface complex which exhibits a ligand-to-metal charge-transfer band in a spectal window of the solid phase and which is photolysed with high quantum yields.

If the surface complex is the chromophore, then the photochemical reductive dissolution occurs as a unimolecular process; alternatively, if the bulk iron(III)(hydr)oxide is the chromophore, then it is a bimolecular process. Irrespective of whether the surface complex or the bulk iron(III)(hydr)oxide acts as the chromophore, the rate of dissolved iron(II) formation depends on the surface concentration of the specifically adsorbed electron donor; e.g. compare Eqs. (10.11) and (10.18). It has been shown experimentally with various electron donors that the rate of dissolved iron(II) formation under the influence of light is a Langmuir-type function of the dissolved electron donor concentration (Waite, 1986).

Also, photolysis of dissolved iron(III) oxalato complexes, which are photolyzed with high quantum yields, has to be taken into account in order to describe this photoredox system. Dissolved iron(III) oxalato complexes are formed via thermal dissolution of hematite where Fe(II) acts as a catalyst as described in Chapter 9. As a result, the rate of dissolved Fe(II) formation is not constant, i.e., zero order with regard to the dissolved oxalate concentration, but increases with time as shown in Fig. 10.12, indicating an autocatalytic process (Cornel and Schindler, 1987; Siffert and Sulzberger, 1991).

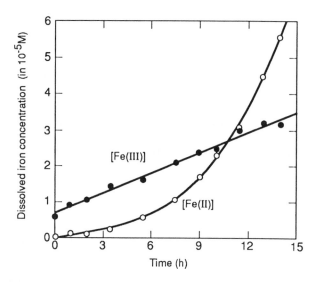

Figure 10.12

Light-induced dissolution of hematite in the presence of oxalate at pH 3. The deaerated hematite suspension was irradiated with light that had passed a monochromator (λ = 375 nm; I_0 = 4 W m^{-2}). Initial oxalate concentration = 3.3 mM.
(From Siffert and Sulzberger, 1991)

Photosynthetic Oxidation of Iron(II)

Photosynthetic processes involving some inorganic iron-bearing minerals (semiconductors) and dissolved Fe-species may be looked at as "primitive" alternatives or precursors to biological photosynthesis.

From the perspective of geochemistry, one can ask the question of whether oxygen evolution occurred in prebiotic times with systems involving iron(III) oxides (see Khader et al., 1987). The question then arises how iron(III) bearing minerals were formed in prebiotic times. The major iron ore sources on the Earth are the banded iron formations. They consist of extensive iron-rich and iron-poor layers within siliceous sedimentary rocks. The banding of some of the Precambrian banded iron formations found in Australia consist of iron mineral-chert varves, averaging about 0.15 mm per annual pair. The origin of the banded iron formation of the Precambrian is the subject of many speculations since the atmosphere in the Precambrian, more than 600 million years ago, was probably mildly reducing. One hypothesis that was proposed by Braterman et al. (1984) for the prebiotic banded iron formation is the photooxidation of ferrous iron under reduction of protons. The evolution of hydrogen in an acid aqueous ferrous solution has been known for a long time:

$$Fe^{2+} + H^+ \xrightarrow{h\nu} Fe^{3+} + \tfrac{1}{2} H_2 \tag{10.32}$$

Braterman and coworkers (1984) have demonstrated that this reaction occurs also at higher pH (up to neutral pH) where ferric iron is then precipitated as iron(III) hydroxide according to the following overall reaction:

$$Fe^{2+} + 2 H_2O \xrightarrow{h\nu} \gamma\text{-}FeOOH + 2 H^+ + \tfrac{1}{2} H_2 \tag{10.33}$$

Also iron(II) bearing minerals such as pyrite may be oxidized photochemically in the absence of oxygen according to the following reaction scheme (Jaegermann and Tributsch, 1988):

$$FeS_2 \xrightarrow{h\nu} FeS_2^* (e^-_{cb}, h^+_{vb}) \tag{10.34}$$

$$15 h^+_{vb} + FeS_2 + 8 H_2O \longrightarrow Fe^{3+} + 2 SO_4^{2-} + 16 H^+ \tag{10.35}$$

$$15 e^-_{cb} + 15 H^+ \longrightarrow 7\tfrac{1}{2} H_2 \tag{10.36}$$

$$\overline{\qquad\qquad\qquad\qquad\qquad\qquad\qquad\qquad\qquad\qquad\qquad}$$

$$FeS_2 + 8 H_2O \xrightarrow{h\nu} Fe^{3+} + 2 SO_4^{2-} + 7\tfrac{1}{2} H_2 + H^+ \tag{10.37}$$

$$OH^- \updownarrow$$

$$Fe(OH)_3(s)$$

Abiotic Photosynthetic Oxygen Evolution

It has been reported (Calzaferri et al., 1984) that oxygen from water is evolved upon irradiation of a silver-loaded zeolite (a group of alumino-silicates) suspension according to the following overall stoichiometry:

$$[(Ag^+)_n \times mH_2O]Z \xrightarrow{h\nu} [Ag_n^{(n-r)+} \times rH^+ \times (m-\tfrac{r}{2})H_2O]Z + \tfrac{r}{4} O_2 \tag{10.38}$$

where Z stands for the zeolite lattice. The authors assume that a ligand-to-metal charge-transfer transition, the lattice oxygen ions of the zeolite being the ligand and the silver ions that are located within the zeolite cavities being the metal centers, is involved in this heterogeneous photoredox reaction. This means that an internal surface complex acts as the chromophore in this case. Upon reduction of silver ions silver clusters are formed within the cavities and act themselves as chromophores. The bigger the clusters become, the more they absorb light in the visible region of the spectrum. Thus, a process which can be called self-sensitization of the photochemical oxygen-evolution takes place. This is shown in Fig. 10.13: The current is a measure of the evolved oxygen concentration, since oxygen was measured with a Clark-electrode. At each wavelength the zeolite-suspension was irradiated for 8 minutes. Before the irradiation at the next wavelength, the suspension was kept in the dark for 22 minutes. Such light-dark cycles at different wavelengths were repeated several times. With increasing cycle number the silver-zeolite becomes more and more active in the visible region for the photochemical evolution of oxygen from water.

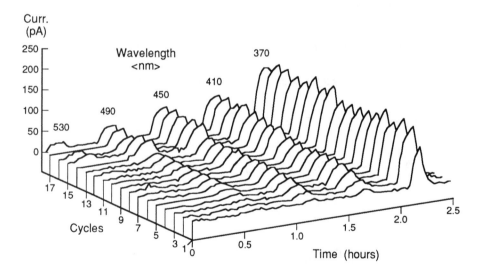

Figure 10.13

Self-sensitization experiment of an aqueous Ag^+ zeolite A suspension. 1 nAs corresponds to 3.94 nl of oxygen.

(From Calzaferri, Hug, Hugentobler, and Sulzberger, 1984)

10.5   The Cycling of Iron in Natural Systems; Some Aspects Based on Heterogeneous Redox Processes

The objectives of this section are to illustrate the relevance of some of the heterogeneous redox processes (that were discussed in Chapter 9) and of the photochemical processes occurring on some inorganic Fe-bearing minerals (this Chapter) to the cycling of iron in natural systems.

The cycling of iron through the various global reservoirs (atmosphere, oceans, soils, sediments) depends on various physical, chemical, and biological processes. In all these reservoirs, the iron bearing solids are present with high surface area to volume ratios. The surface-controlled reactions are of great significance in regulating the composition of the reservoirs and influence the mass fluxes of elements from one reservoir to another. The cycle of iron is interdependent and often kinetically coupled with the cycles of P, S, heavy metals, O_2, and C, and depends on the biota and the energy and intensity of light.

The Importance of Iron in the Global Geochemical Cycling of Elements

Three considerations underline the importance of iron:
1) the quantities involved both in the iron reservoir of sedimentary rocks and the rates of transformations. Approximately 17×10^{20} mol of Fe (ca. 5 – 6 % by weight) are in the sedimentary rocks and ca. 3.5×10^{12} mol yr^{-1} of Fe are transformed from oxidized to reduced reservoirs and vice versa. This corresponds to an equivalent net electron flux. Although this is small in comparison to the electron flux induced by gross photosynthesis (3×10^{16} mol yr^{-1}), the actual flux of electrons passing through Fe ions is much larger because the Fe(II) – Fe(III) "redox wheel" turns locally many times for each net transformation. The ability of Fe-species to become oxidized or reduced and concomitantly to become precipitated and dissolved links the iron cycle to that of oxygen (oxidant) and organic carbon (reductant).
2) The large specific surface areas of the Fe solid phases (Fe(II,III)(hydr)oxides, FeS_2, FeS, Fe-silicates) and their surface chemical reactivities facilitate specific adsorption of various solutes. This is one of the causes for the interdependence of the iron cycle with that of many other elements, above all with heavy metals, some metalloids, and oxyanions such as phosphate.
3) The solid state and the surface chemistry of some of the solid Fe-phases impart to these oxides and sulfides the ability to catalyze redox reactions. Surface complexes and the solid phases themselves acting as semiconductors can participate in photoredox reactions, where light energy is used to drive a thermodynamically unfavorable reaction (heterogeneous photosynthesis) or to catalyze a thermodynamically favorable reaction (heterogeneous photocatalysis).

The (Photo)redox Cycle of Iron

The iron cycle shown in Fig. 10.14 illustrates some redox processes typically observed in soils, sediments and waters, especially at oxic-anoxic boundaries. The cycle includes the reductive dissolution of iron(III)(hydr)oxides by organic ligands, which may also be photocatalyzed in surface waters, and the oxidation of Fe(II) by oxygen, which is catalyzed by surfaces. The oxidation of Fe(II) to Fe(III)(hydr)-oxides is accompanied by the binding of reactive compounds (heavy metals, phosphate, or organic compounds) to the surface, and the reduction of the ferric (hydr)oxides is accompanied by the release of these substances into the water column.

On overall balance the cycle shown in Fig. 10.14 represents a mediation (by iron) of the oxidation of organic matter by oxygen. This oxidation may be important both in the degradation and polymerization of organic matter in soils and waters. The interdependence of the iron cycle with that of other redox cycles is obvious; e.g., Fe(II) can reduce Mn(III,IV) oxides, and HS^- is an efficient reductant of hydrous Fe(III) oxides. Many of the processes mentioned above occur also in soils directly or indirectly mediated by microorganisms. Microorganisms and plants produce a large number of biogenic acids that are effective in ligand promoted dissolution of Fe(III) and other (hydr)oxides. Oxalic, maleic, acetic, succinic, tartaric, ketogluconic and p-hydroxybenzoic acids have been found in top soils with oxalic acid, the most abundant, in concentrations as high as $10^{-5} - 10^{-4}$ M (in soil water). The downward vertical displacement of Al and Fe observed during podzolization of soils can be explained by considering the effect of pH and of complex formers on the solubility of iron and aluminum (hydr)oxides and on their dissolution rates. The same biogenic acids are found typically in waters and sediments and are also produced by fermentation reactions under anoxic conditions. The reductive dissolution of Fe(III) (hydr)oxides is also of importance for iron uptake by higher plants.

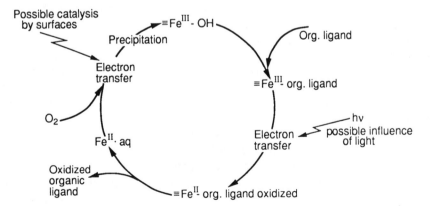

Figure 10.14

Schematic representation of the aquatic (photo)redox cycling of iron. $\equiv$ denotes the lattice surface of an iron(III)(hydr)oxide.

The Iron Cycle in Salt Marsh Sediments. Luther et al. (1991) have proposed that the iron(II) catalyzed dissolution of iron(III)(hydr)oxides plays a significant role in the biogeochemical cycling of iron in salt marsh sediments. No oxygen is present in the porewater of such sediments in Great Marsh, Delaware, and manganese oxides are not likely oxidants in these anaerobic systems. Yet, there is evidence for reaction of dissolved Fe(III) complexes with pyrite at near neutral pH, and thus for the coexistence of dissolved Fe(III) and dissolved Fe(II). The cycling of iron in these systems is shown in Fig. 10.15. The main reactions of this iron cycle are the follwing:

1) Ligand-promoted dissolution of iron(III)(hydr)oxides, where the iron(II) cata- lyzed dissolution plays the major role.
2) Reductive dissolution of iron(III)(hydr)oxides and reduction of dissolved iron(III) species by organic reductants, R, soluble reduced sulfur or solid phase sulfur, S(–II).
3) Oxidation of the sulfide of pyrite by dissolved iron(III).
4) Precipitation of dissolved Fe(II) to form sulfide minerals.

The cycle of iron solubilization will continue as long as bacteria and/or plants pro- duce organic ligands.The cycle will stop when sulfate reduction rates are high and organic ligand production is low. At this point soluble hydrogen sulfide reacts with Fe(II) to form sulfide minerals. The iron cycle shown in Fig. 10.15 for salt marsh sediments may also occur in other marine sedimentary systems.

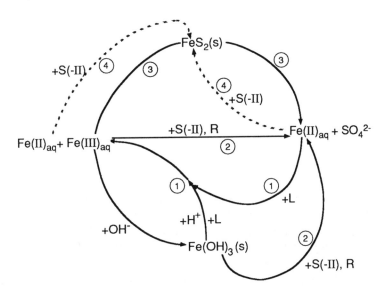

Figure 10.15

Redox cycling of iron in salt marsh sediments. The solid lines and the dashed lines indicate redox reactions and precipitation reactions, respectively.

(Modified from Luther, Kostka, Church, Sulzberger and Stumm, in press)

The Iron Cycle in the Photic Zone of Surface Waters: In the photic zone the formation of iron(II) occurs as a photochemical process. The photochemical iron(II) formation proceeds through different pathways: 1) through the photochemical reductive dissolution of iron(III)(hydr)oxides, and 2) through photolysis of dissolved iron(III) coordination compounds, Fig. 10.16.

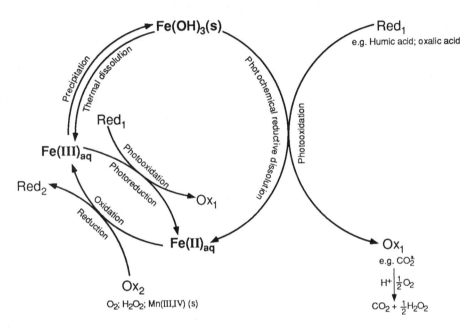

Figure 10.16

Schematic representation of the photoredox cycling of iron in the photic zone of surface waters. The important features are the following:

1) The photochemical reduction of particulate *and* of dissolved iron(III) is coupled to the oxidation of a reductant, i.e., of dissolved organic carbon.
2) The heterogeneous *and* the homogeneous photoredox reactions lead to formation of H_2O_2 through reaction of the primary oxidation product, e.g. $CO_2^{\cdot}$, with oxygen.

Dissolved iron(III) is (i) an intermediate of the oxidative hydrolysis of Fe(II), and (ii) results from the thermal non-reductive dissolution of iron(III)(hydr)oxides, a reaction that is catalyzed by iron(II) as discussed in Chapter 9. Hence, iron(II) formation in the photic zone may occur as an autocatalytic process (see Chapter 10.4). This is also true for the oxidation of iron(II). As has been discussed in Chapter 9.4, the oxidation of iron(II) by oxygen is greatly enhanced if the ferrous iron is adsorbed at a mineral (or biological) surface. Since mineral surfaces are formed via the oxidative hydrolysis of Fe(II), this reaction proceeds as an autocatalytic process (Sung and Morgan, 1980). Both the rate of photochemical iron(II) formation and the rate of oxidation of iron(II) are strongly pH-dependent; the latter increases with increasing pH and the former decreases with increasing pH. The reoxidation of Fe(II) may produce a $Fe(OH)_3(s)$ that is less polymeric and less crystalline than aged Fe(III)(hydr)oxides and thus more soluble and in faster equilibrium with monomeric Fe(III) species, which may control iron uptake by phytoplankton (Rich and Morel, 1990). As shown in Fig. 10.16 the photoredox cycling of iron mediates the oxidation and recycling of organic matter, i.e. of humic and fulvic substances, which may lead to formation of CO in addition to CO_2 (Zepp, private communication; Bartschat and Sulzberger, in preparation). The photochemical oxidation of organic compounds under reduction of iron(III) species is a source of H_2O_2 through the reaction of the primary oxidation product, which is often a radical, with oxygen (see Fig. 10.16). Recently it was suggested (Mopper et al., 1991) that the photochemical degradation pathway is the rate limiting process for the removal of a large fraction of oceanic dissolved organic carbon. The involvement of iron in this photochemical degradation appears likely, although no data are available yet.

As discussed in previous subchapters, the rate of the photochemical reductive dissolution of iron(III)(hydr)oxides depends on the concentration and type of surface complexes present and on the light intensity and its energy. Because the light intensity varies diurnally, also a diurnal variation in the iron(II) concentration can be expected in surface waters. This has been observed in acidic waters (McKnight and Bencala, 1988; Sulzberger et al., 1990). Fig. 10.17 shows such a diurnal variation in the concentration of dissolved Fe(II) in a slightly acidic alpine lake (Lake Cristallina) of Switzerland.

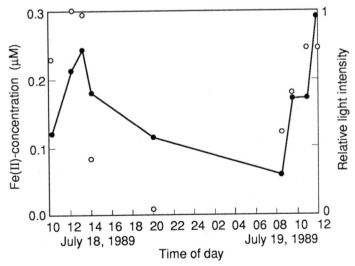

Figure 10.17

Diurnal variation of the concentration of dissolved Fe(II) ● and of the incident light intensity ○ in Lake Cristallina. (The maximal measured light intensity is arbitrarily set to one).

(From Sulzberger, Schnoor, Giovanoli, Hering, Zobrist, 1990)

The net concentration of Fe(II) at any day time reflects the balance of the reductive dissolution and the oxidation/precipitation reactions and parallels the light intenstiy. It is believed that the photochemical reductive dissolution of Fe(III)(hydr)oxides is also of importance for the formation of iron(II) in marine systems; however, around pH = 8 the rate of reductive dissolution is likely to be much smaller than at lower pH-values. Also, in the photic zone of oceanic waters the concentration of H_2O_2 may be sufficiently high that oxidation of Fe(II) by H_2O_2 may compete with oxygenation, which would increase the overall Fe(II) oxidation rate. Since H_2O_2 formation occurs through photochemical processes, e.g., involving dissolved organic carbon (Cooper et al., 1988; Zuo and Hoigné, submitted; Bartschat and Sulzberger, in preparation), iron(II) oxidation by H_2O_2 may also depend on the light intensity.

As discussed in Chapter 10.4, the kinetics of the photochemical reductive dissolution depends strongly on the light energy. This may have consequences for the steady-state concentration of iron(II) at different depths of the photic zone, since longer wavelength-light can penetrate to a greater depth as compared to shorter wavelength-light. The reason for this is that the attenuation of the light due to absorption by dissolved organic carbon decreases with increasing wavelengths (Zepp and Cline, 1977). Wells et al. (1991) have reported that light increases the lability of colloidal iron in seawater of pH 8, and that generally only wavelengths below 400 nm lead to an increase in the lability of iron due to light. From model calculations combining the wavelength-dependence of the attenuation of the light by DOC and the wavelength-dependence of the labilization of colloidal iron the authors predict that the depth-integrated formation of labile iron will be maximal at 390 nm.

The Iron Cycle in Atmospheric Water. The iron cycle is also important in atmospheric water. Although liquid water constitutes less than 10^{-6} % of the troposphere, chemical processes in the aqueous phase (clouds, fog, rain) are of great relevance. Jacob et al. (1989) have suggested that the iron (photo)redox cycle mediates the oxidation of S(IV) by oxygen. Behra and Sigg (1990) have reported that a large fraction of the total iron in fog water is present as dissolved Fe(II). The concentration of Fe(II) increased both with decreasing pH and exposure to light; maximal concentrations of 0.2 mM were reported. These authors assume that Fe(III) (hydr)oxide is reduced by sulfite, organic compounds and also by free radicals formed photochemically during daytime. Not only heterogeneous but also homogeneous photoredox reactions are of importance in atmospheric water. Zuo and Hoigné (submitted) suggest that the concentration of dissolved Fe(III) oxalato complexes in acidic atmospheric water is sufficiently high to make their photolysis a dominant source of in-cloud H_2O_2, $O_2^{\cdot-}$, $H\dot{O}_2$ and $\dot{O}H$ radicals and a major sink of atmospheric oxalic acid. The occurrence of iron cycling in the atmosphere illustrates that redox reactions of Fe(II)/Fe(III) can occur readily also in the absence of microorganisms.

Reading Suggestions

Stumm, W., and B. Sulzberger (1991), "The Cycling of Iron in Natural Environments; Considerations Based on Laboratory Studies of Heterogeneous Redox Processes", *Geochim. Cosmochim. Acta* (in print).

Sulzberger, B. (1990), "Photoredox Reactions at Hydrous Metal Oxide Surfaces; a Surface Coordination Chemistry Approach", in W. Stumm, Ed., *Aquatic Chemical Kinetics*, Wiley-Interscience, New York, pp. 401-429.

Waite, T. D. (1990), "Photo-Redox Processes at the Mineral-Water Interface", in M. F. Hochella Jr. and A. F. White, Eds., *Mineral-Water Interface Geochemistry*, Mineralogical Society of America, Washington, D.C., pp. 559-603.

Chapter 11

Regulation of Trace Elements by the Solid-Water Interface in Surface Waters

11.1 Introduction

The solid-water interface, mostly established by the particles in natural waters and soils, plays a commanding role in regulating the concentrations of most dissolved reactive trace elements in soil and natural water systems and in the coupling of various hydrogeochemical cycles (Fig. 1.1). Usually the concentrations of most trace elements (M or mol kg^{-1}) are much larger in solid or surface phases than in the water phase. Thus, the capacity of particles to bind trace elements (ion exchange, adsorption) must be considered in addition to the effect of solute complex formers in influencing the speciation of the trace metals.

The particles in natural systems are characterized by a great diversity (minerals, including clays and organic particles; organisms – biological debris, humus, macromolecules – and inorganic particles coated with organic matter, etc.) (Table 7.1).

This chapter is organized as follows: We first attempt to discuss, in terms of simplified models, how particles carrying functional groups behave in solutions whose variables are known or controlled. This is followed by observations and interpretations on the concentration of trace elements in rivers and how these trace elements are distributed between particulate and dissolved phase. Then, we will consider the regulation of metal ions and of other reactive elements in lakes; above all, it will be shown that the interaction of these trace elements with biotic and non-biotic particles and the subsequent settling of these particles will be of utmost importance for their removal from the water/column. Finally considerations will be given to inquire to what extent similar interpretations can be given to oceans.

11.2 The Particle Surface as a Carrier of Functional Groups

Abstracting from the complexity of the real systems, there is one common property of all natural particles. Their surfaces contain functional groups which can interact with H^+, OH^- and metal ions and – if Lewis acid sites, e.g., $\equiv Al$ and $\equiv Fe$, are available on the surface – with ligands. Many inorganic solids (oxides and silicates) contain hydroxo groups; carbonates and sulfides expose $-C-OH$, $-C\underset{OH}{\overset{O}{\diagup}}$, MeOH and $-SH$ groups, respectively. While the interaction of alkaline and earth-alkaline ions with clays occurs mostly through ion exchange processes, the adsorption of heavy metals on clays is often dominated by surface complex formation with the single coordinated OH groups on the edge surfaces (aluminol and silanol groups). The surfaces of humic substances is characterized mostly by carboxyl and phenolic – OH groups (some imino and amino as well as some –SH groups may also be present). Biological surfaces contain –COOH, –NH$_2$ and –OH groups. Despite the diversity of these functional groups, they all have the characteristics of surface ligands able to bind protons and metal ions.

Much insight into the surface properties of natural particles is obtained by measuring H$^+$ ion binding as a function of pH, metal ion (or ligand) binding by the surfaces, as a function of pH. For an evaluation of equilibrium quotients and for a generalizable characterization of the pH-dependent interaction we need to determine the maximum protolyzable site concentrations (capacity of functional groups to bind protons) and the metal complexing capacity (and ligand complexing capacity) of the particle surfaces. The proton binding capacity is obtained most readily from alkalimetric titration curves. The complexing properties of metal ions can be determined by titrating the particles with metal ions at constant pH or quite expediently by titrating the metal ions at constant pH with surface ligands, i.e. particles that have been previously separated from the water. It is usually observed that the apparent binding constant may depend somewhat on the ratio of metal bound/surface area because the binding constant becomes – as with natural ligands in solution (Buffle and Altmann, 1987) – weaker when the strongest binding sites become limited. Electrophoretic measurements indicate that in the pH-range of natural waters most particles carry a negative charge; this charge can usually be accounted for by knowing the composition of the particle, its permanent charge, and the cations and anions that have become sorbed to (Chapter 3).

Model Studies. In model studies of adsorption, one deals with simple, well-defined systems, where usually a single well-characterized solid phase is used and the composition of the ionic medium is known, so that reactions competing with the adsorption may be predicted. It is not a trivial problem to compare the results from such model studies with those from field studies, or to use model results for the interpretation of field data. In field studies, a complex mixture of solid phases and dissolved components, whose composition is only poorly known, has to be considered; competitive reactions of major ions and trace metal ions for adsorption may take place, and the speciation of the trace metal ions is often poorly understood. In order to relate field studies to model studies, distribution coefficients of elements between the dissolved and solid phases are useful. These distribution coefficients are of the following form:

$$K_D = \frac{c_s}{c_w} \quad (\ell \, kg^{-1})$$

(11.1)

where c_s is the concentration in the solid particles (mol kg^{-1}) and c_w is the concentration in water (mol ℓ^{-1}).

The distribution coefficients are independent of the concentration of suspended solids in water, which can vary over a wide range; they thus give a better picture than the fraction of metal ions in solution. Such distribution coefficients can be predicted on the basis of the equilibrium constants defining the complexation of metals by surfaces and their complexation by solutes (Table 11.1).

Distribution coefficients based on adsorption equilibria are independent of the total concentrations of metal ions and suspended solids, as long as the metal concentrations are small compared with the concentration of surface groups. Examples of the K_D obtained from calculations for model surfaces are presented in Fig. 11.1. A strong pH dependence of these K_D values is observed. The pH range of natural lake and river waters $(7 - 8.5)$ is in a favorable range for the adsorption of metal ions on hydrous oxides.

Increasing concentrations of a soluble ligand cause a decrease in K_D for the simple case in which the complexation in solution and at the surface are competing with each other (see Example 11.1 and Fig. 11.4).

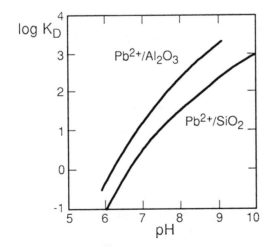

Figure 11.1

Distribution coefficients (ℓ kg^{-1}) calculated for surface complexation with $\equiv$AlOH and $\equiv$SiOH surface group. The following species were taken into account for Pb: Pb^{2+}, $PbOH^+$, $PbCO_3^0$, $Pb(CO_3)_2^{2-}$, $\equiv$Al–O–Pb$^+$, ($\equiv$AlO)$_2$Pb^0, and with SiO$_2$ ($\equiv$SiO)$_2$Pb^0 and $\equiv$SiO–Pb$^+$ {$\equiv$AlOH} = 0.25 mol kg^{-1} {$\equiv$SiOH} = 1.5 mol kg^{-1}.

(Modified from Sigg, 1987)

Table 11.1   Determination of the Distribution Coefficient K_D from Surface Complex Formation (From Schindler, 1984 and Sigg, 1987)

Species at surface: $\equiv S\text{–}O\text{–}M^-$, $(\equiv SO)_2M$

Species in solution: M^{2+}, MOH^-, $M(OH)_m^{(2-m)+}$, ML_1, ML_2, where L_1 and L_2 are known soluble ligands.

$$K_D = \frac{\{\equiv S\text{–}O\text{–}M^-\} + \{(\equiv S\text{–}O)_2M\}}{[M^{2+}] + [MOH^-] + [M(OH)_m] + [ML_1] + [ML_2]} \qquad \left(\frac{mol/kg}{mol/\ell^{-1}}\right)$$

$$K_D = \frac{K_M^s \{\equiv SOH\} / [H^+] + \beta_M^s \{\equiv SOH\}^2 / [H^+]^2}{1 + K_{OH}[OH^-] + \beta_{OH_m}[OH^-]^m + K_1[L] + \beta_2[L]^2} \qquad (\ell/kg)$$

where { } denotes concentration in mol/kg of solid phase.

$$K_M^s = \{\equiv S\text{–}O\text{–}M^+\} [H^+] / \{\equiv SOH\} [M^{2+}]$$
$$\beta_M^s = \{\equiv S\text{–}O)_2 M\} [H^+] / \{\equiv SOH\}^2 [M^{2+}]$$

The K^s values can be corrected for electrostatic effects

$$K_{OH} = [MOH^+] / [M^{2+}] [OH^-]; \qquad \beta_{OH_m} = [M(OH)_m] / [M^{2+}] [OH^-]^m$$
$$K_1 = [ML] / [M^{2+}] [L]; \qquad \beta_2 = [ML_2] / [M^{2+}] [L]^2$$

K_D depends on:
  – pH
  – kind of surface; number of OH groups per surface
  – complexation in solution

Natural Particles in Comparison to Oxides

Many naturally occurring particles – even organic particles, and surfaces of bacteria and algae – interact with metal ions in a similar way as oxides. In Fig. 11.2, the adsorption of Pb(II) on natural particles (isolated from a small river) is compared with that on goethite surfaces. Adsorption isotherms, at a given pH, are compared (data from Müller and Sigg, 1990). The approximate fit of the data of both adsorbents to a Langmuir isotherm (equivalent to a surface complex formation model, cf. Chapter 4.3) is evident from a plot of the reciprocal Langmuir equation.

To normalize the data in reference to the surface parameter, the quantity of mol Pb(II) sorbed is plotted per mol of functional groups present. These were determined experimentally from the adsorption capacity of Pb(II). Perhaps surprising is that the binding of Pb by natural particles at pH = 8 is not significantly different than that by goethite. The slightly stronger affinity of the surface of the natural suspended particles for Pb^{2+} than that of goethite can nearly be accounted for by the fact that the acidity constant of the particle OH-groups have a lower pK value (probably around 5) than the goethite OH-groups (pK around 6.7). The slight deviation of the goethite data from a Langmuir fit, is most likely due to electrostatic effects.

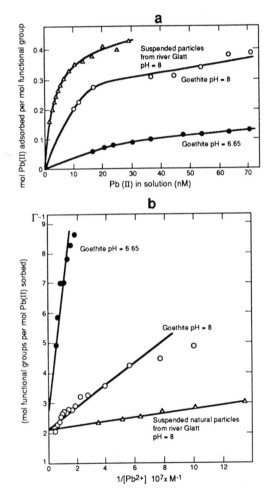

Figure 11.2

Adsorption of heavy metal ions to the surface of goethite and of natural particles

a) Surface complex formation of Pb^{2+} on goethite and the surface of natural particles (Glatt River). The data are interpreted in terms of Langmuir adsorption isotherms (Eq. 4.7).

b) The reciprocal of Eq. (4.7) is plotted (cf. Eq. (4.10c)). Slope and intercept permit to estimate $\Gamma_{M(max)}$ and K_{ads}. Data from Müller (1989) and Müller and Sigg (1990).

The comparison between goethite and natural particles at pH = 8 shows a slightly larger tendency of the natural particles to bind Pb^{2+} than of goethite.

11.3   Titrating Metal Ions with Particles

The extent of adsorption of reactive elements, e.g., metal ions, to particles can be readily determined by titrating a particle suspension or a sample of natural water containing particles with a metal ion (Fig. 11.3a) or to inverse the titration, i.e., to titrate a dilute standard metal solution with particles (Fig. 11.3b).

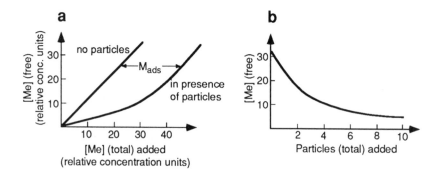

Figure 11.3

a) Schematic representation of the titration of a solution with a standard metal solution at a constant pH. The displacement of the curve (in comparison to a particle-free solution) by the particles reflects the adsorption isotherm and permits the determination of the extent of adsorption.
b) A given metal ion concentration is titrated with a particle suspension. The slope of this curve also reflects the adsorption isotherm and permits to determine the extent of adsorption of metal ions to the particles.

Analytical-operational Difficulties. In order to work close to the conditions in natural waters, very low concentrations of metal ions (in the nanomolar range) and of particles as well as pH values in the neutral range have to be used. Analytical difficulties occur because of undesired adsorption of metal ions to the experimental devices (walls of beakers, glass filtration devices, etc.)[1] and of insufficient separation of the particulate and dissolved phase (particles in the colloidal size range).

Two methodological approaches have been found useful:
1) The voltammetric determination of metal ions in the presence of particles are in principle able to differentiate without prefiltering the water sample between dissolved and labile species, i.e., the metal ions electrochemically available within the diffusion layer and (in addition to other non-labile complexes) those bound to particles and colloids (Gonçalves et al., 1985, 1987).

---

[1] Such metal ion adsorption effects become relatively significant – especially in very dilute solutions at pH-values above 7.

2) The adsorption of metal ions to the walls of the titration vessel can be better controlled by reversing the conventional titration procedure, i.e., instead of titrating the ligands (particles) with metal ions, a given concentration of metal ions is titrated with the ligands (in this case with incremental aliquots of the particles that were previously isolated from the river by centrifugation (Fig. 11.3b). The initial distribution of the metal ion between the walls of the titration vessel and the aqueous phase can be determined and a correction factor can be applied to the results of the subsequent titration. A further advantage of this approach is that a large range of low ratios of metal bound/(surface ligands) may be attained (Müller and Sigg, 1990).

Example 11.1: *Effect of soluble Ligands and of Particles on the Distribution of Zn(II) between particulate and soluble Phases*

In order to evaluate the two important variables which – in addition to pH – affect the residual concentrations of a metal ion, we use a simple equilibrium approach to assess the effect of these two variables. We assume a constant pH and characterize the effect of particle ligands, $\equiv L$, by the surface complex formation equilibrium.

$$K_{\equiv LZn} = [\equiv LZn] / [Zn^{2+}] [\equiv L] \qquad = 10^{8.2} \qquad \text{(i)}$$

and we characterize the equilibrium with the soluble complex former X as

$$K_{ZnX} = [ZnX] / [Zn^{2+}] [X] \qquad = 10^7 \qquad \text{(ii)}$$

For the calculation, we assume the following total concentrations:

$$[Zn_T] = [\equiv LZn] + [Zn^{2+}] + [ZnX] = 10^{-7} \text{ M} \qquad \text{(iii)}$$
$$[\equiv L_T] = [\equiv L] + [\equiv LZn] \qquad\qquad = 10^{-9} - 10^{-6} \text{ M} \qquad \text{(iv)}$$
$$[X_T] = [X] + [ZnX] \qquad\qquad\quad = 10^{-9} - 10^{-6} \text{ M} \qquad \text{(v)}$$

In order to demonstrate the effect we first "titrate" a 10^{-7} M Zn(II) solution in presence of $[X] = 5 \times 10^{-8}$ M (= constant) with particles. The results are given in Fig. 11.4a. Then we "titrate" the solution in the presence of a small (constant) concentration of particles with a soluble complex former X (Fig. 11.4b).

As a consequence of the addition of particles, soluble [Zn(II)] and free $[Zn^{2+}]$ decrease while the particle-borne Zn, $[\equiv LZn]$, increases. The addition of a soluble complex former increases soluble [Zn(II)] but decreases free $[Zn^{2+}]$; the particle-borne Zn $[\equiv LZn]$ decreases with increasing $[X_T]$.

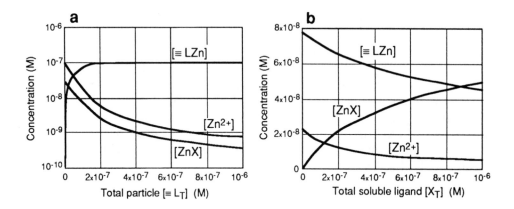

Figure 11.4

Role of Particles in regulating metal ions in rivers

a) In simple calculation (see Example 11.1) a 10^{-7} molar Zn solution is "titrated" with particles $[X_T] = 5 \times 10^{-8}$ M.

b) A 10^{-7} molar Zn-solution in presence of a small particle concentration $[\equiv L_T] = 10^{-7}$ M is "titrated" with a soluble complex former X.

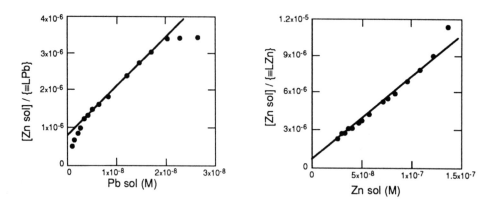

Figure 11.5

Representative examples of linearized titration curves of metal ions with natural particles (isolated from a river) plotted according to Eq. (11.5). Increments of a suspension of the particles were added to a solution of 1.53×10^{-7} M resp. 4.83×10^{-8} M Pb (ionic medium: 10^{-2} M $KNO_3/10^{-3}$ M HCO_3^-, purged with $N_2/0.06$ % CO_2, pH = 8). In the range 3 – 30 (Pb) and 3 – 60 (Zn) mg particles per liter, the experimental points fit a straight line of which the slope and intercept give values for the binding capacity $\equiv L_T$ and K_L. Deviation from this line are observed at low Me/surface ratio (low end of titration curve) where sites with higher affinity are effective, and at high Me/surface ratio, where saturation of the surface occurs.

(From Müller and Sigg, 1990)

Comparing Field Data with Laboratory Experiments. Can the distribution of metal ions between particulate and dissolved phases be explained in terms of simple interactions with surfaces on the basis of the obtained surface parameters? Can the differences in distribution (as reflected by the distribution coefficients) be explained by different binding properties of the particles?

In order to answer these questions, laboratory experiments were compared with field measurements in a local river (Müller and Sigg, 1990). Laboratory experiments with the centrifuged particulate matter, isolated from a river, were used. The results obtained by titrating a metal solution with a suspension of the centrifuged particles were interpreted in terms of binding capacities and conditional stability constants (Fig. 11.5).

Conditional stability constants of the type:

$$K_L = \frac{[\equiv LMe]}{[Me]\,[\equiv L]} \tag{11.2}$$

are derived at constant pH, together with corresponding binding capacities $\equiv L_T$. With:

$$[Me_T] = [\equiv LMe] + [Me_{sol}] = [\equiv LMe] + \alpha_1\,[Me^{n+}] = \text{constant} \tag{11.3}$$

and

$$[\equiv L_T] = [\equiv L] + [\equiv LMe] = \{\equiv L_T\} \times a \tag{11.4}$$

where $\{\equiv L_T\}$ means the total concentration of surface ligands in (mol/kg) of solid material, $[\equiv LMe]$ the concentration of metal bound to the surface ligands L expressed in (mol/ℓ), $[Me_{sol}]$ is the concentration of metal in solution, $\alpha_1 = [Me^{n+}]/[Me_{sol}]$ (α_1 can be calculated for a solution of known composition; in the simplest case only hydroxo and carbonato species are included); and a the concentration of suspended solids in (kg/ℓ). One may derive the following expression (Ruzic, 1987; Van den Berg, 1979):

$$\frac{[Me_{sol}]}{\{\equiv LMe\}} = \frac{\alpha_1}{K_L\{\equiv L_T\}} + \frac{[Me_{sol}]}{\{\equiv L_T\}} \tag{11.5}$$

which allows the calculation of $\equiv L_T$ and K_L from a linear plot of $\dfrac{[Me_{sol}]}{\{\equiv LMe\}}$ versus $[Me_{sol}]$.

This approach is limited to cases where one type of surface group dominates a substantial part of the experimental range and thus gives a straight line. The parameters characterize the solid material of varying composition and allow a comparison of different samples.

With these experimental surface reaction parameters and the speciation program MICROQL the species distribution in the river using metal and particle concentrations, pH and alkalinity measured in the river at the concerned sampling data was calculated. Various assumptions for the complexation in solution may be used, which affect α_1 and thus K_d.

It is surprising that data on natural particles can be fitted over a range of concentrations (representative of those encountered in natural waters) on the basis of a "single-site" surface complex formation model. Apparently similar types of binding groups are predominant and of importance in these particles.

11.4   Regulation of Dissolved Heavy Metals in Rivers

Table 11.2 gives some representative examples of dissolved trace metal concentrations in rivers, especially in large impolluted rivers.

The data show that rivers contain remarkably low concentrations (in some cases concentrations are as low as 10^{-11} M) of dissolved metal ions. (Many data that have been reported in the literature have been based on total particulate *and* dissolved concentrations; furthermore, analytical procedures have often not been able to discriminate against contamination during sampling and sample processing.) The metal concentrations in rivers are a consequence of
1) geochemistry of the rocks in the catchment area (metals released into the water by weathering);
2) anthropogenic pollution (by waste inputs and atmospheric deposition); and
3) river chemistry (adsorption of metal ions to particles and other surfaces and particle deposition into the sediments).

Windom et al. (1991) report that on the average 62, 40, 90, and 80 % of the Cd, Cu, Pb and Zn, respectively, carried by US east coast rivers is on particles. Similar conclusions were reached in an earlier study by Martin and Whitfield (1983). These high proportions of particulate metal ions are representative of large rivers that are often relatively unpolluted and that are characterized by high loads of turbidity (low ionic strength). In many small rivers with anthropogenic metal pollution and of low turbidities (calcareous drainage area) a significant fraction of metal ions may be present in dissolved form (cf. Fig. 11.7). The effect of particles on the residual concentrations is also apparent from the pH dependence. Fig. 11.6 shows an example of this pH dependence. (Similar data on the pH dependence of Cd and Pb were reported by Windom, 1991.) The experimental results show a decrease in [Zn] with increasing pH. Such a dependence is in a first approximation compatible with a reaction of the type

$$S{-}OH + Me^{2+} \rightleftharpoons S{-}OMe^+ + H^+ \tag{11.6}$$

The relatively simple interaction between metal ions and particles as described in Fig. 11.4 permits the application of this model (Eqs. 11.2 – 11.5) to make estimates on changes in distribution between particular and dissolved phase and in metal speciation as a function of variations in river chemistry (e.g., a change in turbidity, i.e., in particle concentration as a consequence of increased runoff, or a change in complex forming capacity by complex forming pollutants or a change in pH).

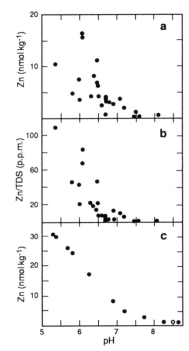

Figure 11.6

Dissolved zinc plotted against pH
a) Zinc in relatively undisturbed major rivers including the Yangtze (Chiang Jiang) and tributaries of the Amazon and Orinoco.
b) Zinc normalized to total dissolved solids for the same set of major rivers.
c) Zinc in pH-adjusted aliquots of Mississippi River water (April 1984, 103 mg ℓ^{-1} suspended load, pH 7.7); the adjusted aliquots were allowed to equilibrate overnight before filtration and analysis.
(From Shiller and Boyle, 1985)

Fig. 11.7 gives examples on the speciation of Pb(II) in a local river (Glatt, Switzerland) in presence of concentrations of EDTA and NTA (typically encountered in this river). EDTA (Ethylene diaminetetracetate) and NTA (nitrilo triacetate) are strong complex formers used in industry and households; they are typically found in receiving waters as pollutants.The data of Müller and Sigg (1990) for natural particles are very similar to those determined for marine sediment particles by Balistieri and Murray (1984). Johnson (1986) has demonstrated heavy metal regulation in a heavily polluted acid-mine drainage river. The concentration dependence could be accounted for by surface complexation with hydrous iron(III)oxide particles.

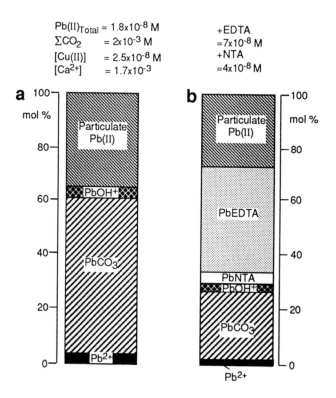

Natural River pH=8

Particles 3.5 mg/ℓ (=5x10^{-8} M sites)

$Pb(II)_{Total}$ = 1.8x10^{-8} M

ΣCO_2 = 2x10^{-3} M

$[Cu(II)]$ = 2.5x10^{-8} M

$[Ca^{2+}]$ = 1.7x10^{-3}

+EDTA

=7x10^{-8} M

+NTA

=4x10^{-8} M

Figure 11.7

Speciation of Pb(II) in Glatt river. The concentrations given for CO_2, Pb(II), Cu(II) and $[Ca^{2+}]$ as well as for the pollutants EDTA and NTA are representative of concentrations encountered in this river. The speciation is calculated from the surface complex formation constants determined with the particles of the river and the stability constants of the hydroxo-, carbonato-, NTA- and EDTA-complexes.The presence of $[Ca^{2+}]$ and $[Cu^{2+}]$ is considered.

(From Müller and Sigg, 1990)

Table 11.2   Examples of Dissolved Trace Metal Concentrations in Rivers
These data are based on advanced instrumentation and sampling methodology, paying attention to the elimination of contamination during sampling, storage, and analysis. See article by Windom et al. (1991) from where the data for this table are taken.

	Concentration nM				
	Cd	Cu	Zn	Pb	Ref
US East Coast Rivers	0.095	17	13.0	0.11	*a)*
Mississippi	0.12	23	3.0		*b)*
Yangtze	<0.01	18–21	0.6–1.2		*c)*
Amazon	0.06	24	0.3–3.8		*c)*
Orinow	0.035	19	2.0		*c)*

a) investigated mean of data from two sampling campaigns by Windom et al. (1991);
b) Shiller and Boyle, Nature *317*, 49–52 (1985);
c) data from Shiller and Boyle (1985), from Boyle et al. (1982) and Edmond et al. (1985).

11.5 Regulation of Trace Elements in Lakes

In Fig. 1.1 we already encountered a scheme of some of the chemical, biological, and physical processes which regulate the concentration of trace elements in the water column of lakes and oceans. When these trace elements are introduced into a lake by riverine and atmospheric input, they interact
1) with solutes (complex formation) and
2) with inorganic and organic (phytoplankton) particles (adsorption and assimilation).
The affinity of the reactive elements to the particles which settle through the water column, determines essentially the relative residence time of these elements and their residual concentrations and ultimate fate. (Fig. 11.8).

As shown in Table 11.3, the concentrations of trace elements in the water column is – despite anthropogenic pollution – extremely small ($10^{-11} - 10^{-7}$ M) illustrating the remarkable efficiency of the continuous "conveyor belt" of the settling adsorbing and scavenging particles. The *sedimentary record* reflects the accumulation of trace elements in sediments and a profile of concentration vs sediment-depth (or age) gives a "memory record" on the loading in the past (Fig. 11.9).

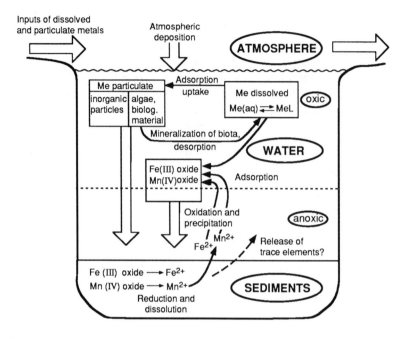

Figure 11.8

Schematic representation of the cycling of trace elements in a lake. Trace elements are removed to the sediments together with settling material, which consists for a large part of biological material.

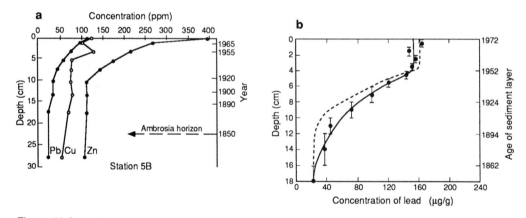

Figure 11.9

Depth-concentration profiles in sediment cores

a) Pb, Cu, Zn in Lake Erie (from Nriagu, Kempe, Wong and Harper, 1979);

b) Pb in Lake Michigan. The dashed line reflects a model fit assuming contributions from gasoline only (from Edgington and Robbins, 1976).

Table 11.3   Examples of concentrations of trace elements in different lakes[1]

Lakes		Cu nM	Zn nM	Cd pM	Pb nM
Lake Zurich	a)	6 − 12	5 − 45	40 − 100	0.05 − 1
Lake Constance	b)	5 − 20	15 − 60	50 − 100	0.2 − 0.5
Lake Michigan	c)	10	9	170	0.25
Lake Erie	d)	9 − 13	0.4 − 0.8	63 − 97	0.008 − 0.15
Lake Ontario	d)	11 − 16	0.04 − 1.7	6 − 83	0.004 − 1.37
Southern Sweden	e)	5 − 15	60 − 382	45 − 1070	0.97 − 5.30
Northern Sweden	e)	4 − 42	<30 − 305	35 − 804	1.45 − 5.30
Lake Baikal	f)	2.4 − 3.9	3.1 − 8.5	22 − 130	−

a) Sigg et al. (1987)
b) Sigg et al. (1983)
c) Shafer and Armstrong (1990)
d) Coale and Flegal (1989)
e) Borg (1983)
f) Falkner et al., (1991)

The data in Lake Erie, Lake Ontario and of the Swedish lakes were measured only in surface waters. The swedish lakes are in crystalline area and have somewhat lower pH values. In these lakes and in Lake Baikal there is no $CaCO_3$ precipitation.

[1] Modified from Sigg (1992)

Role of Settling Particles

Both biogenic organic particles (algae, biological debris) and inorganic particles (e.g., manganese and iron oxides) (cf. Table 11.4) contribute to the binding and transport of reactive elements. The photosynthetic production of algae and their sedimentation is a dominant process, especially in eutrophic lakes. Near the sediment-water interface, anoxic conditions may occur, under which iron and manganese oxides undergo reduction and dissolution. As discussed in Chapter 9, trace elements are affected by these processes in different ways (interaction with iron and manganese oxides, precipitation as and complexation with sulfides).

Table 11.5 shows that sedimentation rates of $0.1 - 2$ g m^{-2} d^{-1} are typically observed in lakes; still higher values are found in very eutrophic lakes. The settling material can be collected in sediment traps; it can then be characterized in terms of chemical composition, morphology, and size distribution of the particles. The composition is subject to seasonal variations caused primarily by different biological activities in the various seasons. Representative examples for Lakes Zurich and Constance are given in Fig. 11.10. These two lakes are prealpine lakes, located in regions of predominantly calcareous rocks, both are under the influence of eutrophication.

Table 11.4 The Role of Settling Particles in Regulating Trace Elements in Lakes

Components of Settling Particles	Characteristics
– *Phytoplankton and biological debris*	– Surface of organisms has strong affinity for heavy metals such as Cu(II), Pb(II), Zn(II), Cd(II), Ni(II) (surface complex formation).
	– Organisms also absorb (assimilate) nutrients (P, N, Si, S etc.) and nutrient metal ions (e.g., Cu(II), Zn(II), Co(II)) and metal ions that are mistaken (by the organisms) as nutrients (e.g., Cd(II), As(V)).
	– Phytoplankton is mineralized in water column and sediments.
– *CaCO₃* (usually precipitated within the lake)	– Heavy metals and phosphate are adsorbed and may become incorporated; CaCO₃ crystals are usually large (d > 5 μm); because small specific surface area (in comparison to other settling material) the effect of CaCO₃ on overall removal of trace elements is small.
– *Fe(III)(hydr)oxides* introduced into the lake and formed within the lake	– Strong affinity (surface complex formation) for heavy metals, phosphates, silicates and oxyanions of As, Se; Fe(III) oxides even if present in small proportions can exert significant removal of trace elements.
	– At the oxic-anoxic boundary of a lake (see Chapter 9.6) Fe(III) oxides may represent a large part of settling particles. Internal cycling of Fe by reductive dissolution and by oxidation-precipitation is coupled to the cycling of metal ions as discussed in Chapter 9.
– *Mn(III,IV) oxides* mostly formed within the lake	– High affinity for heavy metals and high specific surface area. Redox cycling $[MnO_x(s) \rightleftarrows Mn^{2+}(aq)]$ (Chapter 9.6) is usually important in regulating trace element concentrations and transformations in lower portion of lake and sediments.
– *Aluminum silicates* clays, oxides	– Ion exchange; binding of phosphates and metal ions; unless present in large concentrations, overall effect on trace element removal is small.

Table 11.5   Sedimentation rates in different lakes

The ratios Zn/C and Cu/C are given in order to compare the composition of the settling particles in different systems

Lakes	Sed. rate g m^{-2} d^{-1}	C mmol m^{-2} d^{-1}	Fe mmol m^{-2} d^{-1}	Zn μmol m^{-2} d^{-1}	Cu μmol m^{-2} d^{-1}	Pb μmol m^{-2} d^{-1}	Cd μmol m^{-2} d^{-1}	Zn/C μmol/mol	Cu/C μmol/mol	Ref
Constance	2		0.41	5.2	0.8	0.4		530	71	a)
Zurich	2.5	12.5	0.34	5.8	1.2	0.7	0.027	466	95	a)
Michigan	0.48	2.3	0.05	1.3	0.4	0.1	0.007	565	155	b)
Windermere	2.8	28.6	1.29	14.9	12.3	2.3		526	430	c)
Kejimukjik	1.2	13.9		2.9	1.1	1.6		209	80	d)
Mountain	0.45	9.3		2.1	0.7	0.1		230	78	d)

(From Sigg, 1992)

a) Sigg, 1987
b) Shafer and Armstrong, 1990
c) Hamilton-Taylor et al., 1986
d) Nriagu et al., 1989

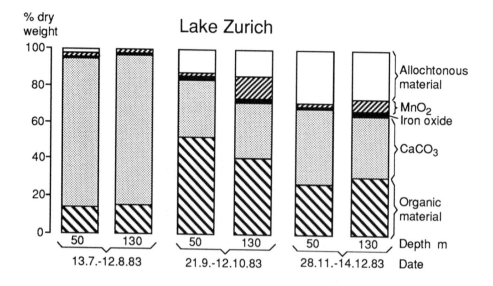

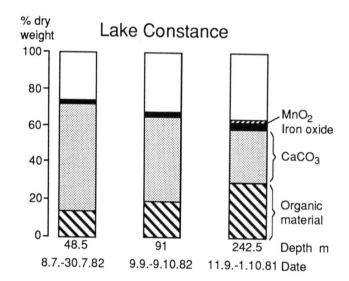

Figure 11.10

Some examples of the composition of settling particles from Lake Zurich and Lake Constance and of their variations with time and depth (% dry weight of organic material, calcium carbonate, iron and manganese oxide). Sediment traps in Lake Zurich were exposed at 50 m, below the productive layer, and at 130 m, that means ~5 meters above the sediment-water interface (maximum depth ≈ 135 m). In Lake Constance, the maximum depth is ~250 m. Settling particles shown here were collected during the productive summer stagnation and at the end of stagnation.

(From Sigg, 1987)

The Binding of Metals to the Settling Particles. The partition of metals between particles and the water depends
1) on the affinity of the metal ions for the solute ligands (complex formation);
2) the affinity for the surfaces of the settling particles, and
3) the extent of uptake by the biota.
Some of the characteristics of the various components of the settling particles and their surfaces is discussed in Table 11.4. The individual components of particles usually do not settle separately; they agglomerate (coagulation) to heterogeneous particles. Algae, calcite crystals and Fe(III)(hydr)oxides assist the agglomeration process so that typically the aggregates form relatively large and relatively rapidly settling agglomerates (see Chapter 7.7). It still is reasonable to assume that the binding and incorporation of metal ions is fast in comparison to the settling.

Since the overall component composition of the settling material (proportion of algae, $CaCO_3$, aluminum silicates etc.) changes seasonally, a corresponding change in element and metal composition of the particles provides clues on the affinity of the various components for metal ions. Investigations on such correlations document the important role of settling phytoplankton in playing a dominant role in the transport of metal ions.

Phytoplankton

Organic material can strongly adsorb metal ions; the functional groups on their surfaces act as ligands (carboxyl, amino groups etc.) for metal ions. All these functional groups favor the surface complex formation with metals; the adsorption reactions are favored at higher pH (Fig. 11.11).

In addition to *ad*sorption processes, phytoplankton can *ab*sorb (assimilate) certain nutrient metal ions (or metal ions that are by the organisms mistaken as nutrients). As with other nutrients, this uptake can occur in stoichiometric proportions. The uptake (and subsequent release upon mineralization) of nutrients in stoichiometric proportions was claimed already 1934 by Redfield. In referring to the atomic proportions C : N : P : Si etc. one refers to the *Redfield Ratios*. This stoichiometry is well established (at least for the conventional nutrients) in oceanic waters; it has also been postulated for lakes (Stumm and Morgan, 1970).

Organic Matter in the settling material originates mostly from phytoplankton. The chemical composition can be compared to the Redfield stoichiometry of algae $\langle(CH_2O)_{106}(NH_3)_{16}(H_3PO_4)\rangle$. In particles of Lake Constance (collected mostly during the summer) a mean composition $C_{113}N_{15}P_1$ was found while the particles from Lake Zurich (collected over a whole year had a mean ratio of C : P of 97 : 1 (N was not measured) (Sigg et al., 1987). In these lakes 15 – 40 weight-% of the settling particles consists of organic matter. This fraction varies during the year due to seasonal variations of primary productivity.

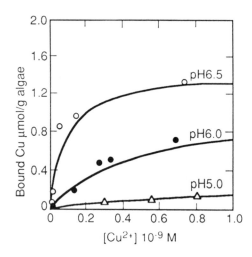

Figure 11.11

Cu(II) binding isotherms for algal surfaces at different pHs. Bound Cu was determind by AAS measurements of extracts of the algae after reaction with Cu at given pH; the reaction was carried out in a
suspension containing 0.01 M KNO_3, $1.89 - 2 \times 10^{-5}$ M NTA, $0.1 - 1.8 \times 10^{-5}$ M TCu and 75 –120 mg
ℓ^{-1} algae dry wt. All isotherms follow a simple Langmuir equation at low coverage. The resulting bind-
ing constants and capacities are log K 8.4, 9.1 and 10.0, Γ_{max} 9×10^{-7}, 1.7×10^{-6} and 1.4×10^{-6} mol g^{-1} for pH 5.0, 6.0 and 6.5, respectively.

(From Xue and Sigg, 1990)

Settling Biota as a major Carrier of heavy Metals. Among the components of settling material mentioned above, several lines of evidence point to the importance of
biological material as a major carrier of trace metal ions. The binding of metal ions
to surface ligands also represents a first step in the uptake of metal ions into the
organisms (Fig. 11.12). Since organisms require a number of essential trace elements, such as Cu and Zn, it may be expected that these elements will be bound in
certain ratios by algae similar to the Redfield ratio for C, N, P. If these elements are
mostly bound to biological material in the settling particles, the ratios found in these
particles should correspond to such a Redfield ratio.

The ratios of different elements to C are listed in Table 11.5 as an attempt to test
this assumption in various systems. For Lake Zurich and Lake Constance the
following tentative ratio was calculated:

$$(CH_2O)_{97}(NH_3)_{16}(H_3PO_4)_1 Cu_{0.006}Zn_{0.03}.$$

In Lake Zurich and Lake Constance, correlations between the contents of different
trace elements in the settling particles and phosphorus, which may be used as an
indicator of biological material, indicate that especially copper and zinc were likely

to be associated with biological material (Fig.11.13). In Lake Zurich, the highest sedimentation fluxes of Cu and Zn were observed during summer, simultaneously with the sediment fluxes of organic carbon and phosphorus (Sigg et al., 1987). Similar tendencies were also found for Cd and Pb, but the data were more scattered, due possibly to the tendency of these elements to adsorb to different types of particles. In Lake Michigan (Shafer and Armstrong, 1990) diatoms appear to represent an important transport phase for several elements; these data listed in Table 11.5 indicate similar ratios of Cu and Zn to C in the settling particles as in Lake Zurich and Constance. An earlier study on seston in Lake Michigan (Parker et al., 1982) also pointed out the role of phytoplankton for the sedimentation of Zn and Cd. The data on Windermere (Hamilton-Taylor et al., 1984) indicate the uptake of Cu by diatoms and the role of their rapid decomposition, although in this case the inputs of detrital material are also important. Data from Mountain Lake (Nriagu et al., 1989), a small acidic lake, also show evidence for the role of biological material, which constitutes a significant fraction of the total settling particles; the somewhat lower ratios of Zn and Cu to C may in this case be related to the lower pH.

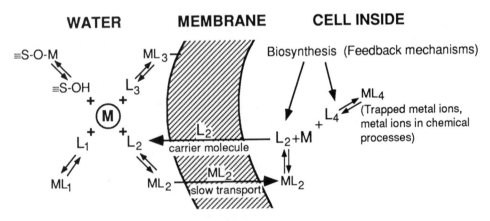

Figure 11.12

Uptake model

In a simplified model, the metal ions equilibrate on the outside of the cell with biologically produced and excreted ligands L_2 or ligands on the cell surface L_3; these reactions are followed by a slow transport step to the inside of the cell. In the cell, the metal ions may be used in biochemical processes or become trapped in inactive forms as a detoxification mechanism. (After Williams, 1983) (cf. Fig. 4.15a).

(From Sigg, 1987)

Mn- and Fe-Oxides

The freshly precipitated manganese and iron oxides which precipitate at an oxic-anoxic boundary within the lake water column or in the top layers of the sediment, form small particles with high surface area; they cause an additional scavenging at

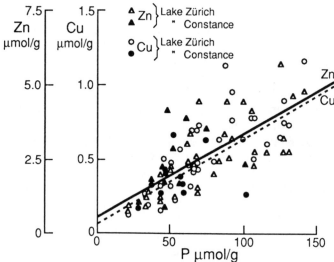

Figure 11.13

Concentrations of Cu and Zn as a function of P in the settling material from Lake Zurich and Lake Constance. P serves as an indicator of biological material. The regression lines for Zn and Cu fall nearly together, indicating that rather constant ratios Cu : Zn : P are found in this material.
(From Sigg et al., 1987)

the depth of the oxic-anoxic boundary. Data from Lake Zurich (Sigg et al., 1987) suggested such a mechanism, since the concentrations of several trace elements were higher in the deeper sediment traps, which were located below the depth of the manganese oxide precipitation and in which manganese oxide accumulated. In other systems, such as more oligotrophic and acidic lakes (Tessier et al., 1989), iron oxides have been shown to be important binding phases for metal ions in the sediments. Binding constants derived from the sediment data were compared to surface complexation constants obtained with single iron hydroxide phases.

Residual Concentrations

Table 11.3 illustrates some data. They are mostly from lakes with high sedimentary fluxes and illustrate that very low concentrations are observed in lakes, in spite of large pollutional inputs into these lakes. The residual concentrations are close to those observed in the open ocean (Bruland, 1983). Even in lakes which are located closer to pollution sources, like Lake Zurich, the concentrations in the water column are at a similarly low level as in the Great Lakes.

The higher concentrations of different trace metals in the Swedish lakes are likely to be related to the lower pH range. Within the two sets of data, increasing concentrations of Zn, Pb and Cd are correlated with decreasing pH. A similar effect was observed for Cd in different lakes from Central Ontario (Borg, 1983). These pH-effects may be understood on the basis of a decrease in the binding of metal ions to particles with decreasing pH.

Steady State Models

Simple steady-state models may be used in order to relate quantitatively the mean concentration in the lake water column and the residence time of metal ions to the removal rate by sedimentation (for a detailed treatment of lake models see Imboden and Schwarzenbach, 1985). In a simple steady-state model, the inputs to the lake equal the removal by sedimentation and by outflow; the water column is considered as fully mixed; mean concentrations and residence times in the water column can be derived from the measured sedimentation fluxes. The binding of metals to the particles is fast in comparison to the settling.

The quantitative relationships and an example are summarized in Table 11.6. Under the assumption of steady-state, the residence time of an element (τ_M) that is removed from the water column by sedimentation and by outflow is given by

$$\tau_M^{-1} = \tau_w^{-1} + \tau_s^{-1}$$

where τ_w is the residence time of water in the lake and τ_s is the residence time of an element with respect to sedimentation. The removal rate by sedimentation τ_s^{-1} can be expressed as: $\tau_s^{-1} = f_p \times k_s$, where f_p is the fraction of the element in particulate form and k_s (d^{-1}) is the rate constant characterizing sedimentation. The fraction of an element in the particulate phase relates to its tendency to bind to particles and depends on the partition coefficient of the element between particulate phase and solution (dependent on the chemical interactions with the particles and in solution), and on the concentrations of particles in the water column. The removal rate by sedimentation can be calculated from the flux of an element to the sediments and its total amount in the water column (Table 11.6). The removal rate by sedimentation is large if both the fraction of an element bound to particles and the sedimentation rate are significant, the residence time of an element in the water column is much smaller than the water residence time. The mean concentration in the water column can be shown to depend on the removal rate by sedimentation and the water residence time according to Eq. (vi) Table 11.6. This means that if the removal rate by sedimentation is high, the mean concentration in the water column turns out to be much lower than the input concentration. Low concentrations of metal ions in the water column can thus be expected to occur if a metal ion binds to a significant extent to particles and if the sedimentation rate is high. The chemical factors which determine the binding to particles (pH, surface ligands, ligands in solution, which have been discussed in Chapters 2 – 4 and the chemical factors such as Ca^{2+} and humic acid concentrations that influence coagulation and, in turn, sedimentation rates, as discussed in Chapter 7.7, affect the distribution of a metal ion between particulate phase and solution and thus the fraction in particulate form

Table 11.6 Removal of Metal Ions from Lake Water Column by Sedimentation (from Sigg, 1992)

Residence time of element M: $\qquad \tau_M^{-1} = \tau_w^{-1} + \tau_s^{-1}$ (i)

where τ_M = residence time of element M [d]
τ_w = residence time of water [d]
τ_s = residence time of element with respect to sedimentation [d]

Removal rate by sedimentation (τ_s^{-1}): $\qquad \tau_s^{-1} = f_p \times k_s = \dfrac{F_M}{h \times [M]_w}$ (ii)

where f_p = fraction of element bound to particles [−]
F_M = sedimentation rate of element M [mol m^{-2} d^{-1}]
$[M]_w$ = concentration of element M in water [mol m^{-3}]
k_s = removal rate of particles [d^{-1}]
h = mean depth of water colum [m]

Example: *Pb in Lake Zurich, summer*

Sedimentation rate of Pb: $\qquad F_{M(Pb)} = 1 \; \mu mol \; m^{-2} \; d^{-1}$

Mean concentration in water column: $[Pb]_w = 8 \cdot 10^{-4} \; \mu M = 0.8 \; \mu mol \; m^{-3}$

Removal rate: $\qquad \dfrac{F_M}{h \times [Pb]_w} = 2.5 \times 10^{-2} \; d^{-1}$

Residence time of water: $\qquad \tau_w = 400 \; d$

With $\tau_w^{-1} = 2.5 \times 10^{-3} \; d^{-1}$:

residence time of Pb: $\qquad \tau_M = 36 \; d$

Steady-state concentration: $[Pb]_w \qquad = \dfrac{\tau_w^{-1}}{\tau_w^{-1} + \tau_s^{-1}} \times c_{in} = 0.09 \; c_{in}$ (iii)

where c_{in} = input concentration

f_p. For a single element under similar chemical conditions in different lakes, the residence times and mean concentrations depend on the sedimentation rates.

Simple steady-state models can only predict mean concentrations. Seasonal variations and concentration depth profiles in the water column of lakes give further insight into the mechanisms governing the removal of metal ions. Data on depth concentration profiles of trace metals in lakes are however still scarce (Sigg, 1985; Sigg et al., 1983; Murray, 1987). In a similar way as in the oceans, it might be expected to observe in lakes different types of profiles for different elements, depending on the predominant removal mechanism (Murray, 1987; Whitfield and Turner, 1987).

11.6   Oceans

The same processes that were discussed for lakes are operative in oceans. Antochtonous particles, above all phytoplankton and other organisms make up the continuous conveyor belt of settling materials. The interaction with biota is reciprocal. Nutrient metal ions coregulate the growth of phytoplankton, but the phytoplankton community has a pronounced effect on the metal ions and their speciation. The bulk of the trace elements, adsorbed most efficiently to the biological surfaces in the surface layers is carried into the deep sea by settling; as the particles fall through the water column, they provide a food source for successive populations of filter feeders so that the material is then repackaged many times en route to the sediment (Whitfield and Turner, 1987).

Large spatial and temporal differences in both trace metal concentrations and chemical speciation in the sea have led to wide variations in biological availability of metals and their effects on phytoplankton (Sunda, 1991). As Sunda points out, trace metals are usually taken up by algae via the formation of coordination complexes with specialized transport ligands in their outer-membranes (similar to the situation given in Fig. 11.12), and metal uptake is determined by the interplay between redox, complexation, or oxide dissolution reactions of metals in seawater and ligand exchange reactions at these sites. Some metals, such as Cu(II) and Zn(II) are heavily chelated by organic ligands in seawater; their biological availability is determined by the concentration of free metal ions.

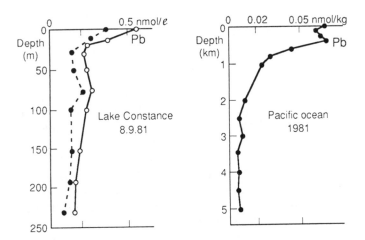

Figure 11.14

Lead profiles in Lake Constance (Summer 81 data: Sigg, 1985) and in the Pacific Ocean (1981 data: Schaule and Patterson, 1981). The similar shape of these profiles despite the difference in length scales (kilometers for the ocean and meters for the lake), illustrates the influence of the atmospheric deposition on the upper layers and the scavenging of Pb(II) by the settling particles.
(Modified from Sigg, 1985)

In Fig. 11.14 the depth profile of the Pb-concentrations of the Central Pacific is compared with that of Lake Constance. In either case, the Pb concentrations of the surface waters are higher than in the deep water; atmospheric transport plays in both cases a significant role in supplying Pb to the surface water. The decrease in the concentration of Pb with depth occurs by particles that scavenge Pb(II) most efficiently. Patterson (e.g., Settle and Patterson, 1980) has used data on the memory record of sediments to compare prehistoric and present-day eolian inputs. These data suggest that the present Pb(II) input is two orders of magnitude larger than that of prehistoric time.

As in lakes, other potential scavenging and metal regeneration cycles operate near the sediment-water interface. Subsequent to early epidiagenesis in the partially anoxic sediments, iron(II) and manganese(II) and other elements depending on redox conditions, are released by diffusion from the sediments to the overlying water, where iron and manganese are oxidized to insoluble iron(III) and manganese(III,IV) oxides. These oxides are also important conveyors of heavy metals near the sediment surface.

Whitfield (1979) and Whitfield and Turner (1987) have shown that the elements in the ocean can be classified according to their oceanic residence times, τ_i:

$$\tau_i = \frac{\text{total number of moles in i in ocean}}{\text{rate of addition or removal (mol time}^{-1})}$$

which are, in turn, a measure of the intensity of their particle-water interaction. Thus, the elements that show the strongest interactions with the particulate phase have very short residence times; those elements that interact little with particles are characterized by long residence times (Fig. 11.15).

As mentioned, the type of concentration-depth profiles observed in oceans should also be observed in lakes. However, the vertical concentration differences in lakes are often not as pronounced as in the ocean. The reason for this is, that the water column in lakes is much shorter; mixing and stagnation in lakes is much more dynamic than in the oceans. Due to the presence of high concentrations of different particles in lakes, the release of trace elements from biogenic particles may not be clearly observed, due to readsorption to other particles. This would mean that low concentrations are observed throughout the water column, but that concentration differences are small. Atmospheric inputs to the upper water layers may also make it more difficult to observe a depletion of certain elements in the epilimnion.

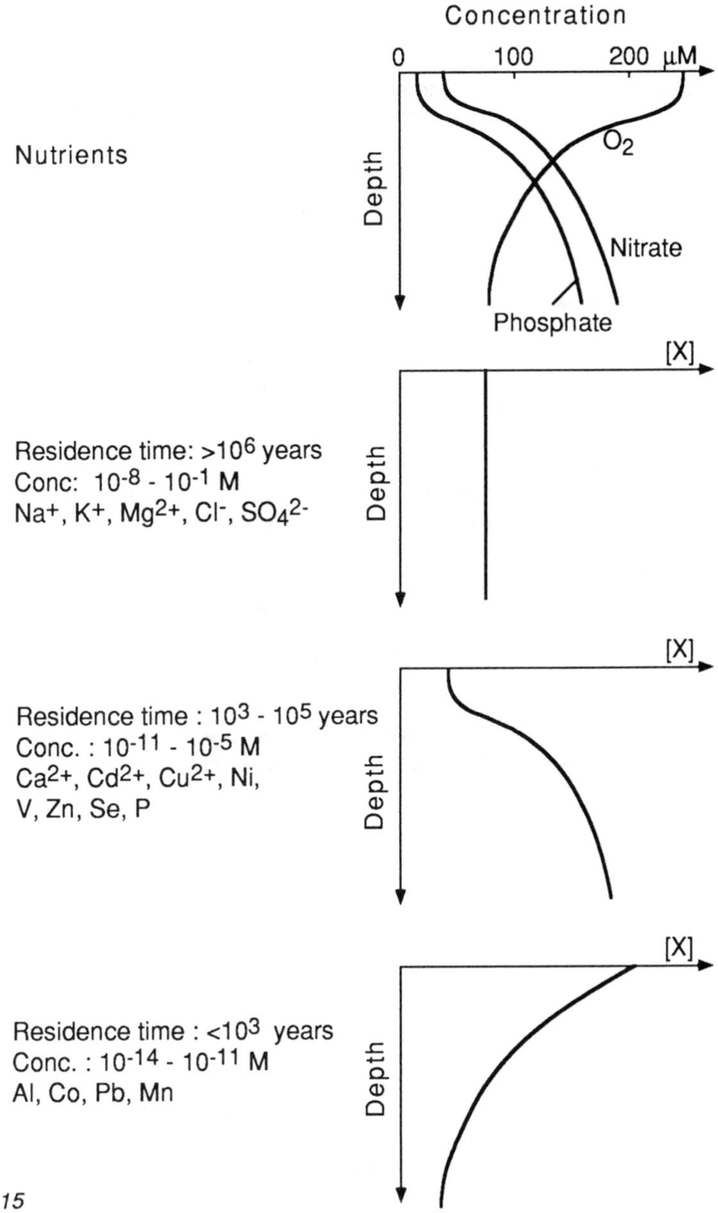

Figure 11.15

Schematic depth ocean profiles for elements. This figure is based on a classification of elements according to their oceanic profiles given by Whitfield and Turner (1987). Uptake of some of the elements, especially the recycled ones, occurs somewhat analogously as that of nutrients. There are some elements such as Cd that are non-essential but may be taken up (perhaps because they mimick essential elements) the same way as nutrients. The concentration ranges given show significant overlap, since the concentrations of the elements also depend on crustal abundance.

(Modified from Whitfield and Turner, 1987)

Reading Suggestions

Broecker, W.S. (1974), "Chemical Oceanography", expecially Chapters 1 (Internal Cycling and Throughput) and 2 (The Sedimentary Imprint), Hartcourt Brace Jovanovich, New York.

Burton, J.D. and P.J. Stakam, (1990), "Trace metals as tracers in the ocean", in H. Charnock, J.E. Lovelock, P.S. Liss and M. Whitfield, Eds., *Tracers in the Ocean*, Princeton University Press, Princeton, N.J.

Morel F.M.M. and R.J.M. Hudson (1985), "The geobiological cycle of trace elements in aquatic systems: Redfield revisited", in W. Stumm, Ed., *Chemical Processes in Lakes*, Wiley-Interscience, New York.

Murray, J.W. (1987), "Mechanisms controlling the distribution of trace elements in oceans and lakes", in R.A. Hites and S.J. Eisenreich, Eds., *Sources and Fates of Aquatic Pollutants*, Adv. Chem. Ser. 216.

Sigg L. (1985), "Metal Transfer Mechanisms in Lakes; the Role of Settling Particles", in W. Stumm, Ed., *Chemical Processes in Lakes*, Wiley-Interscience, New York.

Sigg, L. (1987), "Surface Chemical Aspects of the Distribution and Fate of Metal Ions in Lakes", in W. Stumm, Ed., *Aquatic Surface Chemistry*, Wiley-Interscience, New York.

Sunda, W.G. (1991), "Trace Metal Interactions with Marine Phytoplankton". Biol. Oceanography **6**, 411-442.

Whitfield, M. and D.R. Turner (1987), "The Role of Particles in Regulating the Composition of Seawater", in W. Stumm, Ed., *Aquatic Surface Chemistry*, Wiley-Interscience, New York.

References

Adamson, A. W. (1990), *Physical Chemistry of Surfaces.* 5th Edition, John Wiley and Sons, New York.

Aiken, G. R., D. M. McKnight, R. L. Wershaw, and P. MacCarthy, Eds. (1985), *Humic Substances in Soil, Sediment and Water. Geochemistry, Isolation and Characterization*, John Wiley and Sons, New York, pp. 692.

Al-Ekabi, H., N. Serpone, E. Pelizzetti, C. Minero, M. A. Fox, and R. B. Draper (1989), "Kinetic Studies in Heterogeneous Photocatalysis. 2. TiO_2-Mediated Degradation of 4-Chlorophenol Alone and in a 2,4-Dichlorophenol, and 2,4,5-Trichlorophenol in Air-Equilibrated Aqueous Media," *Langmuir* **5**, 250-255.

Alexander, G. B., W. M. Heston, and R. K. Iler (1954), "The Solubility of Amorphous Silica in Water," *J. Phys. Chem.* **58**, 453-455.

Amrhein, C., and D. L. Suarez (1988), "The Use of a Surface Complexation Model to Describe the Kinetics of Ligand-Promoted Dissolution of Anorthite", *Geochim. Cosmochim. Acta* **52**, 2795-2807.

Anderson, M. A., and A. J. Rubin, Eds. (1981), *Adsorption of Inorganics at Solid-Liquid Interfaces*, Ann Arbor Science Publ., 357 pp.

Archer, M. D., and J. R. Bolton (1992) *Photoconversion of Solar Energy*, to be published by John Wiley and Sons, New York.

Arends, J., J. Christoffersen, M. R. Christoffersen, H. Eckert, B. O. Fowler, J. C. Heughebaert, G. H. Nancollas, J. P. Yesinowski, and S. J. Zawacki (1987), "A Calcium Hydroxyapatite Precipitated from an Aqueous Solution: An International Multimethod Analysis", *J. Crystal Growth* **84**, 515-532.

Arnold, R. G., T. J. di Christina, and M. R. Hoffmann (1986), "Dissimilative Fe(III) Reduction by Pseudomonas sp. 200 – Inhibitor Studies", *Appl. Environ. Microbiol.* **52**, 281-294.

Ballistrieri, L. S., and J. W. Murray (1983), "Metal-Solid Interactions in the Marine Environment: Estimating Apparent Equilibrium Binding Constants", *Geochim. Cosmochim. Acta* **47**, 1091-1098.

Balzani, V., L. Moggi, and F. Scandola (1987), "Towards a Supramolecular Photochemistry: Assembly of Molecular Components to Obtain Photochemical Molecular Devices", in V. Balzani, Ed., *Supramolecular Photochemistry*, NATO ASI Series, Ser. C, Vol. 214, D. Reidel, Dordrecht, Holland.

Bard, A. J. (1980), *Electrochemical Methods; Fundamentals and Application*, chapter 14, John Wiley and Sons, New York.

Bartschat, B. M., and B. Sulzberger, in preparation.

Baur, M. E. (1978), "Thermodynamics of Heterogeneous Iron-Carbon Systems: Implications for the Terrestrial Promotive Reducing Atmosphere", *Chemical Geology* **22**, 189-206.

Beer, R., G. Calzaferri, J. Li, and B. Waldeck (1991), "Towards Artificial Photosynthesis; Experiments with Silver Zeolites", Part 2, *Coord. Chem. Rev.*, in press.

Behra, P., and L. Sigg (1990), "Evidence for Redox Cycling of Iron in Atmospheric Water", *Nature* **344**, 419-421.

Bennett, P. C. (1991), "Quartz Dissolution in Organic-Rich Aqueous Systems", *Geochim. Cosmochim. Acta* **55**, 1781-1797.

Berner, R. A. (1980), *Early Diagenesis*, Princeton University Press, Princeton N.Y., 241 pp.

Berner, R. A. (1986), "Kinetic Approach to Chemical Diagenesis", in F. A. Mumpton, Ed., *Studies in Diagenesis*, U.S. Geol. Surv. 1598, Washington, 13-20.

Berner, R. A., A. C. Lasaga, and R. M. Garrels (1983), "The Carbonate-Silicate Geochemical Cycle and its Effect on Atmospheric Carbon Dioxide Over the Past 100 Million Years", *Amer. J. Sci.* **283**, 641-683.

Berner, R. A., and A. C. Lasaga (1989), "Modeling the Geochemical Carbon Cycle", *Scientific Amer.* **260**, 74-81.

Berner, R. A., and J. W. Morse (1974), "Dissolution Kinetics of Calcium Carbonate in Seawater: IV. Theory of Calcite Dissolution", *Am. J. Sci.* **274**, 108-134.

Bischoff, W. D., F. T. Mackenzie, and F. C. Bishop (1987), "Stabilities of Synthetic Magnesian Calcites in Aqueous Solution: Comparison with Biogenic Materials," *Geochim. Cosmochim. Acta* **51**, 1413-1423.

Blesa, M. A., H. A. Marinovich, E. C. Baumgartner, and A. J. G. Maroto (1987), "Mechanism of Dissolution of Magnetite by Oxalic Acid-Ferrous Ion Solutions", *Inorg. Chem.* **26**, 3713-3717.

Blum, A. E., and A. C. Lasaga (1987), "Monte Carlo Simulations of Surface Reaction Rate Laws", in W. Stumm, Ed., *Aquatic Surface Chemistry*, John Wiley and Sons, New York, 255-292.

Bodine, M. W., H. D. Holland, and M. Borcsik (1965), "Coprecipitation of Manganese and Strontium with Calcite. Symposium: Problems of Postmagmatic Ore Deposition", *Prague* **2**, 401-406.

Borg, H. (1983), "Trace Metals in Swedish Natural Fresh Waters", *Hydrobiol.* **101**, 27-34.

Borkovec, M., B. Buchter, H. Sticher, P. Behra, and M. Sardin (1991), "Chromatographic Methods and Transport of Chemicals in Soils", *Chimia* **45**, 221-227.

Boyle, E. A., S. S. Huested, and B. Grant (1982), "The Chemical Mass Balance of the Amazon Plume – II. Copper, Nickel, and Cadmium", *Deep-Sea Res.* **29**, 1355-1364.

Brady, P. V., and J. V. Walther (1989), "Controls on Silicate Dissolution Rates in Neutral and Basic pH Solutions at 25° C", *Geochim. Cosmochim. Acta* **53**, 2823-2830.

Brady, P. V., and J. V. Walther (1991), "Kinetics of Quartz Dissolution at Low Temperatures", *Chem. Geol.,* in press .

Brady, P. V., and J. V. Walther (1992), in preparation.

Braterman, P. S., A. G. Cairns-Smith, and R. W. Sloper (1984), "Photo-Oxidation of Iron(II) in Water Between pH 7.5 and 4.0", *J. Chem Soc. Dalton Trans.*, 1441-1445.

Broecker, W. S. (1971), "A Kinetic Model for the Chemical Composition of Seawater", *Quaternary Res.* **1**, 188-207.

Brown Jr., G. E. (1990), "Spectroscopic Studies of Chemisorption Reaction Mechanisms at Oxide-Water Interfaces", in M. F. Hochella Jr. and A. F. White, Eds., *Mineral-Water Interface Geochemistry*, 309-363.

Brown Jr., G. E., G. A. Parks, and C. J. Chisholm-Brause (1989), "In-Situ X-Ray Absorption Spectroscopic Studies of Ions at Oxide-Water Interfaces", *Chimia* **43**, 248-256.

Brown, J. C., and J. E. Ambler (1964), "Reductants Released by Roots of Fe Deficient Soybeans", *Agron. J.* **65**, 311.

Brown, W. H., W. S. Fyfe, and F. J. Turner (1962), "Aragonite in California Glaucophane Schists", *J. Petrol* **3**, 566-587.

Bruland, K. W. (1983), "Trace Elements in Seawater", in J. P. Riley and R. Chester, Eds., *Chemical Oceanography*, 157 pp.

Bruno, J., I. Casas, and I. Puigdomènech (1991), "The Kinetics of Dissolution of UO_2 Under Reducing Conditions and the Influence of an Oxidized Surface Layer (UO_{2+x}): Application of a Continuous Flow-through Reactor", *Geochim. Cosmochim. Acta* **55**, 647-658.

Bruno, J., W. Stumm, P. Wersin, and F. Brandberg (1991), "The Influence of Carbonate in Mineral Dissolution. Part 1, The Thermodynamics and Kinetics of Hematite Dissolution in Bicarbonate Solution at T = 25° C", in preparation.

Buffle, J. (1988), *Complexation Reactions in Aquatic Systems. An Analytical Approach*, Ellis Horwood, Chichester.

Buffle, J., D. Perret and M. Newman (1992), "The use of Filtration and Ultrafiltration for Size Fractionation of Aquatic Particles, Colloids and Macromolecules", in J. Buffle and H.P. v. Leeuwen, Eds., *IUPAC Series in Environmental Physical and Analytical Chemistry,* Lewis Publ., Chelsea, MI.

Buffle, J. (1992), Ed., *Chemical and Biological Regulation of Aquatic Processes*, in press.

Buffle, J., and R. S. Altmann (1987), "Interpretation of Metal Complexation by Heterogeneous Complexants", in W. Stumm, Ed., *Aquatic Surface Chemistry. Chemical Processes at the Particle Water Interface*, John Wiley and Sons, New York, 351-383.

Burton, W. K., N. Carbrera, and F. C. Frank (1951), "The Growth of Crystals and the Equilibrium Structure of Their Surfaces", *Phil Trans. Roy. Soc.* **A243**, 299-358.

Busch, P. L., and W. Stumm (1968), "Chemical Interactions in the Aggregation of Bacteria; Bioflocculation in Waste Treatment", *Environ. Sci. Technol.* **2**, 49-53.

Busenberg, E., and C. V. Clemency (1976), "The Dissolution Kinetics of Feldspars at 25° C and 1 Atm. CO_2 Partial Pressure", *Geochim. Cosmochim. Acta* **41**, 41-49.

Busenberg, E., and L. N. Plummer (1986), "A Comparative Study of the Dissolution and Crystal Growth Kinetics of Calcite and Aragonite", in F. A. Mumpton, Ed., *Studies in Diagenesis*, U.S. Geol. Surv. Bull. 1578, 139-168.

Calzaferri, G., S. Hug, T. Hugentobler, and B. Sulzberger (1984), "Self-Sensitization of Photo-Oxygen Evolution in Ag^+ Zeolites: Computer-Controlled Experiments," *J. Photochem.* **26**, 109-118.

Cannon, R. D. (1980), *Electron Transfer Reactions*, Butterworths, London.

Cederberg, G. A., R. L. Street, and J. O. Leckie (1985), "A Groundwater Mass Transport and Equilibrium Chemistry Model for Multicomponent Systems", *Water Resources Research* **21**, 1095-1104.

Chandar, P., P. Somasundaran, and N. J. Turro (1987), "Fluorescence Probe Studies on the Structure of the Adsorbed Layer of Dodecyl Sulfate at the Alumine Water Interface", *J. Coll. Interf. Sci.* **117**, 31-46.

Charlet, L., P. Wersin, and W. Stumm (1990), "Surface Charge of Some Carbonate Minerals", *Geochim. Cosmochim. Acta* **54**, 2329-2336.

Chen, Y. S., J. N. Butler, and W. Stumm (1973), "Kinetic Study of Phosphate Reaction with Aluminum Oxide and Kaolinite", *Environ. Sci. Technol.* **7**, 327-332.

Chiang, P-P., M. D. Donohue, and J. L. Katz (1988), "A Kinetic Approach to Crystallization from Ionic Solution. II. Crystal Nucleation", *J. Colloid Interf. Sci.* **122**, 251-265.

Chisholm-Brause, C. J., G. E. Brown Jr., and G. A. Parks (1989), "EXAFS Investigation of Aqueous Cu(II) Adsorbed on Oxide Surfaces In-Situ", *Physica B* **158**, 646-648.

Chou, L., and R. Wollast (1984), "Study of the Weathering of Albite at Room Temperature and Pressure with a Fluidized Bed Reactor", *Geochim. Cosmochim. Acta* **48**, 2205-2218.

Chou, L., and R. Wollast (1985), "Steady-state Kinetics and Dissolution Mechanisms of Albite", *Am. J. Sci.* **285**, 963-993.

Chou, L., R.M. Garrels, and R. Wollast (1989), "Comparative study of the kinetics and mechanisms of dissolution of carbonate minerals", *Chemical Geology* **78**, 269-282.

Christofferson, J., M. R. Christoffersen, W. Kibalczyc, and W. G. Perdok (1988), "Kinetics of Dissolution and Growth of Calcium Fluoride and Effects of Phosphate", *Acta Odontol Scand.* **46**, 325-336.

Chukhrov, F. V., B. B. Zvyagin, L. P. Ermilova, and A. I. Gorshkov (1973), "New Data on Iron Oxides in the Weathering Zone", in J. M. Serratosa and A. Sanchez, Eds., *Proceedings of the International Clay Conference*, Division de Ciencias C.S.I.C., Madrid, 333-341.

Coale, K. H., and A. R. Flegal (1989), "Copper, Zinc, Cadmium and Lead in Surface Waters of Lakes Erie and Ontario", *Sci. Tot. Env.* **87/88**, 297-304.

Comans, R. N. J. (1987), "Adsorption, Desorption and Isotopic Exchange of Cadmium on Illite: Evidence for Complete Reversibility", *Wat. Res.* **21**, 1573-1576.

Comans, R. N. J. (1990), "Adsorption, Desorption and Isotopic Exchange of Cadmium on Illite: Evidence for Complete Reversibility", in *Sorption of Cadmium and Cesium at Mineral/Water Interfaces*, Ph. D. Thesis, Rijksuniversiteit Utrecht, Netherlands.

Cooper, W. J., R. G. Zika, R. G. Petasne, and J. M. C. Plane (1988), "Photochemical Formation of H_2O_2 in Natural Waters Exposed to Sunlight", *Environ. Sci. Technol.* **22**, 1156-1160.

Cornel, P. R., R. S. Summers, and P. V. Roberts (1986), "Diffusion of Humic Acid in Dilute Aqueous Solutions", *J C I S* **110**, 149.

Cornell, R. M., and P.W. Schindler (1987), "Photochemical Dissolution of Goethite in Acid/Oxalate Solution", *Clays and Clay Minerals* **35**, 347-352.

Cornell, R. M., and U. Schwertmann (1979), "Influence of Organic Anions on the Crystallization of Ferrihydrite", *Clays and Clay Minerals* **27**, 402-410.

Cornell, R. M., R. Giovanoli, and W. Schneider (1989), "Review of the Hydrology of Iron(III) and the Crystallization of Amorphous Iron(III) Hydroxide Hydrate", *J. Chem. Techn. Biology* **46**, 115-134.

Ćosović, B. (1990), "Adsorption Kinetics of the Complex Mixture of Organic Solutes at Model and Natural Phase Boundaries", in W. Stumm, Ed., *Aquatic Chemical Kinetics*, John Wiley and Sons, New York, 291-310.

Cox, P. A. (1987), *The Electronic Structure and Chemistry of Solids*, Oxford University Press, Oxford.

Crumbliss, A. L., and J. M. Garrison (1988), "A Comparison of Some Aspects of the Aqueous Coordination Chemistry of Al(III) and Fe(III)". Comments. *Inorg. Chem.* **8**, 135-141.

Darwent, J. R. (1981), "Hydrogen Production Photosensitized by Aqueous Semiconductor Dispersions", *J. Chem. Soc., Faraday Trans. 2* **77/9**, 1703-1709.

Davies, S. H. R., and J. J. Morgan (1988), "Manganese(II) Oxidation Kinetics on Oxide Surfaces", *J. Colloid Interface Sci.*, **129/1**, 63-77.

Davis, J. A. (1984), "Complexation of Trace Metals by Adsorbed Natural Organic Matter", *Geochim. Cosmochim. Acta* **48**, 679.

Davis, J. A., and D. B. Kent (1990), "Surface Complexation Modeling in Aqueous Geochemistry", in M. F. Hochella Jr. and A. F. White, Eds., *Mineral-Water Interface Geochemistry*, Mineralogical Society of America, 177-260.

Davis, J. A., and J. D. Hem (1989), "The Surface Chemistry of Aluminum Oxides and Hydroxides", in G. A. Sposito, Ed., *The Environmental Chemistry of Aluminum*, CRC Press, Boca Raton, FL, 185-219.

Davis, J. A., and J. O. Leckie (1978a), "Surface Ionization and Complexation at the Oxide/Water Interface. II. Surface Properties of Amorphous Iron Oxyhydroxide and Adsorption of Metal Ions," *J. Colloid Interface Sci.* **67**, 90-107.

Davis, J. A., and J. O. Leckie (1978b), "Effect of Adsorbed Complexing Ligands on Trace Metal Uptake by Hydrous Oxides," *Environ. Sci. Tech.* **12**, 1309-1315.

Davis, J. A., C. C. Fuller, and A. D. Cook (1987), "A Model for Trace Metal Sorption Processes at the Calcite Surface: Adsorption of Cd^{2+} and Subsequent Solid Solution Formation", *Geochim. Cosmochim. Acta* **51**, 1477-1490.

Davison, W. (1985), "Conceptual Models for Transport at a Redox Boundary", in W. Stumm, Ed., *Chemical Processes in Lakes*, John Wiley and Sons, New York, 31-53.

Davison, W., G. W. Grime, J. A. W. Morgan, and K. Clarke (1991), "Distribution of Dissolved Iron in Sediment Pore Waters at Submillimetre Resolution," *Nature* **352**, 323-325.

De Vitre, R. R., D. Perret, and J. Buffle (1992), in J. Buffle, Ed., *Chemical and Biological Regulation of Aquatic Processes*, in press.

Delaney, M., and E. A. Boyle (1987), "Cadmium/Calcium in Late Miocene Benthic Foraminifera and Changes in the Global Organic Carbon Budget", *Nature* **330**, 156-159.

Deng, Y., A. Ruf, and W. Stumm, in preparation.

Deng, Y., and W. Stumm, in preparation.

Di Toro, D. M., J. D. Mahony, P. R. Kirchgraber, A. L. O'Bryne, L. R. Pasquale, and D. C. Piccirilli (1986), "Effects of Nonreversibility, Particle Concentration, and Ionic Strength on Heavy Metal Sorption", *Environ. Sci. Technol.* **20**, 55.

Diem, D., and W. Stumm (1984), "Is Dissolved Mn^{2+} being oxidized by O_2 in the Absence of Mn-Bacteria or Surface Catalysts?", *Geochim. Cosmochim. Acta* **48**, 1571-1573.

Doerner, H. A., and W. M. Hoskins (1925), "Coprecipitation of Radium and Barium Sulfates", *J. Amer. Chem. Soc.* **47**, 662-675.

dos Santos-Afonso, M., and W. Stumm, in preparation.

Drever, J. I. (1988), *The Geochemistry of Natural Waters,* 2nd Ed., Prentice Hall, NJ, 437 pp.

Driscoll, C. T., and G. E. Likens (1982), "Hydrogen Ion Budget of an Aggrading Forested Ecosystem", *Tellus* **34**, 283-292.

Ducker, W. A., T. J. Senden, and R. M. Pashley (1991), "Direct Measurement of Colloidal Forces Using an Atomic Force Microscope", *Nature* **353**, 239-241.

Dzombak, D. A., and F. M. M. Morel (1990), *Surface Complexation Modeling; Hydrous Ferric Oxide*, John Wiley and Sons, New York.

Edgington, D. N., and J. A. Robbins (1976), "Records of Lead Deposition in Lake Michigan Sediments Since 1800," *Env. Sci. Technol.* **10**, 266-274.

Edmond, J. M., A. Spivack, B. C. Grant, H. Ming-Hui, C. Zexiam, C. Sung, and Z. Xiushau (1985),

"Chemical Dynamics of the Changjiang Estuary", *Continental Shelf Research* **4**, 17-36.

Eggleston, C. M., and M. F. Hochella Jr. (1990), "Scanning Tunnelling Microscopy of Sulfide Surfaces", *Geochim. Cosmochim. Acta* **54**, 1511-1517.

Elimelech, M., and Ch. R. O'Melia (1990), "Effect of Particle Size on Collision Efficiency in the Deposition of Brownian Particles with Electrostatic Energy Barriers", *Langmuir* **6/6**, 1153-63.

Falkner, K. K., C. I. Measures, S. E. Heberlin, J. M. Edmond, and R. F. Weiss (1991), "The Major and Minor Element Geochemistry of Lake Baikal", *Limnol. Oceanogr.* **36**, 413-433.

Farley, K. J., D. A. Dzombak, and F. M. M. Morel (1985), "A Surface Precipitation Model for the Sorption of Cations on Metal Oxides", *J. Colloid Interface Sci.* **106**, 226-242.

Faust, B.C., and M. R. Hoffmann (1986), "Photoinduced Reductive Dissolution of α-Fe_2O_3 by Bisulfite," *Environ. Sci. Technol.* **20**, 943-948.

Fisher, N. S., P. Bjerregaard, and S. W. Fowler (1983), "Interactions of Marine Plankton with Transuranic Elements. 1. Biokinetics of Neptunium, Plutonium, Americium, and Valifornium in Phytoplankton", *Limnol. Oceanogr.* **28**, 432.

Fleer, G. J., and J. Lyklema (1983), "Adsorption of Polymers", in G. D. Parfitt and C. H. Rochester, Eds., *Adsorption from Solution at the Solid/Liquid Interface,* Chapter 4, Academic Press, London.

Fokkink, L. G. J. (1987), *Ion Adsorption on Oxides*, Ph.D. Thesis, University of Wageningen, Netherlands.

Fokkink, L. G. J., A. de Keizer, and J. Lyklema (1989), "Temperature Dependence of the Electrical Double Layer on Oxides: Rutile and Hematite", *J. of Colloid and Interface Sci.* **127**, 116-131.

Fowkes, F. M. (1965), In: *Chemistry and Physics at Interfaces*, American Chemical Society, Washington, D.C.

Fowkes, F. M. (1968), "Attractive Forces at Interfaces", *J. Colloid Interf. Sci.* **28**, 493.

Freeman, J. S., and D. L. Rowell (1981), "The Adsorption and Precipitation of Phosphate onto Calcite", *J. Soil Sci.*, **32/1**, 75-84.

Freeze, R. A., and J. A. Cherry (1979), *Groundwater*, Prentice Hall, NJ, 604 pp.

Frei, H., D. J. Fitzmaurice, and M. Grätzel (1990), "Surface Chelation of Semiconductors and Interfacial Electron Transfer", *Langmuir* **6**, 198-206.

Fuller, C. C., and J. A. Davis (1987), "Processes and Kinetics of Cd^{2+} Sorption by a Calcareous Aquifer Sand", *Geochim. Cosmochim. Acta* **51**, 1491-1502.

Furrer, F., and W. Stumm (1986), "The Coordination Chemistry of Weathering, I. Dissolution Kinetics of δ-Al_2O_3 and BeO", *Geochim. Cosmochim. Acta* **50**, 1847-1860.

Fyfe, W. S. (1987), "From Molecules to Planetary Environments: Understanding Global Change", in W. Stumm, Ed., *Aquatic Surface Chemistry*, John Wiley and Sons, New York, 495-508.

Garrels, R. M., and C. Christ (1965), *Minerals Solutions and Equilibria*, Harper and Row, New York.

Garrels, R. M., and E. A. Perry (1974), "Cycling of Carbon, Sulfur, and Oxygen through Geologic Time", in E. D. Goldberg, Ed., *The Sea*, Vol. 5, John Wiley and Sons, New York, 303-336.

Garrels, R. M., and F. T. Mackenzie (1971a), *Evolution of Sedimentary Rocks*, W. W. Norton and Co., New York, 397 pp.

George, P. (1954), "The Oxidation of Ferrous Perchlorate by Molecular Oxygen", *J. Chem. Soc.*, 4349-4359.

Gieseking, J. E. (1975), *Soil Components,* Vol. 2, Springer, New York.

Giovanoli, R., J. L. Schnoor, L. Sigg, W. Stumm, and J. Zobrist (1989), "Chemical Weathering of Crystalline Rocks in the Catchment Area of Acidic Ticino Lakes, Switzerland", *Clays Clay Min.* **36**, 521-529.

Goldberg, S., and G. Sposito (1984), "A Chemical Model of Phosphate Adsorption by Soils: I. Reference Oxide Minerals; II. Noncalcareous Soils", *Soil Sci. Soc. Am.* **48**, 772-783.

Goldschmidt, V. M. (1933), *Fortschr. Mineral. Krist. Petrol.* **17**, 112.

Gonçalves, M. S., L. Sigg, and W. Stumm (1985), "Voltammetric Methods for Distinguishing Between Dissolved and Particulate Metal Ion Concentrations in Presence of Hydrous Oxides", *Env. Sci. Techn.* **19/2**, 141-146.

Gonçalves, M. S., L. Sigg, M. Reutlinger, and W. Stumm (1987), "Metal Ion Binding by Biological Surfaces; Voltammetric Assessment in Presence of Bacteria", *The Sci. Tot. Env.* **60**, 105-119.

Gordon, G., and H. Taube (1962), "Oxygen Tracer Experiments on the Oxidation of Aqueous Uranium(IV) with Oxygen-Containing Oxidizing Agents", *Inorg. Chem.* **1**, 69-75.

Gordon, L., M. L. Salutsky, and H. H. Willard (1959), *Precipitation from Homogeneous Solution*, John Wiley and Sons, New York.

Grandstaff, D. E. (1976), "A Kinetic Study of the Dissolution of Uraninite", *Econ. Geology* **71**, 1493-1506.

Grandstaff, D. E. (1986), "The Dissolution Rate of Forsteritic Olivine from Hawaiian Beach Sand", in S. M. Colman and D. P. Dethier, Eds., *Rates of Chemical Weathering of Rocks and Minerals*, Academic Press, 41-59.

Gratz, A. J., S. Manne, and P. K. Hansma (1991), "Atomic Force Microscopy of Atomic-Scale Ledges and Etch Pits Formed During Dissolution of Quartz", *Science* **251**, 1343-46.

Grauer, R. (1983), "Zur Koordinationschemie der Huminstoffe", *PSI-Report, Würenlingen, Switzerland* **24**, 103.

Grauer, R., and W. Stumm (1982), "Die Koordinationschemie oxidischer Grenzflächen und ihre Auswirkung auf die Auflösungskinetic oxidischer Festphasen in wässrigen Lösungen", *Colloid. Polymer. Sci.* **260**, 959-970.

Grätzel, M. (1989), *Heterogeneous Photochemical Electron Transfer*, CRC Press, Boca Raton, Florida.

Greenland, D. J., and M. H. B. Hayes (1978), *The Chemistry of Soil Constituents*, Wiley, Chichester.

Grenthe, I., W. Stumm, M. Laaksuharju, A. C. Nilsson, and P. Wikberg (1992), "Redox Potentials and Redox Reactions in Deep Groundwater Systems", submitted to *Chemical Geology.*

Grim, R. E. (1968), *Clay Mineralogy,* 2nd Edition, McGraw Hill, New York.

Grütter, A., H. R. von Gunten, M. Kobler, and E. Rössler (1990), "Sorption, Desorption and Exchange of Cs^+ on Glaciofluvial Deposits", *Radiochimica Acta* **50**, 177-184.

Gschwend, Ph. M., and M. D. Reynolds (1987), "Monodisperse Ferrous Phosphate Colloids in an Anoxic Groundwater Plume", *J. Contam. Hydrol.* **1/3**, 309-327.

Guy, C., and J. Schott (1989), "Multisite Surface Reaction Versus Transport Control During the Hydrolysis of a Complex Oxide", *Chem. Geol.* **78**, 181-204.

Hachiya, K., M. Sasaki, Y. Saruta, N. Mikami, and T. Yasanuga (1984), "Static and Kinetic Studies of Adsorption-Desorption of Metal Ions on the γ-Al_2O_3 Surface", *J. Phys. Chem.* **88**, 23-31.

Hamilton-Taylor, J., M. Willis, and C. S. Reynolds (1984), "Depositional Fluxes of Metals and Phytoplankton in Windermere as Measured by Sediment Traps", *Limnol. Oceanogr.* **24**, 695-710.

Harding, I. H., and T. W. Healy (1985), "Electrical Double Layer Properties of Amphoteric Polymer Latex Colloids", *J. Coll. Interf. Sci.* **107**, 382-397.

Hayes, K. F., A. L. Roe, G. E. Jr. Brown, K. O. Hodgson, J. O. Leckie, and G. A. Parks (1987), "In Situ X-Ray Absorption Study of Surface Complexes: Selenium Oxyanions on α-FeOOH", *Science* **238**, 783-786.

Hayes, K. F., and J. O. Leckie (1986), "Mechanism of Lead Ion Adsorption at the Goethite-Water Interface", *J. Am. Chem. Soc.* **323**, 114-141.

Hering, J. G., and F. M. M. Morel (1990), "Kinetics of Trace Metal Complexation: Ligand-Exchange Reactions", *Environm. Sci. Technol.* **24**, 242-252.

Hering, J., and W. Stumm (1990), "Oxidative and Reductive Dissolution of Minerals", in M. F. Hochella and A. F. White, Eds., *Reviews in Mineralogy 23: Mineral-Water Interface Geochemistry*, Mineralogical Society of America, 427-465.

Hering, J., and W. Stumm (1991), "Fluorescence Spectroscopic Evidence for Surface Complex Formation at the Mineral-Water Interface: Elucidation of the Mechanism of Ligand-Promoted Dissolution," *Langmuir* **7**, 1567-1570.

Hiemenz, P. C. (1986), *Principle of Colloid and Surface Chemistry,* 2nd Ed., M. Dekker, New York.

Hiemstra, T., W. H. van Riemsdijk, and H. G. Bolt (1989), "Multisite Proton Adsorption Modeling at the Solid/Solution Interface of (Hydr)Oxides, I. Model Description and Intrinsic Reaction Constants", *J. Colloid Interf. Sci.* **133**, 91-104.

Hiemstra, T., J. C. M. de Wit, and W. O. van Riemsdijk (1989), "Multisite Proton Adsorption Modeling at the Solid/Solution Interface of (Hydr)Oxides, II. Application to Various Important (Hydr)Oxides", *J. Colloid Interf. Sci.* **133**, 105-117.

Hoffmann, M. R. (1990), "Catalysis in Aquatic Environments", in W. Stumm, Ed., *Aquatic Chemical Kinetics; Reaction Rates of Processes in Natural Waters*, John Wiley and Sons, New York.

Hoffmann, R. (1988), *Solids and Surfaces; A Chemist's View of Bonding in Extended Structures*, VCH Publ., New York.

Hohl, H. (1982), "Personal communications", unpublished.

Hohl, H., and W. Stumm (1976), "Interaction of Pb^{2+} with Hydrous γ-Al_2O_3", *J. Colloid and Interface Sci.* **55**, 281-288.

Hohl, H., E. Werth, E. Posch, and R. Giovanoli (1982), presented at ACS-Meeting.

Hohl, H., E. Werth, R. Giovanoli, and E. Posch (1985), "Heterogeneous Nucleation. I. Nucleation of Calcium Fluoride on Cerium(IV) Oxide. Quoted in: Schindler P.W. Grenzflächenchemie oxidischer Mineralien", *Öster. Chem. Z.* **86**, 141-147.

Hohl, H., L. Sigg, and W. Stumm (1980), "Characterization of Surface Chemical Properties of Oxides in Natural Water; The Role of Specific Adsorption Determining the Specific Charge", *Particulates in Water, ACS* **189**, 1-31.

Holdren Jr., G. R., and P. M. Speyer (1987), "Reaction Rate-Surface Area Relationships during the Early Stages of Weathering. II. Data on Eight Additional Feldspars", *Geochim. Cosmochim. Acta* **51/9**, 2311-2318.

Holland, H. D. (1978), *The Chemistry of the Atmosphere and Oceans*, John Wiley and Sons, New York, 351 pp.

Holsom, T. M., E. R. Taylor, Y.-C. Seo, and P. R. Anderson (1991), "Removal of Sparingly Soluble Organic Chemicals from Aqueous Solutions with Surfactant-Coated Terrihydrite", *Environ. Sci. Technol.* **25**, 1585-1589.

Honeyman, B. D., and J. O. Leckie (1986), "Macroscopic Partitioning Coefficient for Metal Ion Adsorption: Proton Stoichiometry at Variable pH and Adsorption Density", in J. A. Davis and K. F. Hayes, Eds., *Geochemical Processes at Mineral Surfaces*, ACS Symposium, Washington, DC.

Huang, C. P., and E. A. Rhoads (1989), "Adsorption of Zn(II) onto Hydrous Alumino-Silicates," *J. Colloid Interf. Science* **131**, 289-306.

Hunter, R. J. (1989), *Foundation of Colloid Science* Vol. 1, Clarendon Press, Oxford.

Imboden, D. M., and R. P. Schwarzenbach (1985), "Spatial and Temporal Distribution of Chemical Substances in Lakes: Modeling Concepts", in W. Stumm, Ed., *Chemical Processes in Lakes*, John Wiley and Sons, New York, 1-30.

Israelachvili, J. N., and G. E. Adams (1978), "Measurement of Forces Between Two Mica Surfaces in Aqueous Electrolyte Solutions in the Range 0 – 100 nm", *J. Chem. Soc., Faraday Trans. 1* **74/4**, 975-1001.

Jacob, D., E. W. Gottlieb, and M. J. Prather (1989), "Chemistry of Polluted Cloudy Boundary Layer", *J. Geophys. Research* **94**, 12975.

Jaegermann, W., and H. Tributsch (1988), "Interfacial Properties of Semiconducting Transition Metal Chalcogenides", in S. G. Davison, Ed., *Progress in Surface Science* **29**, Pergamon Press, New York.

Johnson, C. A. (1986), "The Regulation of Trace Element Concentrations in Rivers and Estuarine Waters Contaminated with Acid Mine Drainage: The Adsorption of Cu and Zn on Amorphous Fe Oxyhydroxides", *Geochim. Cosmochim. Acta* **50**, 2433-2438.

Johnson, C. A., and L. Sigg (1985), "Acidity of Rain and Fog: Conceptual Definitions and Practical Measurements of Acidity", *Chimia* **39**, 59-61.

Karickhoff, S. W., D. S. Brown, and T. A. Scott (1979), "Sorption of Hydrophobic Pollutants on Natural Sediments", *Water Res.* **13**, 241-248.

Khader, M. M., N. N. Lichtin, G. H. Vurens, M. Salmeron, and G. A. Somorjai (1987), "Photoassisted Catalytic Dissociation of H_2O and Reduction of N_2 to NH_3 on Partially Reduced Fe_2O_3," *Langmuir* **3**, 303-304.

Kim, J. I. (1991), "Actinide Colloid Generation in Groundwater," *Radiochim. Acta* **52-53**, 71-81.

Knauss, K. G., and T. J. Wolery (1989), "Muscovite Dissolution Kinetics as a Function of pH and Time at 70° C", *Geochim. Cosmochim. Acta* **53/7**, 1493-1501.

Koike, I., H. Shigemitsu, T. Kazuki, and K. Kazuhiro (1990), "Role of Sub-Micrometre Particles in the Ocean", *Nature* **345**, 242-244.

Kormann, C., D. W. Bahnemann, and M. R. Hoffmann (1988), "Photocatalytic Production of H_2O_2 and Organic Peroxides in Aqueous Suspensions of TiO_2, ZnO, and Desert Sand", *Environ. Sci. Technol.* **22**, 798-806.

Kriegmann, M. R., and M. Reinhard (1989), *Electron Transfer Reactions of Haloaliphatics and Ferrous Iron Bearing Minerals*, Abst. Amer. Chem. Soc., Annual Meeting, Miami, Florida.

Kronberg, B. I., W. S. Fyfe, O. H. Leonardos Jr., and A. M. Santos (1979), "The Chemistry of Some Brazilian Soils: Element Mobility during Intense Weathering", *Chem. Geol.* **24**, 11-229.

Kummert, R., and W. Stumm (1980), "The Surface Complexation of Organic Acids on Hydrous γ-Al_2O_3", *J. Colloid Interface Sci.* **75**, 373-385.

Kung, H. H. (1989), *Transition Metal Oxides: Surface Chemistry and Catalysis*, Elsevier, Amsterdam.

Kunz, B., and W. Stumm (1984), "Kinetik der Bildung und des Wachstums von Calcium-Carbonat", *Vom Wasser* 62, 279-293.

Kurbatov, M. H., G. B. Wood, and J. D. Kurbatov (1951), "Isothermal Adsorption of Cobalt from Dilute Solutions", *J. Phys. Chem.* 55, 1170-1182.

LaMer, V. K., and T. W. Healy (1963), "Adsorption Flocculation Reactions of Macromolecules at the Solid-Liquid Interface", *Rev. Pure Appl. Chem.* 13, 112.

Langmuir, D., and D. O. Whittemore (1971), "Variations in the Stability of Precipitated Ferric Oxy-hydroxides", *Adv. Chem. Ser.* 106, 209-234.

Lasaga, A. C. (1981), "Rate Laws of Chemical Reactions", in A. C. Lasaga and R. J. Kirkpatrick, Eds., *Kinetics of Geochemical Processes*, Soc. Amer. Rev. in Mineral. 8, Washington D.C, 1-68.

Lasaga, A. C., and G. V. Gibbs (1990), "Ab Initio Quantum-Mechanical Calculations of Surface Reactions – A New Era?", in W. Stumm, Ed., *Aquatic Chemical Kinetics*, John Wiley and Sons, New York, 259-290.

Lasaga, A. C., and R. J. Kirkpatrick (1981), *Kinetics of Geochemical Processes*, Mineralogical Society of America.

Lerman, A. (1979), *Geochemical Processes – Water and Sediment Environments*, John Wiley and Sons, New York.

Lerman, A. (1990), "Transport and Kinetics in Surficial Processes", in W. Stumm, Ed., *Aquatic Chemical Kinetics*, John Wiley and Sons, New York, 505-534.

Liang, L., and J. J. Morgan (1990), "Chemical Aspects of Iron Oxide Coagulation in Water: Laboratory Studies and Implications for Natural Systems", *Aquatic Sciences* 52/1, 32-55.

Litter, M. I., and M. A. Blesa (1988), "Photodissolution of Iron Oxides I. Maghemite in EDTA Solutions", *J. Colloid Interface Sci.* 125, 679-687.

Lovley, D. R., and E. J. P. Phillips (1988), "Novel Mode of Microbial Energy Metabolism: Organic Carbon Oxidation Coupled to Dissimilatory Reduction of Iron or Manganese", *Applied and Environ. Microbiology* 54/6, 1472-1480.

Lovley, D. R., D. J. Baedeker, D. J. Lonergan, I. M. Cozanelli, E. J. P. Phillips, and D. I. Siegel (1989), "Oxidation of Aromatic Contaminants Coupled to Microbial Iron Reduction", *Nature* 339, 297-300.

Lovley, D. R., E. J. P. Phillips, Y. A. Gorby, and E. R. Landa (1990), "Microbial Reduction of Uranium", *Nature* 350, 413-416.

Lowenstam, H. A., and S. Weiner (1985), *On Biomineralization*, Oxford University Press, New York.

Luther III, G. W. (1987), "Pyrite Oxidation and Reduction: Molecular Orbital Theory Considerations", *Geochim. Cosmochim. Acta* 51, 3193-3199.

Luther III, G. W. (1990), "The Frontier-Molecular-Orbital Theory Approach in Geochemical Processes", in W. Stumm, Ed., *Aquatic Chemical Kinetics*, John Wiley and Sons, New York, 173-198.

Luther III, G. W., T. M. Church, J. E. Kostka, B. Sulzberger, and W. Stumm (1991), "Seasonal Iron Cycling in the Marine Environment: The Importance of Ligand Complexes with Fe(II) and Fe(III) in the Dissolution of Fe(III) Minerals and Pyrite, Respectively", submitted.

Lyklema, J. (1978), "Surface Chemistry of Colloids in Connection with Stability", in K. J. Ives, Ed., *The*

Scientific Basis of Flocculation, Sijthoff and Noordhoff, The Netherlands, 3-36.

Lyklema, J. (1985), "How Polymers Adsorb and Affect Colloid Stability, Flocculation, Sedimentation, and Consolidation", in B. M. Moudgil and P. Somasundaran, Eds., *Flocculation, Sedimentation and Consolidation: Proceedings of the Engineering Foundation Conference, Sea Island, Georgia,* United Engineering Trustees, Inc., 3-21.

Lyklema, J. (1987), "Electrical Double Layers on Oxides: Disparate Observations and Unifying Principles", *Chemistry and Industry,* 741-747.

Lyklema, J. (1991), *Fundamentals of Interface and Colloid Science,* Vol. 1, Chapter 5, Academic Press, London.

Mann, S. (1988), "Molecular Recognition in Biomineralization", *Nature* **332,** 119-124.

Mann, S., J. Webb, and R. J. P. Williams (1989), *Biomineralization,* VCH Verlag, Weinheim.

Marcus, R. A. (1975), "Electron Transfer in Homogeneous and Heterogeneous Systems", in E. D. Goldberg, Ed., *The Nature of Seawater,* Dahlem Konferenz, Berlin.

Margerum, D. W., G. R. Cayley, D. C. Weatherburn, and G. K. Pagenkopf (1978), "Kinetics and Mechanism of Complex Formation and Ligand Exchange", in A. Martell, Ed., *Coordination Chemistry,* Vol. 2, ACS Symposium Series No. 174, Washington, DC, 1-220.

Martin, J. M., and M. Meybeck (1979), "Elemental Mass-Balance of Material Carried by World Major Rivers", *Marine Chem.* **7,** 173-206.

Martin, J.-M., and M. Whitfield (1983), "The Significance of the River Input of Chemical Elements to the Ocean", in C. S. Wong, E. Boyle, K. W. Bruland, J. D. Burton, and E. D. Goldberg, *Trace Metals in Sea Water,* Plenum Press, New York, 265-296.

Mast, M. A., and J. I. Drever (1987), "The Effect of Oxalate on the Dissolution Rates of Oligoclase and Tremolite", *Geochim. Cosmochim. Acta* **51/9,** 2559-2568.

Matijević, E. (1982), *Conference on the Chemistry of Solid-Liquid Interfaces,* Cavtat, Yugoslavia.

McCarty, P. L., M. Reinhard, and B. E. Rittmann (1981), "Trace Organics in Groundwater", *Environ. Sci. Technol.* **15/1,** 40-51.

McKnight, D. M., and K. E. Bencala (1988), "Dial Variations in Iron Chemistry in an Acidic Stream in the Colorado Rocky Mountains", *Arctic and Alpine Research* **20,** 492-500.

Meybeck, M. (1979b), "Pathways of Major Elements from Land to Ocean through Rivers", in *Review and Workshop on River Inputs to Ocean Systems,* FAO, Rome, 26-10.

Mickers, M. A. H., and H. P. van Leeuwen (1991), "Dynamic Aspects of Metal Speciation in Aquatic Colloidal Systems", submitted.

Millero, F. (1985), "The Effect of Ionic Interactions on the Oxidation of Metals in Natural Waters", *Geochim. Cosmochim. Acta* **49,** 547-553.

Millero, F., S. Sotolongo, and M. Izaguirre (1987), "The Oxidation Kinetics of Fe(II) in Seawater", *Geochim. Cosmochim. Acta* **51,** 793-801.

Modi, H. J., and D. W. Fuerstenau (1957), "Streaming Potential Studies on Corundum in Aqueous Solutions of Inorganic Electrolytes", *J. Phys. Chem.* **61,** 640-643.

Mohr, E. C. J., F. A. Van Baren, and J. Van Schuylenborgh (1972), *Tropical Soils,* 3rd Edition, Mouton-Ichtiar Baru-Van Hoeve, The Hague, Netherlands.

Mopper, K., X. Zhou, R. J. Kieber, D. J. Kieber, R. J. Sikorski, and R. D. Jones (1991), "Photochemical

Degradation of Dissolved Organic Carbon and its Impact on the Oceanic Carbon Cycle", *Nature* **353**, 60-62.

Morel, F. M. M. (1983), *Principles of Aquatic Chemistry*, John Wiley and Sons, New York.

Morel, F. M. M., and P. M. Gschwend (1987), "The Role of Colloids in the Partitioning of Solutes in Natural Waters", in W. Stumm, Ed., *Aquatic Surface Chemistry*, John Wiley and Sons, New York, 405-422.

Morel, F. M. M., and R. J. M. Hudson (1985), "The Geobiological Cycle of Trace Elements in Aquatic Systems: Redfield Revisited", in W. Stumm, Ed., *Chemical Processes in Lakes*, John Wiley and Sons, New York, 251-281.

Morse, J. W, and F. J. Mackenzie (1990), *Geochemistry of Sedimentary Carbonates*, Elsevier, Amsterdam.

Morse, J. W., and R. A. Berner (1972), "Dissolution of Calcium Carbonate in Seawater II. A Kinetic Origin for the Lysocline", *Amer. J. Sci.* **272**, 840-851.

Motschi, H. (1985), "Cu(II) EPR: A Complementary Method for the Thermodynamic Description of Surface Complexation", *Adsorption Sci. Technol.* **2**, 39.

Motschi, H. (1987), "Aspects of the Molecular Structure in Surface Complexes; Spectroscopic Investigations", in W. Stumm, Ed., *Aquatic Surface Chemistry*, John Wiley and Sons, New York, 111-124.

Mouvet, Ch., and A. C. M. Bourg (1983), "Speciation (Including Adsorbed Species) of Copper, Lead, Nickel and Zinc in the Meuse River", *Water Res.* **17/6**, 641-49.

Murad, E., and W. R. Fisher (1988), "The Geochemical Cycle of Iron", in J. W. Stucky B. A. Goodman and U. Schwertmann, Eds., *Iron in Soils*, D. Reidel, Dordrecht.

Murray, J. W. (1987), "Mechanisms Controlling the Distribution of Trace Elements in Oceans and Lakes", in R. A. Hites and S. J. Eisenreich, Eds., *Adv. Chem. Ser.*, **216**.

Müller, B. (1989), *Über die Adsorption von Metallionen an Oberflächen aquatischer Partikel*, Ph. D. Thesis No. 8988, ETH Zurich, Switzerland.

Müller, B., and L. Sigg (1990), "Interaction of Trace Metals with Natural Particle Surfaces: Comparison between Adsorption Experiments and Field Measurements", *Aquatic Sciences* **52/1**, 75-92.

Müller, G. (1977), "Heavy Metals and PAH in a Sediment Core from Lake Constance", *Naturwissenschaften* **64**, 427-431.

Nancollas, G. H. (1989), "In Vitro Studies of Calcium Phosphate Crystallization", in S. Mann, J. Webb, and J. P. Williams, Eds., *VCH Verlagsgesellschaft*, Weinheim, 157-187.

Nancollas, G. H., and M. M. Reddy (1974), "Crystal Growth Kinetics of Minerals Encountered in Water Treatment Processes", *Aqueous-Environ. Chem. Met.*, 219-253.

Nancollas, G. H., and M. M. Reddy (1974), "Kinetics of Crystallization of Scale-Forming Minerals", *Soc. Petrol. Eng. J.* **14/2**, 117-126.

Nealson, K. H. (1982), "Microbiological Oxidation and Reduction of Iron", in H. D. Holland and M. Schidlowski, Eds., *Mineral Deposits and Evolution of the Biosphere; Dahlem Konferenzen*, Springer Verlag, New York, 51-56.

Nealson, K. H., and C. R. Myers (1990), "Iron Reduction by Bacteria: A Potential Role in the Genesis of Brandes Iron Formation", *Amer. J. of Science* **290**, 35-45.

Neihof, R. A., and G. I. Loeb (1972), "The Surface Charge of Particulate Matter in Seawater," *Limnol. Oceanogr.* **17**, 7-16.

Newman, A. C. D. (1987), "The Interaction of Water with Clay Mineral Surfaces" in A. C. D. Newman, Ed., *Chemistry of Clays and Clay Minerals*, Longman, Essex, England, 237-274.

Nielsen, A. E. (1964), *Kinetics of precipitation*, Pergamon Press, Oxford, 151 p.

Nielsen, A. E. (1981), "Theory of Electrolyte Crystal Growth. The Parabolic Rate Law", *Pure Appl. Chem.* **53**, 2025-2039.

Nielsen, A. E. (1986), "Mechanisms and Rate Laws in Electrolyte Crystal Growth from Aqueous Solution", in J. A. Davis and K. F. Hayes, Eds., *Geochemical Processes at Mineral Surfaces*, Amer. Chem. Soc. Symposium Series 323, 600-614.

Nielsen, A. E., and O. Söhnel (1971), "Interfacial Tension Electrolyte Crystal-Aqueous Solution, from Nucleation Data", *J. Cryst. Growth* **11**, 233-242.

Nriagu, J. O., A. L. W. Kemp, H. K. T. Wong, and N. Harper (1979), "Sedimentary Records of Heavy Metal Pollution in Lake Erie", *Geochim. Cosmochim. Acta* **43**, 247-258.

Nriagu, J. O., and H. K. T. Wong (1989), "Dynamics of Particulate Trace Metals in the Lakes of Kejimkujik National Park, Nova Scotia, Canada", *Sci. Tot. Env.* **87/88**, 315-328.

O'Melia, C. R. (1972), "Coagulation and Flocculation", in W. J. Weber Jr., Ed., *Processes for Water Quality Control*, John Wiley and Sons, New York, 61-110.

O'Melia, Ch. R. (1972), "Water and Waste-Water Filtration", *Air Water Pollut.*, **6**, 177-198.

O'Melia, C. R. (1987), "Particle-Particle Interactions", in W. Stumm, Ed., *Aquatic Surface Chemistry*, John Wiley and Sons, 385-403.

O'Melia, C. R. (1989), "Particle-Particle Interactions in Aquatic Systems", *Colloids Surf.* **39**, 255-271.

O'Melia, C. R. (1990), "Kinetics of Colloid Chemical Processes in Aquatic Systems", in W. Stumm, Ed., *Aquatic Chemical Kinetics, Reaction Rates of Processes in Natural Waters*, John Wiley and Sons, New York, 447-474.

Ochial, E.-I. (1991), "Biomineralization (Part V)", *J. Chem. Ed.* **68**, 827-830.

Ochs, M. (1991), *Humic Substances at Aquatic Interfaces: A Comparison of Hydrophobic vs Coordinative Adsorption and Subsequent Effects on Mineral Weathering*, Ph.D. Thesis, ETH Zurich, Switzerland.

Ochs, M., B. Ćosović and W. Stumm (1992), in preparation.

Overbeek, J. Th. G. (1966), "Colloid Stability in Aqueous and Non-Aqueous Media", *Discus. Faraday Soc.* **42**, 7.

Overbeek, J. Th. G. (1976), "Polyelectrolytes, Past, Present and Future", *Pure Appl. Chem.* **46**, 91-101.

Paces, T. (1983), "Rate Constants of Dissolution Derived from the Measurements of Mass Balance in Hydrological Catchments", *Geochim. Cosmochim. Acta* **47**, 1855-1863.

Parker, C.A. (1968), *Photoluminescence of Solutions*, Elsevier, Amsterdam, 208-214.

Parker, J. L., K. A. Stanlaw, J. S. Marshall, and C. W. Kennedy (1982), "Sorption and Sedimentation of Zn and Cd by Seston in Southern Lake Michigan", *J. Great Lakes Res.* **8**, 520-531.

Parks, G. A. (1967), "Aqueous Surface Chemistry of Oxides and Complex Oxide Minerals; Isoelectric Point and Zero Point of Charge," in *Equilibrium Concepts in Natural Water Systems*, Advances in Chemistry Series, No. 67, American Chemical Society, Washington, DC.

Parks, G. A. (1984), "Surface and Interfacial Free Energies of Quartz", *J. of Geophysical Research* **89**, 3997-4008.

Parks, G. A. (1990), "Surface Energy and Adsorption at Mineral/Water Interfaces: An Introduction", in M. F. Hochella Jr. and A. F. White, Eds., *Mineral-Water Interface Geochemistry*, Mineralogical Society of America, 133-175.

Parsons, R. (1987), "The Electric Double Layer at the Solid-Solution Interface", in W. Stumm, Ed., *Aquatic Surface Chemistry*, John Wiley and Sons, New York, 33-47.

Peri, J. B. (1965), "A Model for the Surface of γ-Alumina", *J. Phys. Chem.* **69**, 220-230.

Pichat, P., and Fox, M. A. (1988), "Photocatalysis on Semiconductors", in M. A. Fox and M. Chanon, M., Eds., *Photoinduced Electron Transfer*, Part D, Elsevier, Amsterdam.

Plummer, L. N., and F. T. Mackenzie (1974), "Predicting Mineral Solubility from Rate Data: Application to the Dissolution of Magnesian Calcites", *Amer. J. Sci.* **274**, 61-83.

Plummer, L. N., T. M. L. Wigley, and D. L. Parkhurst (1978), "The Kinetics of Calcite Dissolution in CO_2-Water Systems at 5° to 60° C and 0.0 to 1.0 atm CO_2", *Amer. J. Sci.* **278**, 179-216.

Postma, D., C. Boesen, H. Kristiansen, and F. Larsen (1991), "Nitrate Reduction in an Unconfined Sandy Aquifer: Water Chemistry, Reduction Processes, and Geochemical Modeling", *Water Resources Research* 27/8, 2027-2045.

Pou, T. E., P. J. Murphy, V. Young, and J. O. M. Bockris (1984), "Passive Films on Iron: The Mechanism of Breakdown in Chloride Containing Solutions", *J. Electrochem. Soc.* **131**, 1243.

Price, N. M., and F. M. M. Morel (1990), "Role of Extracellular Enzymatic Reactions in Natural Waters", in W. Stumm, Ed., *Aquatic Chemical Kinetics*, John Wiley and Sons, New York, 235-257.

Purcell, K. F., and J. C. Kotz (1980), *An Introduction to Inorganic Chemistry*, Saunders, Philadelphia, 637 pp.

Regazzoni, A. E., M. A. Blesa, and A. J. G. Maroto (1983), "Interfacial Properties of Zirconium Dioxide and Magnetite in Water", *J. Coll. Interf. Sci.* **91**, 560-570.

Rich, H. W., and F. M. M. Morel, "Availability of Well-Defined Iron Colloids to the Marine Diatom Thalassiosira Weissflogii", *Limnol. Oceanogr.*, accepted.

Roberts, A. L., and P. M. Gschwend (1991), "Mechanism of Pentachloroethane Dehydrochlorination to Tetrachloroethylene", *Environ. Sci. Technol.* **25**, 76-86.

Robie, R. A., and D. R. Waldbaum (1968), *Thermodynamic Properties of Minerals and Related Substances at 298.15° K (25.0° C) and One Atmosphere (1.013 Bars) Pressure and at Higher Temperatures*, U.S Government Printing Office, Washington DC.

Robie, R. A., B. S. Hemingway, and J. R. Fisher (1979), *Thermodynamic Properties of Minerals and Related Substances at 298.15° K and 1 Bar (10² Pascals) Pressure and at Higher Temperatures*, Geol. Survey Bull. 1452. U.S. Government Printing Office, Washington, DC., 456 pp.

Rönngren, L., S. Sjöberg, Z. Sun, W. Forsling, and P. W. Schindler (1991), "Surface Reactions in Aqueous Metal Sulfide Systems", *J. Coll. Interf. Sci.* **145**, 396-404.

Ruf, A. (1992), Ph.D. Thesis, ETH Zurich, in preparation.

Ruzić, I. (1987), "Time Dependence of Adsorption at Solid-Liquid Interfaces", *Croat. Chem. Acta* **60/3**, 457-475.

Ruzić, I., H. J. Ulrich, and B. Ćosović, (1988), "Time Dependence of the Adsorption of Valeric Acid at the Mercury-Sodium Chloride Interface", *J. Colloid Interface Sci.* **126**, 525-536.

Ryan, J. N., and P. M. Gschwend (1990), "Colloid Mobilization in Two Atlantic Coastal Plain Aquifers: Field Studies", *Water Resources Research* **26**, 307-322.

Ryan, J. N., and P. M. Gschwend (1991), "Extraction of Iron Oxides from Sediments Using Reductive Dissolution by Titanium(III)," *Clays and Clay Minerals*, in press.

Sayles, F. L., and P. C. Mangelsdorf Jr. (1979), "Cation-Exchange Characteristics of Amazon River Suspended Sediment and its Reaction with Seawater", *Geochim. Cosmochim. Acta* **43**, 767-779.

Schaule, B. K., and C. C. Patterson (1981), "Lead Concentrations in the Northeast Pacific: Evidence for Global Anthropogenic Perturbations", *Earth Planet. Sci. Lett.* **54**, 97.

Scheutjens, J. M. H. M., and G. J. Fleer (1980), "Statistical Theory of the Adsorption of Interacting Chain Molecules. 2. Train, Loop, and Tail Size Distribution", *J. Phys. Chem.* **84/2**, 178-90.

Schindler, P. (1967), *Advances in Chemistry Series. In: Equilibrium Concepts in Natural Water Systems*, American Chemical Society No. **67**, p. 196, Washington DC.

Schindler, P. W. (1984), "Surface Complexation", in H. Sigel, Ed., *Metal Ions in Biological Systems*, Vol. 18, M. Dekker Inc., New York.

Schindler, P. W. (1985), "Grenzflächenchemie oxidischer Mineralien", *Österreichische Chemie-Zeitschrift* **86**, 141-146.

Schindler, P. W. (1990), "Co-Adsorption of Metal Ions and Organic Ligands: Formation of Ternary Surface Complexes", in M. F. Jr.. Hochella and A. F. White, Eds., *Mineral-Water Interface Geochemistry*, Mineralogical Soc. of America, Washington, DC, 281-307.

Schindler, P. W., and W. Stumm (1987), "The Surface Chemistry of Oxides, Hydroxides, and Oxide Minerals", in W. Stumm, Ed., *Aquatic Surface Chemistry*, John Wiley and Sons, New York, 83-110.

Schindler, P. W., P. Liechti, and J. C. Westall (1987), "Adsorption of Copper, Cadmium and Lead from Aqueous Solutions to the Kaolinite-Water Interface", *Netherlands J. of Agricultural Sci.* **35**, 219.

Schneider, W. (1988), "Iron Hydrolysis and the Biochemistry of Iron – The Interplay of Hydroxide and Biogenic Ligands", *Chimia* **42**, 9-20.

Schneider, W. (1990), *Roche Symposium Aspects of Bioinorganic Chemistry*.

Schneider, W., and B. Schwyn (1987), "The Hydrolysis of Iron in Synthetic, Biological and Aquatic Media", in W. Stumm, Ed., *Aquatic Surface Chemistry*, John Wiley and Sons, New York, 167-196.

Schnoor, J. L. (1990), "Kinetics of Chemical Weathering: A Comparison of Laboratory and Field Weathering Rates", in W. Stumm, Ed., *Aquatic Chemical Kinetics*, John Wiley and Sons, 475-504.

Schnoor, J. L., and W. Stumm (1985), "Acidification of Aquatic and Terrestrial Systems", in W. Stumm, Ed., *Chemical Processes in Lakes*, John Wiley and Sons, New York, 311-338.

Schnoor, J. L., and W. Stumm (1986), "The Role of Chemical Weathering in the Neutralization of Acidic Deposition", *Schweiz. Z. f. Hydrologie* **48/2**, 171-195.

Schott, J., and R. A. Berner (1985), "Dissolution Mechanism of Pyroxenes and Olivines During Weathering", in J.I. Drever, Ed, *The Chemistry of Weathering*, NATO ASI SERIES C **149**, 35-53.

Schwarzenbach, R. P., R. Stierli, and J. Zeyer (1990), "Quinone and Iron Porphyrine Mediated Reduction of Nitroaromatic Compounds in Homogeneous Aqueous Solution", *Env. Sci. Technol.* **24/10**, 1566-1574.

Schwarzenbach, R. P., W. Giger, E. Hoehn, and J. K. Schneider (1983), "Behavior of Organic Compounds During Infiltration of River Water to Groundwater. Field Studies", *Environ. Sci. Technol.* **17**, 472-479.

Schwarzenbach, R. P., and J. Westall (1981), "Transport of Nonpolar Organic Compounds from Surface Water to Groundwater. Laboratory Sorption Studies", *Environ. Sci. Technol.* **15**, 1360-1367.

Schwertmann, U., and R. M. Cornell (1991), *Iron Oxides in the Laboratory*, VCH Verlagsges., Weinheim, 132 p.

Segal, M. G., and R. Sellers (1984), "Redox Reactions at Solid-Liquid Interfaces", *Advances in Inorganic and Bioinorganic Mechanisms* **3**, 97-129.

Settle, D., and C. C. Patterson (1980), "Lead in Albacore: Quide to Lead Pollution in Americans", *Science* **207**, 1167-1176.

Shafer, M. M., and D. E. Armstrong (1990), "Trace Element Cycling in Southern Lake Michigan: Role of Water Column Particle Components", *Abstract ACS Meeting*, 273-277.

Shiao, S. Y. (1979), "Ion Exchange Equilibria Between Montmorillonite and Solutions of Moderate to High Ionic Strength", in *Radioactive Water in Geological Storage*, ACS Symposium 100, Amer. Chem. Soc.

Shiller, A. M., and E. Boyle (1985), "Dissolved Zinc in Rivers", *Nature* **317**, 49-52.

Siegel, D. I., and H. O. Pfannkuch (1984), "Silicate Dissolution Influence on Filson Creek Chemistry, Northeastern Minnesota", *Geol. Soc. Am. Bull.* **95**, 1446-1453.

Siever, R. (1968), "Sedimentological Consequences of a Steady-State Ocean-Atmosphere", *Sedimentology* **11**, 5.

Siffert, Ch., and B. Sulzberger (1991), "Light-Induced Dissolution of Hematite in the Presence of Oxalate: A Case Study", *Langmuir* **7**, 1627-1634.

Sigg, L. (1985), "Metal Transfer Mechanisms in Lakes; The Role of Settling Particles", in W. Stumm, Ed., *Chemical Processes in Lakes*, John Wiley and Sons, New York, 283-310.

Sigg, L. (1987), "Surface Chemical Aspects of the Distribution and Fate of Metal Ions in Lakes", in W. Stumm, Ed., *Aquatic Surface Chemistry*, John Wiley and Sons, New York, 319-349.

Sigg, L. (1992), "Regulation of Trace Elements in Lakes", in J. Buffle, Ed., *Chemical and Biological Regulation of Aquatic Processes*, in press.

Sigg, L., and W. Stumm (1981), "The Interaction of Anions and Weak Acids with the Hydrous Goethite (α-FeOOH) Surface", *Colloids and Surfaces* **2**, 101-117.

Sigg, L., and W. Stumm (1991), *Aquatische Chemie*, 2nd. Edition, Teubner Verlag, Stuttgart, Germany.

Sigg, L., M. Sturm, and D. Kistler (1987), "Vertical Transport of Heavy Metals by Settling Particles in Lake Zürich", *Limnol. Oceanogr.* **32**, 112-130.

Sigg, L., M. Sturm, J. Davis, and W. Stumm (1982), "Metal Transfer Mechanisms in Lakes", *Thalassia Jugoslavica* **18**, 293-311.

Sigg, L., W. Stumm, and B. Zinder (1984), "Chemical Processes at the Particle/Water Interface; Implications Concerning the Form of Occurrence of Solute and Adsorbed Species", in C. J. H. Kramer and J. Duinker, Eds., *Complexation of Trace Metals in Natural Waters*, Dr. W. Junk Publishers, The Hague, Netherlands, 251-266.

Singer, Ph. C., and W. Stumm (1970), "Acidic Mine Drainage – the Rate-determining Step", *Science* **167**, 1121-1123.

Singer, Ph. C., and W. Stumm (1970), "The Solubility of Ferrous Iron in Carbonate-Bearing Waters", *J. AWWA* **62**, 198-202.

Smith, R. W., and J. D. Hem (1972), "Effect of Aging on Aluminum Hydroxide Complexes in Dilute Aqueous Solutions", *U.S. Geological Survey Water-Supply Paper* **1827-D**, 51.

Somasundaran, P., and D. W. Fuerstenau (1966), "Mechanisms of Alkyl Sulfonate Adsorption at the Alumina-Water Interface", *J. of Physical Chemistry* **70**, 90-96.

Somasundaran, P., and G. E. Agar (1967), "Zero Point of Charge of Calcite", *J. Colloid Interface Sci.* **24/4**, 433-440.

Somasundaran, P., T. W. Healy, and D. W. Fuerstenau (1964), "Surfactant Adsorption at the Solid-Liquid Interface – Dependence", *J. of Physical Chemistry* **68**, 3562-3566.

Sparks, D. L. (1989), *Kinetics of Soil Chemical Processes*, Academic Press, San Diego, 210 pp.

Sposito, G. (1984), *The Surface Chemistry of Soils*, Oxford University Press, New York.

Sposito, G. (1986), "Sorption of Trace Metals by Humic Materials in Soils and Natural Waters", *CRC Crit. Rev. Environ. Control* **16**, 193-229.

Sposito, G. (1989), "Surface Reactions in Natural Aqueous Colloidal Systems", *Chimia* **43**, 169-176.

Sposito, G. (1989), *The Chemistry of Soils*, Oxford University Press.

Steefel, C. I., and P. Van Cappellen (1990), "A New Kinetic Approval to Modelling Water Rock Interaction: The Role of Nucleation, Precursors, and Ostwald Ripening", *Geochim. Cosmochim. Acta* **54**, 2657.

Stipp, S. L., and M. F. Hochella Jr. (1991), "Structure and Bonding Environments at the Calcite Surface as Observed with X-Ray Photoelectron Spectroscopy (XPS) and Low Energy Electron Diffraction (LEED)", *Geochim. Cosmochim. Acta* **55/6**, 1723-1736.

Stone, A. T. (1987), "Microbial Metabolites and the Reductive Dissolution of Manganese Oxides: Oxalate and Pyruvate", *Geochim. Cosmochim. Acta* **51**, 919-925.

Stone, A. T. (1987), "Reductive Dissolution of Manganese(III/IV)Oxides by Substitute Phenols", *Environ. Sci. Technol.* **21**, 979-988.

Stone, A. T., and J. J. Morgan (1987), "Reductive Dissolution of Metal Oxides", in W. Stumm, Ed., *Aquatic Surface Chemistry*, John Wiley and Sons, New York, 221-252.

Stucki, J. W., B. A. Goodman, and U. Schwertmann (1985), Eds., *Iron in Soil and Clay Minerals*, Reidel Dordrecht.

Stumm, W. (1977), "Chemical Interaction in Particle Separation", *Environ. Sci. Technol.* **11**, 1066-1070.

Stumm, W. (1991), "Atmospheric Depositions; Impact of Acids on Lakes", in A. Lerman, Ed., accepted.

Stumm, W., and C.. R. O'Melia (1968), "Stoichiometry of Coagulation", *J. AWWA* **60**, 514-539.

Stumm, W., and E. Wieland (1990), "Dissolution of Oxide and Silicate Minerals: Rates Depend on Surface Speciation", in W. Stumm, Ed., *Aquatic Chemical Kinetics*, John Wiley and Sons, New York, 367-400.

Stumm, W., and G. F. Lee (1961), "Oxygenation of Ferrous Iron", *Industrial and Engr. Chem.* **53**, 143-146.

Stumm, W., and J. J. Morgan (1970), *Aquatic Chemistry*, John Wiley and Sons, New York, 583 pp.

Stumm, W., and J. J. Morgan (1981), *Aquatic Chemistry,* 2nd Ed., John Wiley and Sons, New York, 796 pp.

Stumm, W., and L. Sigg (1979), "Kolloidchemische Grundlagen der Phosphor-Elimination in Fällung, Flockung und Filtration", *Z. f. Wasser- u. Abwasser-Forschung* **12**, 73-83.

Stumm, W., and R. Wollast (1990), "Coordination Chemistry of Weathering: Kinetics of the Surface-Controlled Dissolution of Oxide Minerals", *Reviews of Geophysics* **28/1**, 53-69.

Stumm, W., C. P. Huang, and S. R. Jenkins (1970), "Specific Chemical Interaction Affecting the Stability of Dispersed Systems", *Croat. Chem. Acta* **42**, 223-245.

Stumm, W., Ed. (1990), *Aquatic Chemical Kinetics, Reaction Rates of Processes in Natural Waters,* John Wiley and Sons, New York, 545 pp.

Stumm, W., F. Furrer, and B. Kunz (1983), "The Role of Surface Coordination in Precipitation (Heterogeneous Nucleation) and Dissolution of Mineral Phases", *Croat. Chem. Acta* **56**, 593-611.

Stumm, W., H. Hohl, and F. Dalang (1976), "Interaction of Metal Ions with Hydrous Oxide Surface", *Croat. Chem. Acta* **48**, 491.

Stumm-Zollinger, E. (1972), "Die bakterielle Oxidation von Pyrit", *Arch. Mikrob.* **83**, 110-119.

Suatoni, J. C., R. E. Snyder, and R. O. Clark (1961), "Voltammetric Studies of Phenol and Aniline Ring Substitution," *Anal. Chem.* **33**, 1894-1897.

Sulzberger, B., D. Suter, C. Siffert, S. Banwart, and W. Stumm (1989), "Dissolution of Fe(III)(Hydr) Oxides in Natural Waters; Laboratory Assessment on the Kinetics Controlled by Surface Coordination", *Marine Chemistry* **28**, 127-144.

Sulzberger, B., J. L. Schnoor, R. Giovanoli, J. G. Hering, J. Zobrist, (1990), "Biogeochemistry of Iron in an Acidic Lake," *Aquatic Siences* **52**, 56-74.

Sunda, W. (1991), "Trace Metal Interactions with Marine Phytoplankton", *Biol. Oceanogr.* **6**, 411-442.

Sunda, W. G., and R. R. L. Guillard (1976), "The Relationship Between Cupric Ion Activity and the Toxicity of Copper to Phytoplankton", *J. Marine Res.* **34**, 511.

Sung, W., and J. J. Morgan (1980), "Kinetics and Product of Ferrous Iron Oxygenation in Aqueous Systems," *Env. Sci. Technol.* **14**, 561-568.

Suter, D., C. Siffert, B. Sulzberger, and W. Stumm (1988), "Catalytic Dissolution of Iron(III)(Hydr) Oxides by Oxalic Acid in the Presence of Fe(II)", *Naturwissenschaften* **75**, 571-573.

Suter, D., S. Banwart, and W. Stumm (1991), "Dissolution of Hydrous Iron(III); Oxides by Reductive Mechanisms", *Langmuir* **7**, 809-813.

Svendrup, H. A. (1990), *The Kinetics of Base Cation Release due to Chemical Weathering,* Lund University Press, Lund, Sweden.

Svetlicić, V., V. Zutić, and J. Tomaić (1990), "Estuarine Transformation of Organic Matter: Single Coalescence Events of Estuarine Surface Active Particles", *Marine Chemistry* **32**, 253-267.

Tamura, H., K. Goto, and M. Nagayama (1976), "The Effect of Ferric Hydroxide on the Oxygenation of Ferrous Ions in Neutral Solutions", *Corrosion Sci.* **26**, 197-207.

Tamura, H., S. Kawamura, and M. Nagayama (1980), "Acceleration on the Oxidation of Fe^{2+} Ions by Fe(III) Oxyhydroxides", *Corrosion Sci.* **20**, 963-971.

Tanford, Ch. (1980), *The Hydrophobic Effect: Formation of Micelles and Biological Membranes,* 2nd Ed., John Wiley and Sons, New York, 233 pp.

Tessier, A., R. Carignan, B. Dubreuil, and F. Rapin (1989), "Partitioning of Zinc Between the Water Column and the Oxic Sediments in Lakes", *Geochim. Cosmochim. Acta* **53**, 1511-1522.

Thompson, D. W., and P. G. Pownall (1989), "Surface Electrical Properties of Calcite," *J. Colloid Interface Sci.* **131/1**, 74-82.

Thurman, E. M. (1985), *Organic Geochemistry of Natural Waters*, Nijhoff, Boston.

Tipping, E., and D. Cooke (1982), "The Effect of Adsorbed Humic Substances on the Surface Charge of Goethite in Freshwater", *Geochim. Cosmochim. Acta* **56**, 75.

Tole, M. P., A. C. Lasaga, C. Pantano, and W. B. White (1986), "The Kinetics of Dissolution of Nepheline ($NaAlSiO_4$)", *Geochim. Cosmochim. Acta* **50**, 379-392.

Tomaić, J., and V. Žutić (1988), "Humic Material Polydispersity in Adsorption at Hydrous Alumina Seawater Interface", *J. Colloid and Interface Sci.* **126**, 482-492.

Torrents, A., and A. T. Stone (1991), "Hydrolysis of Phenyl Picolinate at the Mineral/Water Interface", *Env. Sci. Technol.* **25**, 143-149.

Towe, K. M., and W. F. Bradley (1967), "Mineralogical Constitution of Colloidal Hydrous Ferric Oxides", *J. Colloid Interface Sci.* **24**, 384-392.

Troelstra, S. A., and H. R. Kruyt (1943), "Extinktiometrische Untersuchung der Koagulation", in W. Ostwald, Ed., *Kolloid-Beihefte*, Verlag Th. Steinkopf, Dresden, 225-261.

Ulrich, H.-J., W. Stumm, and B. Ćosović (1988), "Adsorption of Aliphatic Fatty Acids on Aquatic Interfaces. Comparison Between 2 Model Surfaces: The Mercury Electrode and δ-Al_2O_3 Colloids", *Environ. Sci. Techn.* **22**, 37-41.

Ulrich, H.-J., and C. Degueldre (1987), "The Sorption of ^{210}Pb, [210]Bi and [210]Po on Montmorillonite: A Study with Emphasis on Reversibility Aspects and on the Effect of the Radioactive Decay of Adsorbed Nuclides", submitted.

Valette, G., and A. Hamelin (1973), "Structure et propriétés de la couche double electrochimique à l'interphase argent/solution aqueuse de fluorure de sodium", *J. Electroanal. Chem.* **45**, 301-319.

Valverde, N., and C. Wagner (1976), "Considerations on the Kinetics and the Mechanism of the Dissolution of Metal Oxides in Acidic Solutions", *Ber. Bunsenges. Physik. Chem.* **80**, 330-333.

Van Cappellen, P. (1991), *Personal Communication*, in preparation.

Van Cappellen, P. (1991), *The Formation of Marine Apatite; A Kinetic Study*, Ph. D. Thesis, Yale University, USA.

Van der Schee, H. A., and J. Lyklema (1984), "A Lattice Theory of Polyelectrolyte Adsorption", *J. Phys. Chem* **88**, 6661-7.

Van der Zee, S. E. A. T. M., and W. H. Van Riemsdijk (1986), "Sorption Kinetics and Transport of Phosphate in Sandy Soils", *Geoderma* **38**, 293-309.

Van Hoof, P. L., and A. W. Andren (1989), "Partitioning and Transport of ^{210}Pb in Lake Michigan", *J. Great Lakes Res.* **15**, 498-509.

Van den Berg, C. M. G., and J. R. Kramer (1979), "Conditional Stability Constants for Copper Ions with Ligands in Natural Waters", in E. Jenne, Ed., *On Chemical Modeling: Speciation, Sorption, Solubility and Kinetics in Aqueous Systems*, ACS Symp. Series.

Van Loosdrecht, M. C. M., W. Norde, J. Lyklema, and A. J. B. Zehnder (1990), "Hydrophobic and Electrostatic Parameters in Bacterial Adhesion", *Aquatic Sciences* **52/1**, 103-114.

Van Olphen, H. (1977), *An Introduction to Clay Colloid Chemistry,* 2nd Ed., John Wiley and Sons, New York.

Vaslow, F., and G. E. Boyd (1952), "Thermodynamics of Coprecipitation: Dilute Solid Solutions of AgBr in AgCl[1]", *J. Amer. Chem. Soc.* **74**, 4691.

Velbel, M. A. (1985), "Geochemical Mass Balances and Weathering Rates in Forested Watersheds of the Southern Blue Ridge", *Am. J. Sci.* **285**, 904-930.

Vlachopoulos, N., P. Liska, A. J. McEvoy, and M. Grätzel (1987), "Efficient Spectral Sensitization of Polycrystalline Titanium Dioxide Photoelectrodes", *Surface-Science* **189/190**, 823-831.

Von Gunten, U., and W. Schneider (1991), "Primary Products of the Oxygenation of Iron(II) at an Oxic-Anoxic Boundary: Nucleation, Aggregation, and Aging", *J. Colloid and Interface Science* **145**, 127-139.

Waite, T. D., and F. M. M. Morel (1984), "Photoreductive Dissolution of Colloidal Iron Oxide: Effect of Citrate," *J. Colloid Interface Sci.* **102**, 121-137.

Waite, T.D. (1986), "Photoredox Chemistry of Colloidal Metal Oxides", in: J.A. Davis and K.F. Hayes, Eds., *Geochemical Processes at Mineral Surfaces*, Washington, ACS Symposium Ser. No. 323.

Walter, L. M., and J. W. Morse (1984a), "Magnesian Calcite Solubilities: A Reevaluation", *Geochim. Cosmochim. Acta* **48**, 1059-1069.

Walter, L. M., and J. W. Morse (1984b), "Reactive Surface Area of Skeletal Carbonates During Dissolution: Effect of Grain Size", *J. Sediment. Petrol* **54**, 1081-1090.

Wang, Z.-J., and W. Stumm (1987), "Heavy Metal Complexation by Surfaces and Humic Acids: A Brief Discourse on Assessment by Acidimetric Titration", *Netherlands J. of Agricult. Sci.* **35**, 231-240.

Wehrli, B. (1989), "Monte Carlo Simulations of Surface Morphologies During Mineral Dissolution", *J. Coll. Int. Sci.* **132**, 230-242.

Wehrli, B. (1990), "Redox Reactions of Metal Ions at Mineral Surfaces", in W. Stumm, Ed., *Aquatic Chemical Kinetics*, John Wiley and Sons, New York, 311-336.

Wehrli, B., and W. Stumm (1989), "Vanadyl in Natural Waters: Adsorption and Hydrolysis Promote Oxygenation", *Geochim. Cosmochim. Acta* **53**, 69-77.

Wehrli, B., B. Sulzberger, and W. Stumm (1989), "Redox Processes Catalyzed by Hydrous Oxide Surfaces", *Chemical Geology* **78**, 167-179.

Wehrli, B., E. Wieland, and G. Furrer (1990), "Chemical Mechanisms in the Dissolution Kinetics of Minerals; the Aspect of Active Sites", *Aquatic Sciences* **52**, 1-114.

Wehrli, B., S. Ibrić, and W. Stumm (1990), "Adsorption Kinetics of Vanadyl(IV) and Chromium(III) to Aluminum Oxide: Evidence for a Two-step Mechanism", *Colloids and Surfaces* **51**, 77-88.

Weilenmann, U., C. R. O'Melia, and W. Stumm (1989), "Particle Transport in Lakes: Models and Measurements", *Limnology & Oceanography* **34**, 1-18.

Weiss, A. (1958), "Die innerkristalline Quellung als allgemeines Modell für Quellvorgänge", *Chem. Ber.* **91**, 487.

Weissbuch, I., L. Addadi, M. Lahav, and L. Leiserowitz (1991), "Molecular Recognition at Crystal Interfaces", *Science* **253**, 637-645.

Wells, A. F. (1984), *Structural Inorganic Chemistry,* 5th Edition, Oxford University Press, Oxford, England.

Wells, M. L., and E. D. Goldberg (1991), "Occurence of Small Colloids in Sea Water", *Nature* **353**, 342-344.

Wells, M. L., L. M. Mayer, O. F. X. Donard, M. M. de Souza Sierra, and S. G. Ackelson, (1991), "The Photolysis of Colloidal Iron in the Oceans", *Nature* **353**, 248-250.

Wersin, P., L. Charlet, R. Karthein, and W. Stumm (1989), "From Adsorption to Precipitation; Sorption of Mn^{2+} on $FeCO_3(s)$", *Geochim. Cosmochim. Acta* **53**, 2787-2796.

West, A. R. (1984), *Solid State Chemistry and its Applications*, John Wiley and Sons, New York.

Westall, J. (1979), "MICROQL – I: A Chemical Equilibrium Program in BASIC", *EAWAG Report, Dübendorf, Switzerland.*

Westall, J. (1987), "Adsorption Mechanisms in Aquatic Surface Chemistry", in W. Stumm, Ed., *Aquatic Surface Chemistry*, John Wiley and Sons, New York, 3-32.

Westall, J., and H. Hohl (1980), "A Comparison of Electrostatic Models for the Oxide Solution Interface", *Adv. Colloid Interface Sci.* **12**, 265-294.

White, A. F. (1990), "Heterogeneous Electrochemical Reactions Associated with Oxidation of Ferrous Oxide and Silicate Surfaces", *Reviews in Mineralogy* **23**, 467-509.

White, A. F., and A. Yee (1985), "Aqueous Oxidation-Reduction Kinetics Associated with Coupled Electroncation Transfer from Iron-containing Silicates at 25° C", *Geochim. Cosmochim. Acta* **49**, 1263-1275.

Whitfield, M. (1979), "The Mean Oceanic Residence Time (Mort) Concept. – A Rationalization", *Marine Chemistry* **8**, 101-123.

Whitfield, M., and D. R. Turner (1987), "The Role of Particles in Regulating the Composition of Natural Waters", in W. Stumm, Ed., *Aquatic Surface Chemistry*, John Wiley and Sons, New York, 457-493.

Wieland, E. (1988), *Die Verwitterung schwerlöslicher Mineralien – ein koordinationschemischer Ansatz zur Beschreibung der Auflösungskinetik*, Ph.D. Thesis, ETH Zurich, Switzerland.

Wieland, E., and W. Stumm (1991), "Dissolution Kinetics of Kaolinite in Acid Aqueous Solutions at 25° C", submitted to *Geochim. Cosmochim. Acta.*

Wieland, W., B. Wehrli, and W. Stumm (1988), "The Coordination Chemistry of Weathering: III. A Generalization on the Dissolution Rates of Minerals", *Geochim. Cosmochim. Acta* **52**, 1969-1981.

Williams, R. J. P. (1983), "The Symbiosis of Metal Ion and Protein Chemistry", *Pure Appl. Chem.* **55**, 35.

Windom, H. L., J. T. Byrd, R. G. Smith, and F. Huan (1991), "Inadequacy of NASQUAN Data for Assessing Metal Trends in the Nation's Rivers", *Environ. Sci. Technol.* **25**, 1137-1142.

Wirth, G. S., and J. M. Gieskes (1979), "The Initial Kinetics of the Dissolution of Vitreous Silica in Aqueous Media", *J. Coll. Int. Sci.* **68**, 492-500.

Wollast, R. (1974), "The Silica Problem", in E. D. Goldberg, Ed., *The Sea* Vol. 5, John Wiley and Sons, New York, 359-392.

Wollast, R. (1990), "Rate and Mechanism of Dissolution of Carbonates in the System $CaCO_3–MgCO_3$", in W. Stumm, Ed., *Aquatic Chemical Kinetics, Reaction Rates of Processes in Natural Waters*, John Wiley and Sons, New York, 431-445.

Wollast, R., and F. T. Mackenzie (1983), "Global Cycle of Silica", in S. R. Aston, Ed., *Silicon Geochemistry and Biogeochemistry*, Academic Press, New York, 39-76.

Wollast, R., and L. Chou (1985), "Kinetic Study of the Dissolution of Albite with a Continuous Flow-through Fluidized Bed Reactor", in J. I. Drever, Ed., *The Chemistry of Weathering*, Reidel, Dordrecht, 75-96.

Wu, L., W. Forsling, and P. W. Schindler (1991), "Surface Complexation of Calcium Minerals in Aqueous Solution. 1. Surface Protonation at Fluorapatite-Water Interfaces," *J. Colloid Interface Sci.* **147/1**, 178-185.

Xue, H., and L. Sigg (1988), "The Binding of Heavy Metals to Algal Surfaces", *Wat. Res.* **22/7**, 917-926.

Xue, H., and L. Sigg (1990), "Binding of Cu(II) to Algae in a Metal Buffer", *Water Res.*, **24/9**, 1129-1136.

Yao, K. M., T. Habibian, and C. R. O'Melia (1971), "Water and Waste Water Filtration, Concepts and Applications", *Environm. Sci. and Tech.* **5**, 1105.

Yates, D. E., and T. W. Healy (1975), "Mechanism of Anion Adsorption at the Ferric and Chromic Oxide/ Water Interfaces", *J. Colloid Interface Sci.* **52**, 222-228.

Yates, D. E., S. Levine, and T. W. Healy (1974), "Site-binding Model of the Electrical Double Layer at the Oxide/Water Interface", *J. Chem. Soc. Faraday Trans.* **70**, 1807.

Zachara, J. M., J. A. Kittrick, and J. B. Harsh (1988), "The Mechanism of Zn^{2+} Adsorption on Calcite", *Geochim. Cosmochim. Acta* **52**, 2281-2291.

Zeltner, W. A., E. C. Yost, M. L. Machesky, M. I. Tejedor-Tejedor, and M. A. Anderson (1986), "Characterization of Anion Binding on Goethite Using Titration Calorimetry and Cylindrical Internal Reflection-Fourier Transform Infrared Spectroscopy", in J. A. Davis and K. F. Hayes, Eds., *Geochemical Processes at Mineral Surfaces*, Am. Chem. Soc., Washington, 142-161.

Zepp, R. G., and D. M. Cline (1977), "Rates of Direct Photolysis in Aquatic Environment," *Current Research* **11**, 359-366.

Zhang, J. W., and G. H. Nancollas (1990), "Mechanism of Growth and Dissolution of Sparingly Soluble Salts", in M. F. Hochella and A. F. White, Eds., *Reviews in Mineralogy 23*, Washington DC., 365-396.

Zinder, B., G. Furrer, and W. Stumm (1986), "The Coordination Chemistry of Weathering. II. Dissolution of Fe(III)Oxides", *Geochim. Cosmochim. Acta* **50**, 1861-1869.

Zobrist, J., and J. I. Drever (1990), "Weathering Processes in Alpine Watersheds Sensitive to Acidification", in M. Johannessen, R. Mosello, and H. Barth, Eds., *Air Pollution Research Report 20 "Acidification Processes in Remote Mountain Lake"*, CEC, Brussels, 149-161.

Zobrist, J., and W. Stumm (1979), "Chemical Dynamics of the Rhine Catchment Area in Switzerland; Extrapolation to the "Pristine" Rhine River Input into the Ocean", *Proc. Review and Workshop on River Inputs to Ocean Systems (RIOS) FAO, Rome*.

Zuo, Y., and J. Hoigné (1992), "Formation of Hydrogen Peroxide and Depletion of Oxalic Acid in Atmospheric Water by Photolysis", *Env. Sci. Technol.*, submitted.

Zutić, V., and J. Tomaić (1988), "On the Formation of Organic Coatings on Marine Particles: Interactions of Organic Matter at Hydrous Alumina/Seawater Interfaces", *Mar. Chem.* **23**, 51-67.

Zutić, V., and W. Stumm (1984), "Effect of Organic Acids and Fluoride on the Dissolution Kinetics of Hydrous Alumina. A Model Study Using the Rotation Disc Electrode", *Geochim. Cosmochim. Acta* **48**, 1493-1503.

Index

hydrophobic substance
    partition coefficient 117
    sorption 116-120

I

igneous rocks
    weathering 189
incongruent dissolution 188
inhibition
    of Cu(II) on δ-Al$_2$O$_3$ 4
    of dissolution 199-205
    inner-sphere 8
inner-sphere complex 21-23
    and dissolution 162
    in Lake Constance 296
    in lake water and in oceans 295
    on hematite and rutile 76
    in soils 120
inner-spheric surface complex 309
interfacial energy 221
    nucleation process 219-222
interfacial tension
    and adsorption 88
    angle of contact 142
    of liquids and solids 145
ion exchange 129-134
    of clays 132
ion exchange capacity
    measurement of 129
ion exchange equilibria
    relative affinity 131
iron
    cycling of 361, 362
    global geochemical cycling 361
    in atmospheric water 366
    passive film 204
    photoredox cycling 364
iron(II)
    photosynthetic oxidation 358
iron(III) hydrolysis 82
iron oxide 82
isomorphic substitution
    octahedral sheet 62
    tetrahedral sheet 62

J

joule 9

K

kaolinite 62, 64, 65
    dissolution 178-183

kinetics
    adsorption of metal ions 98
    crystal growth 233-235
    dissolution of Al-minerals 179
    heterogeneous photoredox reactions 347
    nucleation 211-222
    of anion adsorption 103
kink 170
Kurbatov plot 33

L

lake
    steady state model 391
Lake Cristallina 197
    weathering processes 198
Langmuir equation
    and surface complex formation constant 91
    and Frumkin equation 95
Langmuir isotherm 90-92
ledge 170
Lewis acid 13
Lewis acid site 24
ligand
    bidentate 25
ligand adsorption
    surface protonation 166
ligand binding
    by a hydrous oxide 30, 72
ligand exchange 14, 24-30
    kinetics 103
ligand promoted dissolution 165-169
    light-induced 344
linear free energy relation 26
lipophilicity of a substance 116
Lippmann equation 150-152

M

magnesian calcite 301-303
    solubility 301
magnesite
    dissolution rate 291
manganese(III,IV)oxide
    reduction by phenols 323-328
mean field statistic
    and steady state 169
mechanical weathering 195
membrane filter
    size distributions 284
metal
    binding on clays 141
    binding to the settling particles 387